D0169327

SOLID WASTE ENGINEERING

P. AARNE VESILIND
Bucknell University

WILLIAM A. WORRELL
San Luis Obispo County
Integrated Waste Management Authority

DEBRA R. REINHART
University of Central Florida

BROOKS/COLE

™

THOMSON LEARNING

Australia • Canada • Mexico • Singapore • Spain • United Kingdom • United States

BROOKS/COLE

THOMSON LEARNING

Publisher: *Bill Stenquist*
Marketing Team: *Sam Cabaluna, Chris Kelly, and Mona Weltmer*
Editorial Coordinator: *Valerie Boyajian*
Project Coordinator: *Mary Vezilich*
Production Service: *Matrix Productions*
Manuscript Editor: *Cathy Baehler*
Permissions Editor: *Bob Kauser*

Cover Design: *Denise Davidson*
Text Design: *Roy Neuhaus*
Cover Photo: *ICL/ImageState*
Photo Researcher: *Sue Howard*
Print Buyer: *Vena Dyer*
Typesetting: *Argosy*
Printing and Binding: *R. R. Donnelley & Sons, Inc.—Crawfordsville*

COPYRIGHT © 2002 Brooks/Cole. Brooks/Cole is an imprint of the Wadsworth Group, a division of Thomson Learning, Inc.
Thomson Learning™ is a trademark used herein under license.

For more information about this or any other Brooks/Cole product, contact:
BROOKS/COLE
511 Forest Lodge Road
Pacific Grove, CA 93950 USA
www.brookscole.com
1-800-423-0563 (Thomson Learning Academic Resource Center)

All rights reserved. No part of this work covered by the copyright hereon may be reproduced or used in any form or by any means—graphic, electronic, or mechanical, including photocopying, recording, taping, Web distribution, or information storage and retrieval systems—without the prior written permission of the publisher.

For permission to use material from this work, contact us by
web: www.thomsonrights.com
fax: 1-800-730-2215
phone: 1-800-730-2214

Printed in the United States of America

10 9 8 7 6 5 4 3 2

Library of Congress Cataloging-in-Publication Data

Vesilind, P. Aarne.
  Solid waste engineering / P. Aarne Vesilind, William A. Worrell, Debra R. Reinhart.
    p. cm.
  Includes bibliographical references and index.
  ISBN 0-534-37814-5
  1. Refuse and refuse disposal. I. Worrell, William A.   II. Reinhart, Debra R.   III. Title.

TD791.V47 2002
628.4'4—dc21

2001035223

*This book is dedicated
to the memory of
Richard Ian Stessel.*

Rich Stessel was born in New York City on December 31, 1956. He received his B.A. in Physics from Harvard University in June 1979, then went on to receive an M.A. in Public Policy Sciences and an M.S. and Ph.D. in Environmental Engineering, all from Duke University. He joined the University of South Florida in 1986 as an Assistant Professor in the Civil Engineering and Mechanics Department and spent several years as a member of Columbia University's School of Mines before returning to USF. His research and teaching interests included accelerated decomposition in landfills and the mechanical processing of waste to recover resources and energy. He was one of the premier researchers and teachers of solid waste management in the world. In August of 2001, he was tragically killed by a lightning bolt while walking on the USF campus.

Richard Stessel touched the lives of all three of the authors of this book—as a colleague, a former student, and as a friend. His humor, kindness, and grace will be missed by all who knew him.

CREDITS

This page constitutes an extension of the copyright page. We have made every effort to trace the ownership of all copyrighted materials and to secure permission from copyright holders. In the event of any question arising as to the use of any material, we will be pleased to make the necessary corrections in future printings. Thanks are due to the following authors, publishers, and agents for permission to use the material indicated.

P. xviii, Newsday, 1987/Dan Sheehan

Chapter 1
P. 15, By permission of Johnny Hart and Creators Syndicate ©2000;
P. 24, DILBERT by Scott Adams ©1998 United Feature Syndicate, Inc. Used with permission.

Chapter 3
Fig. 3-7, Courtesy of Volvo;
Fig. 3-13, Coutesy Dempster Dumpster;
Fig. 3-15, Courtesy of Volvo.

Chapter 5
Fig. 5-4, Courtesy Keith Walking Floor;
Fig. 5-11a, Courtesy Shred-Tech;
Fig. 5-11b, Courtesy Shred-Tech.

Chapter 6
Fig. 6-11b, Courtesy of Bulk Handling Systems Inc.;
Fig. 6-30, Courtesy O. S. Walker Co.

Chapter 8
Fig. 8-3, Courtesy Scat Engineering.

Contents

FOREWORD by John Skinner   xi

PREFACE   xvii

CHAPTER ONE

INTEGRATED SOLID WASTE MANAGEMENT   **1**

Solid waste in history   1

Economics and solid waste   4

Legislation and regulations   6

Materials flow   8

Reduction   10

Reuse   11

Recycling   12

Recovery   15

Disposal of solid waste in landfills   17

Energy conversion   21

The need for integrated solid waste management   23

Special wastes   24

Final thoughts   25

Problems   27

CHAPTER TWO

MUNICIPAL SOLID WASTE CHARACTERISTICS AND QUANTITIES   **30**

Definitions   30

Municipal solid waste generation   33

Municipal solid waste characteristics   38
    Composition by identifiable items   39
    Moisture content   45
    Particle size   48
    Chemical composition   49
    Heat value   49
    Bulk and material density   54
    Mechanical properties   56
    Biodegradability   57

Final thoughts   58

Appendix: Measuring particle size   59

Problems   64

CHAPTER THREE

COLLECTION   **67**

Refuse collection systems   67
    Phase 1: house to can   68
    Phase 2: can to truck   69
    Phase 3: truck from house to house   72
    Phase 4: truck routing   76
    Phase 5: truck to disposal   78

Commercial wastes   80

Transfer stations   82

Collection of recyclable materials   85

Litter and street cleanliness   91

Final thoughts   96

Appendix: Design of collection systems   96

Problems   103

CHAPTER FOUR

LANDFILLS   **109**

Planning, siting, and permitting of landfills   110
    Planning   110
    Siting   115
    Permitting   117

Landfill processes   118
   Biological degradation   118
   Leachate production   121
   Gas production   128

Landfill design   132
   Liners   132
   Leachate collection, treatment, and disposal   133
   Landfill gas collection and use   143
   Geotechnical aspects of landfill design   151
   Stormwater management   151
   Landfill cap   152

Landfill operations   155
   Landfill equipment   155
   Filling sequences   155
   Daily cover   158
   Monitoring   158

Post-closure care and use of old landfills   160

Landfill mining   163

Final thoughts   164

Problems   169

CHAPTER FIVE

PROCESSING OF MUNICIPAL SOLID WASTE   173

Refuse physical characteristics   173

Storing MSW   175

Conveying   176

Compacting   182

Shredding   183
   Use of shredders in solid waste processing   183
   Types of shredders used for solid waste processing   185
   Describing shredder performance by changes in particle-
      size distribution   191
   Power requirements of shredders   197
   Health and safety   200
   Hammer wear and maintenance   201
   Shredder design   202

Pulping   206

Roll crushing   207

Granulating   210

Final thoughts   210

Appendix: The pi breakage theorem   210

Problems   221

CHAPTER SIX

MATERIALS SEPARATION   224

General expressions for materials separation   224
   Binary separators   224
   Polynary separators   226
   Effectiveness of separation   227

Picking (hand sorting)   229

Screens   230
   Trommel screens   231
   Reciprocating and disc screens   236

Float/sink separators   239
   Theory of operation   240
   Jigs   246
   Air classifiers   247
   Other float/sink devices   259

Magnets and electromechanical separators   260
   Magnets   260
   Eddy current separators   263
   Electrostatic separation processes   265

Other devices for materials separation   265

Materials separation systems   269
   Performance of materials recovery facilities   272

Final thoughts   275

Problems   277

CHAPTER SEVEN

COMBUSTION AND ENERGY RECOVERY   283

Heat value of refuse   283
   Ultimate analysis   283
   Compositional analysis   285
   Proximate analysis   288
   Calorimetry   288

Materials and thermal balances   293
    Combustion air   293
    Efficiency   295
    Thermal balance on a waste-to-energy combustor   297

Combustion hardware used for MSW   300
    Waste-to-energy combustors   300
    Modular starved air combustors   307
    Pyrolysis   307
    Mass burn versus RDF   310

Undesirable effects of combustion   314
    Waste heat   314
    Ash   316
    Air pollutants   319
    Dioxin   329

Final thoughts   332

Problems   334

CHAPTER EIGHT

BIOCHEMICAL PROCESSES   336

Methane generation by anaerobic digestion   337
    Anaerobic decomposition in mixed digesters   338
    Potential for application of anaerobic digesters   342
    Methane extraction from landfills   343
    Potential for the application of methane extraction from landfills   344

Composting   344
    Fundamentals of composting   344
    Composting municipal solid waste   349
    Potential for composting municipal solid waste   351
    Composting wastes other than refuse   354

Other biochemical processes   355
    Glucose production by acid and enzymatic hydrolysis   355
    Other bacterial fermentation processes   357

Final thoughts   358

Problems   362

CHAPTER NINE

CURRENT ISSUES IN SOLID WASTE MANAGEMENT   **363**

Life cycle analysis and management   363
  Life cycle analysis   363
  Life cycle management   366

Flow control   368

Public or private ownership and operation   369

Contracting for solid waste services   370

Financing solid waste facilities   373
  Calculating annual cost   375
  Calculating present worth   376
  Calculating sinking funds   377
  Calculating capital plus O&M costs   378
  Comparing alternatives   378

Hazardous materials   378

The role of the solid waste engineer   380

Final thoughts   381

Epilogue   383

Problems   385

APPENDIX A   387

APPENDIX B   404

APPENDIX C   416

AUTHOR INDEX   420

SUBJECT INDEX   423

Foreword

When Aarne Vesilind told me that he, William Worrell, and Debra Reinhart were writing a new college textbook on municipal solid waste management, I initially wondered why a new book was necessary. Aarne's response was that this was about solid waste *engineering*. It wasn't until I reviewed the manuscript that I understood the true significance of that one word.

Having been trained as an engineer, but not having practiced engineering for many years, I had almost forgotten that engineers approach things as problems to be solved. In order to solve a problem, you must first be able to break it down into its basic elements, characterize it, describe it mathematically, formulate the physics, chemistry and biology, quantify it, and ultimately measure the results. This is what makes *Solid Waste Engineering* different from other texts. It provides a unique and extremely valuable contribution to the solid waste field.

This is not a text that looks at solid waste management from a solely historical or social perspective. And it is certainly not for the math- or physics-phobic. It is a text that provides engineers with a basic framework in order to use their special skills and education to meaningfully participate in the implementation of improved solid waste management systems.

Solid Waste Engineering should have wide-scale value in improving environmental engineering education around the world. In my service with the United Nations Environmental Programme and as President of the International Solid Waste Association, I am acutely aware of the severe problems caused by improper management of solid wastes worldwide, especially in developing countries. In many instances, officials and practitioners in those countries simply don't know where to start. This text provides a starting point so that university engineering departments and engineering consulting firms in these countries will have on hand the basic tools that have been successfully used in the developed world. As they use this text, they will be able to analyze and evaluate their current state of design practices and begin to implement solutions.

The Preface introduces the reader to the tale of two barges: an ocean-dumping barge that routinely dumped solid wastes off the New York Bight at the turn of the 20th century, and the infamous *Mobro* barge, that cruised up and down the East coast in 1987 searching for a community that would let it dock and off-load its wastes. The ocean-dumping barge shows us how far we have come in developing sound technologies to manage our solid waste. Today in North America, ocean dumping and open dumps have been eliminated and replaced with sophisticated recycling, recovery and disposal systems that are fully capable of managing solid wastes in an environmentally sound manner. However, as the hapless voyage of the *Mobro* demonstrates, solid wastes are often viewed as someone else's unwanted discards and application of even the best engineering solutions will quite often face formidable public opposition. The tale of the two barges gives students who are contemplating a career in solid waste engineering a preliminary glimpse of the social and political arena in which they will be applying their engineering skills.

Chapter 1, "Integrated Solid Waste Management," illustrates how wastes have been a consequence of life for all creatures since the beginning of time. The chapter traces the management of solid wastes from early civilizations to modern society and presents the economic theories governing waste practices from Adam Smith to the Club of Rome. It describes the U.S. federal solid waste legislation from the *Rivers and Harbors Act of 1899,* which prohibited dumping in navigable waters, to the Superfund and RCRA laws of the 1980s and 1990s that were designed to protect human health and the environment from improper management of solid and hazardous waste. The chapter concludes with a description of different elements of an integrated solid waste management system and introduces the reader to the four R's of the solid waste mantra: Reduce, Reuse, Recycle and Recover.

Chapter 2, entitled "Municipal Solid Waste Characteristics and Quantities," provides a thorough description of the composition, and ultimate and proximate chemical analysis of MSW. The chapter provides useful definitions, generation rates and the composition of MSW from both material and product basis. It discusses the need for and process of sampling MSW. Finally, it introduces the reader to all the essential scientific parameters that determine waste recovery potential: density, particle size, moisture content, heating value, mechanical properties and biodegradability.

Chapter 3 on "Collection of Municipal Solid Waste" provides the tools and methodologies to calculate the number of containers on a MSW collection route and the number of collection trucks needed. The chapter includes an introduction to the various truck types and body designs and provides a primer on the fundamentals of truck routing.

Chapter 4 is entitled "Landfills." Here is where the student will learn equations to calculate the landfill volume after compaction, estimate the leachate production rate and time, and calculate the landfill gas production rate. The chapter sets forth the basic equations for designing leachate collection systems and landfill gas collections systems.

Chapter 5 on the "Processing of Municipal Solid Waste" shows the difficulty of processing MSW compared to other materials because solid waste is a very heterogeneous material that is extremely difficult to predict in terms of particle size and physical parameters. The chapter provides the analytical tools that will enable the student to begin to unravel the performance of various material processes when applied to solid waste. It covers all the traditional material conveying technologies, including conveyors, live-bottom hoppers, pneumatic techniques, vibrating feeders, screw conveyors and drag chains. It describes various size reduction processes such as shredders, pulpers, roll crushers and granulators and describes shredder performance due to changes in particle size distribution. For the purists, the chapter contains an elegant discussion of the Pi Breakage Theorem. You really need to get out your scientific calculators to tackle the problems at the end of this chapter.

What do panning for gold and winnowing wheat have in common with materials recovery facilities (MRFs)? Chapter 6 answers this question by describing the basic physics of "Materials Separation." Not for the fainthearted in physics and math, this chapter takes the basic materials separation processes used in the mining and agricultural industries and applies them to the sorting of trash and recyclables. Using Newton's law, Stoke's equation and other formulations, the chapter analyzes the performance of the various unit processes used in many MRFs, including trommel screens (operating in the cascading, cataracting and centrifuging modes), float/sink separators, air classifiers, winnowing and flotation devices, and magnetic and eddy current separators. The last part of the chapter combines various separation unit processes in different sequences, and provides an excellent analysis of the rejection and extraction ratios for several integrated separated systems. Using the techniques and methodologies provided, MRF designers can gain insight into what works and what doesn't, and can identify approaches that would improve the recovery rate and purity of materials separated for recovery.

In Chapter 7, "Combustion and Energy Recovery," the reader first learns how to determine the heating value of refuse from ultimate analysis, compositional analysis, and calorimetry. Next, the chapter explains how to calculate the stochiometric air requirements for combustion, the thermal efficiency of the combustion process, and the stack gas temperature. Reviews are presented for these types of combustion devices: refractory-lined furnaces, water-wall combustors, modular starved air combustors, pyrolytic units, and refuse derived fuel (RDF) combustors. Finally, the text summarizes control of air pollutants from waste combustion including particulate control through fabric filters and electrostatic precipitators, control of dioxin emissions, and control of gaseous pollutants with wet and dry scrubbers.

Chapter 8, "Biochemical Processes," explains how to calculate the amount of methane produced from organic matter and provides an approximation of municipal solid waste's chemical composition to derive the yield of methane per kg of wet refuse. The chapter also shows how to calculate the residence time for a biochemical reactor and the quantity of gas produced. The composting process is described along with the role of the carbon/nitrogen ratio in limiting the composting rate. Finally, there is a discussion of hydrolysis, a biochemical process that is not widely used on solid waste, but which has the important potential to convert the cellulose fraction of the waste into glucose, ethanol, acetone, and other chemicals.

Chapter 9 covers "Current Issues in Solid Waste Management," including life-cycle assessment, flow control, and public and private service provision. Important issues are introduced regarding facility financing, including how to calculate present worth, use of a sinking fund, and how to determine the capital recovery factor. There is a very useful section on contracting out, which includes a list of things that a contract or franchise should cover.

Each chapter is well documented with a lengthy list of references. The problems at the conclusion of every chapter challenge the student to apply the techniques and methodologies which have been covered.

One aspect of the book that is most enjoyable is the section "Final Thoughts." This is where the authors pull the reader back from the science and mathematics and put solid waste in a more contemplative context through a series of vignettes loaded with insightful statements. A few of my favorites include:

"Waste is a consequence of our everyday life; if you don't manage it, it gets dumped."

"More people recycle than vote."

"We do not consume materials; we merely use them and ultimately return them to the environment, often in an altered state."

"Sometimes the engineer is accused of appeasing the establishment and destroying our environment. Some engineers confronted with such harsh criticism withdraw into their professional shells. But this is a classic cop-out."

"A truly professional engineer will infuse ethics into his or her decision making...environmental ethics will play an ever-increasing role in the engineer's professional responsibilities to society."

"The regulator must balance two primary moral values—do not deprive liberty and do no harm. Setting strict regulations would result in unwarranted reduction in liberty, while the absence of adequate regulations can damage public health."

"...if engineering is to maintain its professional autonomy, the public has to trust engineers ... Engineers, as all professionals, must work to maintain public trust."

"If the engineer cannot communicate information to the public, including decision makers such as public officials, then projects will not be implemented."

"It's not easy being a green engineer."

"Engineering is fun."

"The solid waste engineer is responsible for transforming the industry into a professional field with best practices."

Solid Waste Engineering reinforces the notion that engineers are optimists in the art of the possible; they like to look at a situation and make it better. But the book itself admits that the engineering challenges of managing our solid wastes are minor compared to the regulatory, social and political challenges of establishing solid waste management facilities. We have made excellent progress in improving our solid waste management practices over the past 20 years, but there is still so much to be done. Communities must be able to develop solid waste management strategies to meet their local needs. Full consideration of history, politics, economics, and community values are important in developing solutions. Engineers need to be able to establish meaningful dialogue and communicate with local political leaders and community groups and bring their unique engineering judgement into play in a complex social interaction. This is where solid waste engineers make their contributions and really earn their salaries.

While on the one hand, *Solid Waste Engineering* is sympathetic to an engineer's aspirations, it also imparts a forewarning:

"Engineers see themselves as performing a public service. Engineers build civilizations. Engineers serve the public's needs....

Engineers tend to be utilitarian. They look at the overall and aggregate net benefit, diminishing the importance of harm to the individual."

To be successful in responding to socially important problems like management of solid waste, engineers must develop a moral compass that will help to guide them through the uncertainties inherent in the public decision-making process. Solid waste engineers must be able to address and deal with the philosophical perspectives of different stakeholder groups that are often at odds with each other. *Solid Waste Engineering* is a valuable contribution to education literature in the field because it not only provides students with the analytical tools to design engineering solutions, but it also provides insight into the philosophical and social context in which those solutions must be applied.

John H. Skinner, Executive Director and CEO
The Solid Waste Association of North America

Preface: A Tale of Two Barges

Not so very long ago, as the coastal cities of the young United States grew to metropolitan regions, the disposal of municipal refuse was expediently achieved by simply loading up large barges, transporting them some distance from shore, and shoveling the garbage into the water. One such scow, operated out of New York City during the turn of the 20th century, is pictured below. Few complained when some of the refuse floated back to the shore. It was simply the way things were done.

A different story can be told about another barge, named the *Mobro*, pictured on the next page. The year was 1987. The *Mobro* had been loaded in New York with municipal solid waste and found itself with nowhere to discharge the load, and ocean disposal was now illegal. The barge was towed from port to port, with six

states and three countries rejecting the captain's pleas to offload its unwanted cargo.

The media picked up on this unfortunate incident and trumpeted the "garbage crisis" to anyone who would listen. Reporters honed their finest hyperbole, claiming that the barge could not unload because all our landfills were full and that the United States would soon be covered by solid waste from coast to coast. Unless we did something soon, they claimed, we could all be strangled in garbage.

The difference between the two barges, almost 100 years apart in time, is striking. In 1900 there were few laws restricting refuse disposal and thus solid waste disposal practices resulted in severe and permanent detrimental effects to the environment. There is no doubt that much of the refuse being shoveled off the barge 100 years ago is still on the bottom of the New York Bight and will remain there indefinitely as an embarrassment to future generations.

A hundred years later the public is acutely aware of the problem of solid waste disposal, and today we have achieved a degree of technological sophistication in our management of solid waste. Our landfills are constructed with almost no detrimental environmental effect, our solid waste combustors emit essentially no pollutants, and the public is increasingly participating in recycling programs. We have the problem under control, and yet the public perception is exactly opposite of the reality, as exemplified by the *Mobro* incident.

The story of the hapless *Mobro* is actually a story of an entrepreneurial enterprise gone sour. An Alabama businessman, Lowell Harrelson, wanted to construct a facility for converting municipal refuse to methane gas. He recognized that baled refuse would be the best form of refuse for that purpose. He purchased the bales of municipal solid waste from New York City and was going to find a landfill somewhere on the East Coast or in the Caribbean where he could deposit the bales and start making methane. Unfortunately, he did not get the proper permits for bringing refuse into various municipalities, and the barge was refused permission to offload its cargo. As the journey continued, the press coverage grew, and no local politicians would agree to allow the garbage to enter their ports. Harrelson finally had to burn his investment in a Brooklyn incinerator.

The barge *Mobro* is a poor metaphor for the state of municipal solid waste management in the United States because today we manage the discards of society with engineering skill at reasonable cost and at minimal risk to the public. This is not to say that we cannot do things better. Yes, we have to be concerned about non-replenishable resources. Yes, we care about the use of land for refuse storage—land that could be used for other purposes. Yes, we can design better packaging for our consumer goods. Yes, we can initiate programs that promote litter-free roads. Yes, we can design better devices that more effectively separate the various constituents of refuse. And yes, we can do many good things to improve the solid waste collection, treatment, and disposal process. But we should also be proud of the accomplishments of solid waste engineers in managing the collection, recovery, and disposal of municipal refuse. It is this positive theme that we wish to impress on the readers of this book.

This book is written for the engineering student who wants to learn about solid waste engineering, a subset of environmental engineering. Environmental engineering developed during the last 50 years as a major engineering discipline and is now established as an equal alongside such major engineering fields as civil, chemical, mechanical, and electrical engineering. The emergence of environmental engineering is driven in great part by societal need to control the pollution of our environment. Jobs for environmental engineers continue to increase exponentially, and there is no sign that this will slow down.

Using this book as part of a graduate or advanced undergraduate course in solid waste engineering will help to prepare the student to enter the field as an engineer-in-training. Much of the knowledge in solid waste engineering is gained by actual experience while working with experienced engineers in the field, and it is impossible to include all this experience in this book. What we hope is that the student will, at the conclusion of this course, be able to enter into meaningful conversations with experienced engineers and eventually put the basic principles learned in this course to beneficial use.

The course, as taught by the authors at different universities, usually takes one semester. Some of us have taught this course by going through the book in sequence, assigning homework problems as appropriate. One of us has eschewed homework problems completely and used only the design problem, found in Appendix A, requiring students to work in groups and individually write weekly chapters. The structure of the course is probably not as important as the education of the students in the fundamentals of solid waste management. It is not enough to train students to solve certain types of problems. It *is* important for them to emerge from this course being able to think reflectively and logically about the problems and solutions in solid waste engineering.

We believe that the material in this book represents a valuable first course in solid waste engineering. For us, the "proof of the pudding" has been the wide acceptance of our students by the practicing engineering community. We hope that others will be able to use this book to launch exciting and productive careers in solid waste engineering.

Finally, we would like to thank our reviewers: Michael J. Barcelona, University of Michigan; Morton A. Barlaz, North Carolina State University; Simeon J. Komisar, Renesselaer Polytechnic Institute; Robert E. Miller, University of Alaska, Anchorage; Sayed R. Qasim, University of Texas at Arlington; Raymond W. Regan, Pennsylvania State University; William F. Ritter, University of Delaware; Thomas G. Sanders, Colorado State University; and Berrin Tansel, Florida International University.

P. Aarne Vesilind
William A. Worrell
Debra R. Reinhart

1

Integrated Solid Waste Management

All creatures, humans included, constantly make decisions about what to use and what to throw away. A chimpanzee knows that the inside of the banana is good, and that the peel is not, and throws it away. A paramecium uses certain high-energy organic molecules and discharges its products after having extracted the energy in the carbon–carbon or carbon–hydrogen bonds. And humans buy a can of soft drink with the full understanding that the can will become waste. Waste is a consequence of everyday life—of all creatures.

SOLID WASTE IN HISTORY

In this book we consider a special kind of waste created by humans, the so-called *solid waste*, to distinguish it from the waste we emit into the atmosphere or the waste we discharge into the sewerage system. As with all other creatures, humans have been producing solid waste forever as part of life.

When humans abandoned nomadic life at around 10,000 BC, they began to live in communities, resulting in the production of solid waste. Waste piled up, and people wallowed in the offal—a characteristic that seems to be unique to the human animal. There were exceptions, of course. In the Indus valley the city of Mahenjo-Daro had houses with rubbish chutes and probably had waste collection systems. Other towns on the Indian subcontinent—Harappa and Punjab—had toilets and drains, and by 2100 BC the cities on the island of Crete had trunk sewers connecting homes.[1] The sanitary laws written by Moses in 1600 BC still survive in part. By 800 BC old Jerusalem had sewers and a primitive water supply. By 200 BC the cities in China had "sanitary police" whose job it was to enforce waste disposal laws.

But for the most part people in cities lived among waste and squalor. Only when the social discards became dangerous for defense was action taken. In Athens, in 500 BC, a law was passed to require all waste material to be deposited more than a mile out of town because the piles of rubbish next to the city walls provided an opportunity for invaders to scale up and over the walls.[2] Rome had similar problems, and eventually developed a waste collection program in 14 AD.

The cities in the Middle Ages in Europe were characterized by unimaginable filth. Pigs and other animals roamed the streets, and wastewater was dumped out of windows onto unsuspecting passersby. In 1300 the Black Death, which was to a great degree a result of the filth, reduced the populations in cities and alleviated the waste problems until the industrial revolution in the mid-1800s brought people back to the cities.

The living conditions of the working poor in 19th century European cities have been graphically chronicled by Charles Dickens and other writers of that period. Industrial production and the massing of wealth governed society, and human conditions were of secondary importance. Water supply and wastewater disposal were, by modern standards, totally inadequate. For example, Manchester, England, had on average one toilet per 200 people. About one-sixth of the people lived in cellars, often with walls oozing human waste from adjacent cesspools. People often lived around small courtyards where human waste was piled, and which also served as the children's playground.

The Great Sanitary Awakening in the 1840s was spearheaded by a lawyer, Edwin Chadwick (1800–1890), who argued that there was a connection between disease and filth. The germ theory was not, however, widely accepted until the famous incident with the pump handle on Broad Street in London. The public health physician, John Snow (1813–1858), suspected that the water supply from the Broad Street pump was contaminated and was the cause of the cholera epidemic. He removed the handle and prevented people from drinking the contaminated water, thus stopping a cholera epidemic and ushering in the public health revolution.

In the United States the conditions in many of the cities were appalling. Waste was disposed of by the judicious method of throwing it into streets where rag pickers would try to salvage what had secondary value. Animals would devour foodstuff. In 1834 Charleston, West Virginia, enacted a law protecting garbage-eating vultures from hunters.

Recycling in the late 1800s was by individuals who scoured the streets and the trash piles looking for material of value. The first organized municipal recycling program was attempted in 1874 in Baltimore, but it did not succeed.[3]

As early as 1657 the residents of New Amsterdam (present New York) forbade the throwing of garbage into streets, but the cleanliness of streets was still the responsibility of the individual homeowner.[3] Finally, in 1866, two hundred years after the first attempt, the Metropolitan Board of Health in New York declared war on trash, forbidding the throwing of garbage or dead animals into streets. The first incinerator in the New World was built in 1887 on Governor's Island in New York. In 1895 the garbage problem finally became a factor in politics, and great effort was made politically to clean up the cities. Municipal collection systems were created, the most famous and best organized one being in New York City by Col. George Waring. The method of disposal in New York in those days was to carry the waste by barge into the bight area and dump it into the water. Before special barges were built for this purpose, the waste was simply shoveled off the barge by workers. Waring started a comprehensive materials recovery system, using pickers to separate out and then sell recoverable materials. His plan included the collection of separated materials: ashes, garbage, and other material. After several years of operation the public opposition to his schemes was so strong that the materials recovery system was terminated.[3]

Much of the unwanted material in cities was removed by scavengers, who collected over 2000 yd^3 (1530 m^3) daily in the City of Chicago.[1] Partly because of scavengers and partly because of the lifestyle, in 1916 the municipal collection crews collected only about 0.5 lb of refuse per capita per day, compared to about ten times that today. The generation rates were no doubt much higher, however. Open horse-drawn wagons (Figure 1-1) were used to collect the waste, and the horses fouled the streets while the refuse was being collected.

The fouling of beaches forced the passage of federal legislation in 1934 making the dumping of municipal refuse into the sea illegal. Industries and commercial establishments were not covered, however, and continued unabated dumping into offshore waters. The first hole-in-the-ground that was periodically covered with dirt, a precursor of today's modern landfill, was started in 1935 in California. Ironically, the site today is on the U.S. Environmental Protection Agency's (EPA's) Superfund list as containing highly hazardous materials.[3] The American Society of Civil Engineers in 1959 published the first engineering guide to sanitary landfilling, which included the compaction of the refuse and the placement of a daily cover to reduce odor and rodents.

Figure 1-1 *A typical horse-drawn solid waste collection vehicle, used well into the 1920s in many U.S. cities.*

As the management of municipal refuse has changed over the years, so has the composition of the waste. Here are some significant events that changed the characteristics of residential and commercial municipal solid waste.[3]

1908	Paper cups replace tin cups in vending machines
1913	Corrugated cardboard becomes popular as packaging
1924	Kleenex facial tissues are first marketed
1935	First beer can is manufactured
1944	Dow Chemical invents Styrofoam
1953	Swanson introduces the TV dinner
1960	Pop-top beer cans are invented
1963	Aluminum beer cans are developed
1977	PETE soda bottles begin to replace glass

In today's cities solid waste is removed and either is sent to disposal or is reprocessed for subsequent use. This change in thinking from simply getting the stuff out of town to its use for some purpose represents the first paradigm shift in solid waste engineering in nearly 2000 years. Following rapidly on the move to recover materials is the Waste Reduction Revolution—the idea that it is bad to create waste in the first place. These changes have occurred because of both economics and a change in public attitude.

ECONOMICS AND SOLID WASTE

The emergence of the industrial age fostered the science of economics and prompted many leading thinkers to attempt to bring rational order to the seemingly chaotic world around them. The rationalism that resulted led to the common belief that trends could be understood and decisions made best on the basis of numbers. This substitution of the quantitative for the qualitative still pervades modern society and influences our entire set of attitudes toward resources and how they should be distributed.

Adam Smith (1723–1790), through his concept of *the invisible hand*, introduced an element of positive faith and optimism. However, his efforts were overshadowed by a number of pessimistic analysts who predicted continuing misery, poverty, exploitation, and class discrimination. David Ricardo (1772–1823), with his *iron law of wages*, held that wages for the working people would always remain at the poverty level, since any increase in wages would result in a commensurate increase in population, and this would once again drive wages down.

Equally pessimistic was the view held by Thomas Malthus (1766–1834), who in 1798 reasoned that since population growth is geometric and the increased production of food is arithmetic, a famine must result. This *law of population* was part of the *laissez-faire* school of economic liberalism, and was in great part responsible for the earned reputation of economics as a "dismal science." Malthus held that overpopulation can be prevented only by two types of checks: positive and preventive. Numbered among the former are wars, plagues, and similar disasters. Preventive checks include abstention from marriage, limitations on the number of children, and the like. Although the latter is clearly preferable, Malthus had little hope for the world, and insisted that the poor were "authors of their own poverty," simply because they failed to use the preventive checks on population growth.

This thesis was widely believed for many years and held as basic economic dogma. But as populations grew, widespread famine and deprivation were avoided, and Malthus' writings fell from favor. Economists began to think of Malthus as an economic anachronism to be studied, but only in the historical context. Technology, the new god, was able to preserve order, avert disaster, and lead us into the promised land.

This optimism was widely shared during the 19th and well into the 20th century, with only a few disquieting voices. Henry David Thoreau's (1817–1862) distrust of things technical was tolerated with bemusement as the ramblings of an ungrateful crackpot. During the 1950s and 1960s a few more voices in the wilderness became audible. Paul Erlich, with his grand overstatements and predictions of doomsday, seemed strangely reminiscent of Malthus. Barry Commoner became the first public ecologist and helped promote the feeling of disquiet. Slowly, through the 1960s, the public became convinced that there may indeed be something to this doomsday talk.

The most respected and well-publicized voice of pessimism came from an interdisciplinary group of scientists at Massachusetts Institute of Technology. Funded by the Club of Rome, a consortium of concerned industrialists, this group of talented scientists and engineers developed a computer model of the world, based on projections of pollution, agricultural production, availability of natural resources, industrial production, and population. Their ambitious undertaking, led by Dennis Meadows and Jay Forrester, resulted in the publication of a final report that indicated that even our most optimistic projections will eventually lead to the onset of famine, wars, and the destruction of our economic system.[4] It was, in short, a dismal outlook. Malthus would have been pleased.

The Club of Rome report has been criticized for inaccuracies and misinterpretations, and some of these accusations appear to be valid. Indeed, a revised model has shown an increased chance for world survival[5] and more accurate data would seem to reduce the level of pessimism. Nevertheless, the dismal outlook of Malthus is generally reaffirmed by such studies, and we are beginning to realize that our planet is finite and that it has only limited resources and living space. The scarcity of land and nonrenewable resources could indeed have the ultimate devastating effect envisioned by Malthus and now once again suggested by predictive studies. At the very least, the concern is real, and we should begin to seek alternative life systems in order to have more assurance that these disasters can be avoided.

One (of many) possible potentially beneficial alternatives toward global stability is to eliminate solid wastes generated by our materialistic society that are now deposited on increasingly scarce land. The recovery of these resources from solid waste would be a positive step toward establishing a balanced world system where society is no longer dependent on extraction of scarce natural ores and fuels. It seems quite clear that society has to adapt, using less technology in some instances, more in others, to achieve this balance. The technology and philosophy necessary for the development of a solid waste policy for sustained use of the earth's resources is the topic of this text.

LEGISLATION AND REGULATIONS

The United States, like other now independent countries with British judicial traditions, operates on the concept of *common law*. Common law is derived from the principle of fair play, or justice—a purely British invention. Under common law, if a person is wronged, the perpetrator is convicted and sentenced on the basis of *precedence*. That is, if a similar wrongful act occurred previously, then the only right and fair thing to do is to treat the next person in a similar manner. Common law is not written down, except as cases that define the precedent for new cases.

The genius of common law is that it is fair to all, but at the same time is able to change as the needs and values of the people change. It is no longer a crime, for example, to be a witch. But common law changes very slowly since most courts are loathe to make new law. Common law is also ineffective in protecting the environment and in correcting environmental ills because under common law a person (not a forest or a river or any other nonhuman entity) has to be wronged before relief can be sought in the courts. The assumption is that all of nature is owned by humans and it is only the wealth and welfare of these humans that common law protects. Because of the lack of responsiveness of common law to environmental ills, the Congress of the United States has resorted to passing environmental legislation. Such *legislated law* in effect establishes new common law by setting artificial and immediate precedents.

Prior to the 1960s the only federal legislation that addressed solid waste was the 1899 *Rivers and Harbors Act*, which prohibited the dumping of large objects into navigable waterways. The federal government was not involved in solid waste matters, except as a major producer of solid waste, much of which was managed by the Department of the Interior. Municipal solid waste was commonly thrown into unlined open dumps, which were intentionally set on fire to reduce volume. In larger communities solid waste was sent to incinerators, which had minimal air emission controls and which did a poor job of reducing the volume of waste. In Durham, North Carolina, for example, the city incinerator in the 1950s was called the "Durham Toaster" in the newspapers. Apparently the organic matter emerged from the incinerator barely singed.

The first federal legislation intended to assist in the management of solid waste was the 1965 *Solid Waste Disposal Act*, which provided technical assistance to the states through the U.S. Public Health Service. The emphasis in this legislation was the development of more efficient methods of disposal, and not the protection of human health.

On 1 January, 1970, President Nixon signed the *National Environmental Policy Act* (NEPA), which led to the creation of the Environmental Protection Agency (EPA). The most significant part of this mostly policy statement is Section 102, which requires all federal agencies (except those engaged in national defense work) to write *environmental impact statements* (EISs) whenever there is significant effect on the environment. When this law was enacted, many thought that the EISs would be simple one-paragraph boilerplate statements, much like our present antidiscrimination requirement, that agencies would simply attach to all plans and contracts stating that in their opinion there is no significant effect on the environment.

But environmentalists took these first meager environmental impact statements to court and the courts forced the federal agencies to write meaningful EISs. Since that time, most states have passed environmental impact legislation, and now whenever there is public money involved, such as for all landfills, combustors, and other solid waste management facilities, an environmental impact statement is almost certainly required.

In 1976 the Congress of the United States passed the *Resource Conservation and Recovery Act* (RCRA). With its 1984 amendments RCRA is a strong piece of legislation that mainly addresses the problem with hazardous waste, but also specifies guidelines for nonhazardous solid waste disposal. Subtitle D in this act is the municipal solid waste section, and landfills that fall under these requirements are commonly called *Subtitle D landfills*. In 1991 under Subpart D, the EPA adopted regulations to establish minimum national landfill criteria for all solid waste landfills. A key component of the standard was to require landfills to install composite liner systems consisting of a plastic liner on top of compacted clay. Specifications for the composite liner were also included in the regulation. Existing landfills had two years to comply with these standards.

In response to such hazardous waste disasters as Love Canal in Niagara, New York, and the Valley of the Drums in Tennessee, Congress passed the 1980 *Comprehensive Environmental Response, Compensation, and Liability Act* (CERCLA), which is commonly known by its nickname—the *Superfund* act. CERCLA created a financial means of cleaning up old hazardous waste sites by tapping into the coffers of present chemical companies, who most likely were not guilty of anything other than association with the real perpetrators.

The combustion of solid waste is controlled by the 1970 *Clean Air Act* (with subsequent amendments). With this legislation began the process of closing burning dumps and uncontrolled incinerators. In every case the federal agencies involved (mostly the EPA) required the individual states to set up local guidelines that adhere to the federal standards, which are then approved by the EPA. The actual enforcement of those requirements is then left to the states.

The siting of solid waste facilities is further complicated by local opposition, which often drags out the process. In some states it now takes over ten years to site a new landfill, even in the absence of local opposition. Municipalities that have traditionally managed their solid waste locally are now shipping waste across state borders to remote regional landfills just to avoid having to go through the process of developing a local solution to the problem. Communities are increasingly unwilling to go through the process of siting a new landfill, transfer station, or combustor. In the end the engineering challenges are minor compared to the regulatory and political challenges of siting new solid waste facilities.

Most states have passed strong legislation encouraging and promoting recycling. In the 1990s over 40 states established recycling goals. For example, California mandated that 25 percent of the waste be diverted from landfills by 1995 and that 50 percent be diverted by 2000. In Pennsylvania every community over 5000 population is mandated to set up a recycling program. Often local problems in siting new landfills drive the recycling effort. If a substantial amount of solid waste can be diverted from the landfill, then the facility will last longer and a new one will not have to be built.

MATERIALS FLOW

The flow of materials in our society may be illustrated by the schematic diagram shown in Figure 1-2. This diagram emphasizes the fact that we do not *consume* materials; we merely use them and ultimately return them, often in an altered state, to the environment. The production of useful goods for eventual use by those people called (incorrectly) *consumers* requires an input of materials. These materials originate from one of three sources: raw materials, which are gleaned from the face of the earth and used for the manufacture of products; scrap materials produced in the manufacturing operation; and materials recovered after the product has been used. Industrial operations are not totally efficient, and thus produce some waste that must be disposed of. The resulting processed goods are sold to the users of the products, who, in turn, have three options after use: to dispose of this material, to collect the material in sufficient quantities either to use it for energy production or to recycle it back into the industrial sector, or to reuse the material for the same or a different purpose without remanufacture.

This is a closed system, with only one input and one output, emphasizing again the finite nature of our world. At steady state the materials injected into the process must equal the materials disposal back into the environment. This process applies to the sum of all materials as well as to certain specific materials. For example, the manufacture of aluminum beverage containers involves the use of a raw material—

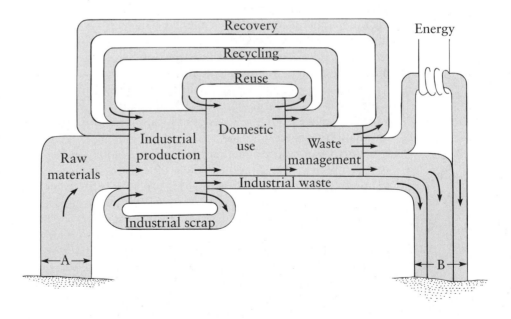

Figure 1-2   Materials flow through society.

bauxite ore—which is refined to produce aluminum. The finished products then are sold to consumers. Some of these cans are defective or for other reasons unfit for consumer use and are recycled as industrial scrap. The consumer uses the cans, and the empty containers or other products are disposed of in the usual manner. Some of the aluminum is returned to the industrial sector (for remanufacture) and some of it might be used for other purposes in the home. The aluminum that is recovered and returned to the manufacturing process gets there only by a conscious effort by the community or other organizations that collect and recycle the material through the system. For many materials this is at a financial loss.

The interaction of the materials flow with the environment is at the input of raw materials and the deposition of wastes. In Figure 1-2 these two interfaces are denoted by the letter **A** for raw materials and by the letter **B** for the materials returned to the environment.

We can argue that both **A** and **B** should be as large as possible since there are many benefits to be gained by increasing these values. For example, a large quantity of raw materials injected into the manufacturing process represents a high rate of employment in the raw materials industry, and this can have a residual effect of creating cheaper raw materials and reducing the cost of manufacturing. A large **B** component is also beneficial in the sense that the waste disposal industry (which includes people as diverse as the local trash collector and the president of a large firm that manufactures heavy equipment for landfills) has a key interest in the quantity of materials that people dispose of. Thus a large **B** component would mean more jobs in this industry.

However, large **A** and **B** components also have detrimental effects. A large raw material input means that great quantities of nonreplenishable raw materials are extracted (often using something less than environmentally sensitive methods, as exemplified by strip mining). Similarly, large quantities of waste can have a significant detrimental effect, such as land areas used for waste disposal, or air pollution from the burning of waste in combustors.

A high rate of raw material extraction can eventually lead to a problem in the depletion of natural resources. At the present time in the United States we have already exhausted our domestic supplies of some nonreplenishable materials, such as copper, zinc, and tin, and are importing a substantial fraction of these materials.[6] If the rest of the world were to attain the standard of living that the developed nations have at the present, the raw materials supply would not be adequate to meet the demand. Our present lifestyle is based on obtaining these materials from concentrated sources (ores), and in using them we are distributing the products over a wide land area. Such a distribution obviously makes recovery and reuse difficult.

Finally, the question of national security for each country is predicated on the nation's ability to obtain reliable supplies of raw materials. In the 1970s we experienced the problems that can be created by relying on other countries for such necessities as oil. There is little doubt that cartels will be developed by nations that have large deposits of other nonreplenishable materials, and that in the future the cost of such products as aluminum, tin, and rubber will increase substantially. We have ample justification for reducing the wastes disposed of into the environment to the smallest quantities practical, and we should redesign our economic system to achieve this end.

As shown in Figure 1-2, if the system is in steady state, the input must equal the output. Hence a reduction of either **A** or **B** necessarily results in a concomitant reduction in the other. In other words, it is possible to attack the problem in two ways.

Looking first at the **A** component, a reduction in raw materials demand could be achieved by increasing the amount of industrial scrap reprocessed, by decreasing the amount of manufactured goods, or by increasing the amount of recovered materials from the post-consumer waste stream. Increasing industrial scrap would involve increasing either *home scrap* (waste material reused within an industrial plant) or *prompt industrial scrap* (clean, segregated industrial waste material used immediately by another company). But scrap represents inefficiency, and an ultimate goal of industry is to produce as little scrap as possible. Clearly, decreasing the demand for raw materials will require another approach.

One possibility for achieving a low use of raw materials is to decrease the amount of manufactured goods. This *reduction* will necessitate a redesign of products in such a way as to use less material. The federal government can legislate a lower rate of material use by placing taxes on excessive packaging, initiating a package charge (e.g., so many cents per pound of packaging), requiring mandatory longer life of manufactured products, and other options. In addition it is possible for consumers to buy fewer manufactured goods or simply to buy products to consciously minimize waste.

A second means of reducing waste is to *reuse* the products. Often this is done without much thought, such as refillable soda and beer bottles, or coffee cans to hold nails, or paper bags for taking out the garbage. In addition, repairing an item instead of discarding it and buying a replacement is an example of the tradeoff between a labor-intensive society and a consumer-based society.

A third means of reducing the waste destined for disposal is to separate out materials that have some economic value, collect these separately, and use them as a source of raw materials. This process is called *recycling* and involves the active participation of the product user.

The fourth means is to process the solid waste so as to recover useful material from the mixed waste. *Recovery* can also include the recovery of energy from the solid waste. For example, a waste-to-energy plant or a landfill gas recovery system is recovering the energy value of the solid waste through a transformation process.

In summary, the feasible options for achieving reduced material use and waste generation are known as the four Rs:

1. Reduction
2. Reuse
3. Recycling
4. Recovery

Reduction

Waste reduction can be achieved in three basic ways: (1) reducing the amount of material used per product without sacrificing the utility of that product, (2) increasing the lifetime of a product, and/or (3) eliminating the need for the product.

Waste reduction in industry is called *pollution prevention*—an attractive concept to industry because in many cases the cost of treating waste is greater than the cost of changing the process so that the waste is not produced in the first place.[7] Any manufacturing operation produces waste. As long as this waste can be readily disposed of, there is little incentive to change the operation. If, however, the cost of waste disposal is great, the company has an incentive to seek improved manufacturing techniques that reduce the amount of waste. Pollution prevention as a corporate concept was pioneered by such companies as 3M and DuPont, and has as its driving force the objective of reducing cost (and hence increasing the competitive advantage of the manufactured goods in the market place). For example, automobile manufacturers for years painted new cars using spray enamel paint. The cars were then dried in special ovens that gave them a glossy finish. Unfortunately, such operations produced large amounts of volatile organic compounds (VOCs) that had to be controlled, and control measures were increasingly expensive. The manufacturers then developed a new method of painting, using dry powders applied under great pressure. Not only did this result in better finishes, but it all but eliminated the problem with the VOCs. Pollution prevention is the process of changing the operation in such a manner that pollutants are not even emitted.

Reduction of waste on the household level is called *waste reduction* (sometimes referred to as *source reduction* by the EPA).[8, 9] Typical alternative actions that result in a reduction of the amount of municipal solid waste being produced include refusing bags at stores, using laundry detergent refills instead of purchasing new containers, bringing one's own bags to grocery stores, stopping junk mail deliveries, and using cloth diapers.[10] Unfortunately, the level of participation in source reduction is low compared to recycling activities. Even though source reduction is the first solid waste alternative for EPA, and eight states have source reduction goals that range from no net increase of waste per capita to 10% reductions, few people participate in such programs. There is some evidence, however, that where communities have initiated disposal fees based on volume or weight of refuse generated, the amount of refuse is reduced by anywhere between 10% and 30%.[11, 12] Public information programs can significantly help in reducing the amount of waste generated. A study of 250 homes in Greensboro, North Carolina, found that a 10% waste reduction can be achieved following a public information program.[13] Advice on how to "shop smart" offered by one municipality is shown in Figure 1-3.[14]

Reuse

Reuse is an integral part of society, from church rummage sales to passing down children's clothing between siblings. Many of our products are reused without much thought given to ethical considerations. These products simply have utility and value for more than one purpose. For example, paper bags obtained in the supermarket are often used to pack refuse for transport from the house to the trash can or to haul recyclables to the curb for pickup. Newspapers are rolled up to make fireplace logs, and coffee cans are used to hold bolts and screws. All of these are examples of reuse.

Figure 1-3   Advice on how to reduce the amount of refuse generated.
Source: (14)

Recycling

The process of recycling requires that the owner of the waste material first separate out the useful fraction so that it can be collected separately from the rest of the solid waste. Many of the components of municipal solid waste can be recycled for remanufacturing and subsequent use, the most important being paper, steel, aluminum, plastic, glass, and yard waste.

Theoretically, vast amounts of materials can be recycled from refuse, but this is not an easy task regardless of how it is approached. In recycling, a person about to discard an item must first identify it by some characteristics and then manually segregate it into a separate bin. The separation relies on some readily identifiable characteristic or property of the specific material that distinguishes it from all others. This characteristic is known as a *code*, and this code is used to separate the material from the rest of the mixed refuse using a *switch*.

In recycling, the code is simple and visual. Anyone can distinguish newspapers from aluminum cans. But sometimes confusion can occur, such as identifying aluminum cans from steel cans, for example, or newsprint from glossy magazines, especially if the glossy magazines are intermingled with the Sunday paper. The most difficult operation in recycling is the identification and separation of plastics. Because mixed plastic has few uses, plastic recycling is more economical if the different types of plastic are separated from each other. Most people, however, cannot distinguish one type of plastic from another. The plastics industry has responded by marking most consumer products with a code that identifies the type of plastic, as shown in Figure 1-4. Plastics that can be recycled are all common products used in everyday life, some of which are listed in Table 1-1.

Theoretically, all a person about to discard an unwanted plastic item has to do is to look at the code and separate the various types of plastic accordingly. In fact, there is almost no chance that a domestic household will have seven different waste receptacles for plastics, nor are there enough of each of these plastics to be economically collected and used. Typically, only the most common types of plastic are recycled, including PETE (polyethylene terephthalate), the material out of which the 2-liter soft drink bottles are made, and HDPE (high-density polyethylene), the white translucent plastic used for milk bottles.

Taking into account transportation and processing charges, it still appears that the economics for curbside recycling and materials recovery facilities in metropolitan areas (close proximity to refuse and markets) are quite favorable. The proof of this, of course, is the impressive number of materials recovery facilities (MRFs) in operation or under construction. In 1997 about 136 million people were served by almost 9,000 curbside recycling programs that fed the MRFs, and communities had created nearly 3,500 composting facilities.[15]

The success of recycling programs has been in spite of the severe obstacles that our present economic system places on the use of secondary materials. Some of these obstacles are identified here:

Location of wastes. The transportation costs of the waste may prohibit the implementation of recycling and recovery. Secondary materials have to be shipped to market, and if the source is too far away, the cost of the transport can be prohibitive.

PETE   HDPE   PVC   LDPE   PP   PS   OTHER

Figure 1-4 *Plastic recycling symbols.*

Table 1-1   Common Types Of Plastics That May Be Recycled

Code Number	Chemical name	Abbreviation	Typical uses
1	Polyethylene terephthalate	PETE	Soft drink bottles
2	High-density polyethylene	HDPE	Milk cartons
3	Polyvinyl chloride	PVC	Food packaging, wire insulation, and pipe
4	Low-density polyethylene	LDPE	Plastic film used for food wrapping, trash bags, grocery bags, and baby diapers
5	Polypropylene	PP	Automobile battery casings and bottle caps
6	Polystyrene	PS	Food packaging, foam cups and plates, and eating utensils
7	Mixed plastic		Fence posts, benches, and pallets

Low value of material. The reason that an item is considered waste is that the material (even when pure) has little value. For example, the price paid for secondary polystyrene has fluctuated from a positive payment of $100 per ton to a negative payment of $300 per ton.

Uncertainty of supply. The production of solid waste depends on the willingness of collectors to transport it, the cooperation of consumers to throw things away according to a predictable pattern, and the economics of marketing and product substitution, which may significantly influence the availability of a material. Conversion from aluminum to plastic beverage containers—whether by legislation, marketing options, or consumer preferences—will significantly change the available aluminum in solid waste. The replacement of a high-value material, aluminum, by a low-value material, plastic, will adversely affect recycling. Potential solid waste processors thus have little control over their raw materials.

Administrative and institutional constraints. Some communities are unwilling to pay the additional cost to implement curbside recycling programs. The costs of these programs typically are in the $1 to $4 per month per household range. Other cities may have labor or contractual restrictions preventing the implementation of resource recovery projects. For example, many yard waste composting facilities are prohibited by land use ordinances from accepting sludge or food waste, both of which would increase the value of the compost.

Legal restrictions. Some cities, such as San Diego, are prohibited from charging their residents for solid waste service, thus making it difficult to implement curbside recycling.

Uncertain markets. Recovery facilities must depend on the willingness of customers to purchase the end products—materials or energy. Often such markets are fickle, being either small, fragile operations or large, vertically integrated corporations that purchase the products on margin so as to satisfy unusually heavy short-duration demand.

Used by permission of Johnny Hart and Creators Syndicate, Inc.

Recovery

Recovery is defined as the process in which the refuse is collected without prior separation, and the desired materials are separated at a central facility. A typical materials recovery facility (MRF) is shown in Figure 1-5.

The various recovery operations in a MRF have a chance of succeeding if the material presented for separation is clearly identified by a code and if the switch is then sensitive to that code. Currently, no such technology exists. It is impossible, for example, to mechanically identify and separate all of the PETE soft drink bottles from refuse. In fact, most recovery operations employ *pickers*, human beings who identify the most readily separable materials—such as corrugated cardboard and HDPE milk bottles—before the refuse is mechanically processed.

Most items in refuse are not made of a single material, and to be able to use mechanical separation, these items must be separated into discrete pieces consisting of a single material. A common "tin can," for example, contains steel in its body, zinc on the seam, a paper wrapper on the outside, and perhaps an aluminum top. Other common items in refuse provide equally challenging problems in separation.

One means of producing single-material pieces and thereby assisting in the separation process is to decrease the particle size of refuse by grinding up the larger pieces. Grinding will increase the number of particles and achieve many clean (single-material) particles. The size-reduction step, although not strictly materials separation, is employed in some materials recovery facilities, especially if refuse-derived fuel is produced. Size reduction is followed by various other processes, such as air classification (which separates the light paper and plastics) and magnetic separation (for the iron and steel) (see Chapter 6).

The recovery of materials, although it sounds terribly attractive, is still a marginal option. The most difficult problem faced by engineers designing such facilities is the availability of firm markets for the recovered product. Occasionally, the markets are quite volatile, and secondary material prices can fluctuate wildly. One example is the secondary paper market.

Paper industry companies are *vertically integrated*, meaning that the company owns and operates all of the steps in the papermaking process. They own the lands on which the forests are grown, they do their own logging, and they take the logs

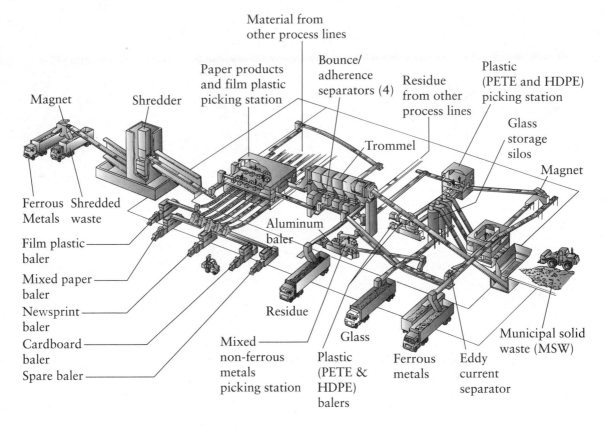

Figure 1-5 Typical materials recovery facility (MRF).

to their own paper mill. Finally, the company markets the finished paper to the public. This is schematically shown in Figure 1-6.

Suppose a paper company finds that it has an annual base demand of 100 million tons of paper. It then adjusts the logging and pulp and paper operations to meet this demand. Now suppose there is a short-term fluctuation of 5 million tons that has to be met. There is no way the paper company can plant the trees necessary to meet this immediate demand, nor are they able to increase the capacity of the pulp and paper mills on such short notice. What they do then is to go to the secondary paper market and purchase the secondary fiber to meet the incremental demand. If several large paper companies find that they have an increased demand, they will all try to purchase the secondary paper, and suddenly the demand will shoot the price of waste paper up. When either the demand decreases or the paper company has been able to expand its capacity, it no longer needs the secondary paper, and the price of waste paper plummets. Because paper companies purchase waste paper *on the margin*, secondary paper dealers are always in either a boom or bust situation, and the price is highly variable. When paper companies that use only secondary paper to produce consumer products increase their production (due to the demand for recycled or recovered paper), these extreme fluctuations are dampened

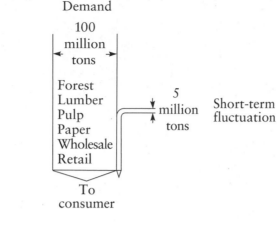

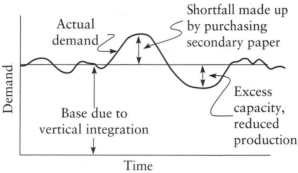

Figure 1-6 *Vertical integration of the paper industry.*

out. Figure 1-7 shows the price fluctuations of old newsprint over the past few years. Such spikes in the cost curve play havoc with long-term planning for secondary materials production and use.

Disposal of Solid Waste in Landfills

The disposal of solid wastes is a misnomer. Our present practices amount to nothing more than hiding waste well enough so it cannot be readily found. The only two realistic options for storing waste on a long-term basis are in the oceans (or other large bodies of water) and on land. The former is forbidden by federal law and is becoming similarly illegal in most other developed nations. Little else needs to be said of ocean disposal, except perhaps that its use was a less than glorious chapter in the annals of public health and environmental engineering.

The placement of solid waste on land is called a *dump* in the USA and a *tip* in Great Britain (as in *tipping*). The dump is by far the least expensive means of solid waste disposal, and thus was the original method of choice for almost all inland communities. The operation of a dump is simple and involves nothing more than making sure that the trucks empty at the proper spot. Volume is often reduced by setting the dumps on fire, thus prolonging dump life.

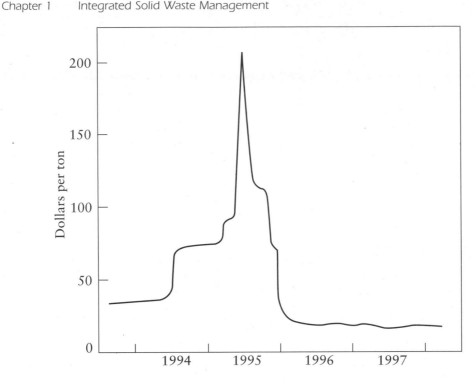

Figure 1-7 Price fluctuations of old newsprint. Source: (12)

Rodents, odor, air pollution, and insects at the dump, however, can result in serious public health and aesthetic problems, and alternate methods for disposal were necessary. Larger communities can afford to use a combustor for volume reduction, but smaller towns cannot afford such capital investment. This has led to the development of the *sanitary landfill*. The sanitary landfill differs markedly from open dumps in that the latter were simply places to dump wastes, while sanitary landfills are engineered operations, designed and operated according to acceptable standards. The basic principle of a landfill operation is to prepare a site with liners to deter pollution of groundwater, deposit the refuse in the pit, compact it with specially built heavy machinery with huge steel wheels, and cover the material with earth at the conclusion of each day's operation (Figure 1-8). Siting and developing a proper landfill require planning and engineering design skills.

Even though the tipping fees paid for the use of landfills are charged on the basis of weight of refuse accepted, landfill capacity is measured in terms of volume, not weight. Engineers designing the landfills first estimate the total volume available to them and then estimate the density of the refuse as it is deposited and compacted in the landfill. The density of refuse increases markedly as it is first generated in the kitchen and then finally placed into the landfill. The *as generated* bulk density of municipal solid waste (MSW) is perhaps 100 to 300 lb/yd^3 (60 to 180 kg/m^3), while the compacted waste in a landfill exceeds 1200 lb/yd^3 (700 kg/m^3). Landfills are required to have daily dirt covers, and the more dirt that is placed on the refuse, the less volume that is available for the refuse itself.

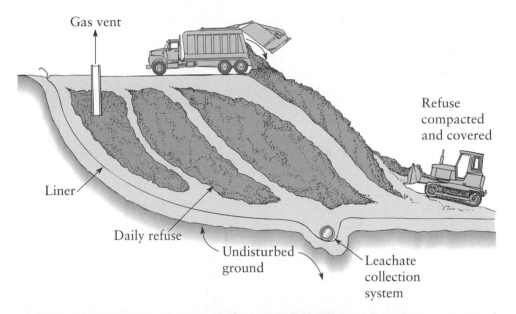

Gas vent

Refuse
compacted
and covered

Liner

Daily refuse

Undisturbed
ground

Leachate
collection
system

Figure 1-8 A typical sanitary landfill.

Commonly, engineers estimate that the volume occupied by the cover dirt is one-fourth of the total landfill volume.

EXAMPLE
1-1

Imagine a town where 10,000 households each fill one 80-gallon container of refuse per week. What volume would this refuse occupy in a landfill? Assume that 10% of the volume is occupied by the cover dirt.

This problem is solved using a mass balance. Imagine the landfill as a black box, and the refuse goes from the households to the landfill.

(mass out) = (mass in)

$$V_L\, D_L = V_P\, D_P$$

where V and D are the volume and density of the refuse, and subscripts L and P denote loose and packed refuse. Assume the density of the refuse when collected is 200 lb/yd^3 and is 1200 lb/yd^3 in the landfill.

$$((10{,}000 \text{ households})(80 \text{ gal/household})(0.00495 \text{ yd}^3/\text{gal}))(200 \text{ lb/yd}^3)$$
$$= (1200 \text{ lb/yd}^3)V_P$$

$$V_P = 660 \text{ yd}^3$$

If 10% of the total volume occupied is taken up by the cover dirt, then the total landfill volume necessary to dispose of this waste is

$$T = 660 + 0.10T$$

where T is the total volume. Thus T = 733 cubic yards.

Sanitary landfills are not inert. The buried organic material decomposes anaerobically, producing various gases, such as methane and carbon dioxide, and liquids that have extremely high pollutional capacity when they enter the groundwater. Liners made of either impervious clay or synthetic materials such as plastic are used to try to prevent the movement of this liquid, called *leachate*, into the groundwater. Figure 1-9 shows how a synthetic landfill liner is installed in a prepared pit. The seams have to be carefully sealed, and a layer of soil placed on the liner to prevent landfill vehicles from puncturing it.

Synthetic landfill liners are useful in capturing most of the leachate, but they are never perfect. No landfill is sufficiently tight that groundwater contamination by leachate is totally avoided. Wells have to be drilled around the landfill to check for groundwater contamination from leaking liners, and if such contamination is found, remedial action is necessary. And, of course, the landfill never disappears—it will be there for many years to come, limiting the use of the land for other purposes.

Modern landfills also require the gases generated by the decomposition of the organic materials to be collected and either burned or vented to the atmosphere. The gases are about 50% carbon dioxide and 50% methane, both of which are greenhouse gases. In the past, when gas control in landfills was not practiced, the gases have been known to cause problems with odor, soil productivity, and even explosions. Larger landfills use the gases for running turbines for the production of

Figure 1-9 *Placement of a synthetic liner in a modern landfill.*

electricity for sale to the power company, while smaller landfills simply vent the gases to the atmosphere. The fact that landfills produce methane gas has been known for a long time, and many accidental explosions have occurred when the gas has seeped into basements and other enclosed areas where it could form explosive mixtures with oxygen. Modern landfills are required to collect the gases produced in a landfill and either flare them or collect them for subsequent beneficial use.

The first system for collecting and using landfill gas was placed into operation in California after a number of gas-extraction wells were driven around a deep landfill to prevent lateral migration of gas. These wells burned off about 1000 ft³/min (44 m³/min) of gas without the need for auxiliary fuel.[16] Based on this experience, the capture and use of this gas instead of just wastefully burning it off seemed a reasonable alternative. Huge landfills (such as Palos Verdes, California, 15 million tons in place) could produce gas with a total heating value of 10 trillion Btu.

Energy Conversion

An alternative to allowing refuse to biodegrade and form a useful fuel is the combustion of refuse and energy recovered as heat. The potential for energy recovery from solid waste is significant. Of the 217 million tons (196 million tonnes) of waste generated annually in the United States, over 80% is combustible, yielding a heat value equivalent to about 1 million barrels of oil per day. This figure is equivalent to about 4.6% of all the fuel consumed by all utilities, 10% of all the coal consumed by all utilities, and about 20% of the electrical energy demand of the private sector of a municipality. Although these figures must obviously be adjusted to reflect the losses incurred in producing electricity, the use of refuse as a source of energy clearly has tremendous potential.

Refuse can be burned as is (the so-called *mass burn combustors*) or processed to produce *a refuse-derived fuel*. A picture of a waste-to-energy facility is shown in Figure 1-10. The refuse is dumped from the collection trucks into a pit that serves to mix and equalize the flow over the 24-hour period since such facilities must operate around the clock. A crane lifts the refuse from the pit and places it in a chute that feeds the furnace. The grate mechanism moves the refuse, tumbling it and forcing in air from the bottom as well as the top as the combustion takes place. The hot gases produced from the burning refuse are cooled by a bank of tubes filled with water. The hot water goes to a boiler where steam is produced. This steam can be used for heating and cooling, or for producing electricity in a turbine. The cooled gases are then cleaned by dry scrubbers, baghouses, electrostatic precipitators, and/or other control devices and discharged through a stack.

The more the solid waste is processed prior to its combustion, the better is its heat value and usefulness as a substitute for a fossil fuel. Such processing removes much of the noncombustible materials, such as glass and metals, and reduces the size of the paper and plastic particles so they burn more evenly. Refuse that has been so processed is called a *refuse-derived fuel* (RDF). The simplest form of RDF is shredding the solid waste to produce a more homogeneous fuel, without removing any of the metals or other noncombustibles. If the metals and glass are removed, the fuel is improved in its heat value and handling. This organic fraction can be further processed into pellets that are an excellent addition to coal in coal-fired furnaces.

One reason waste-to-energy facilities have not found greater favor is the concern with the emissions, but this concern seems to be misplaced. Studies have shown that the risks of MSW combustion facilities on human health are minimal, and with modern air pollution control equipment there should be no measurable effect of either the gaseous or particulate emissions.

Figure 1-10 Typical waste-to-energy facility for combusting municipal solid waste.

The refuse combustion option is, however, complicated by the problem of ash disposal. Although the volume of the refuse is reduced by over 90% in waste-to-energy facilities, the remaining 10% still has to be disposed of. Many special wastes, such as old refrigerators, cannot be incinerated and must either be processed to recover the metals or be landfilled. A landfill is therefore necessary even if the refuse is combusted, and a waste-to-energy plant is not an ultimate disposal facility. Refuse combustion ash concentrates the inorganic hazardous materials in refuse, and the combustion process may create other chemicals that may be toxic. Most solid waste ash is placed in municipal landfills, but other facilities have constructed their own landfills exclusively for ash disposal (so-called *monofills*). Ash has been increasingly used as a raw material for the production of useful aggregate for producing building materials, as an aggregate for road construction, or in other applications where a porous, predictable (and inexpensive) material is useful.[17]

THE NEED FOR INTEGRATED SOLID WASTE MANAGEMENT

Solid waste professionals recognize that issues related to managing solid waste must be addressed using a holistic approach. For example, if more waste is recycled, this can have a negative financial impact on the landfill because less refuse is landfilled. Since many landfill costs are fixed (there is a cost regardless if any refuse is landfilled), a drop in the incoming refuse can have severe economic ramifications. The various methods of solid waste management are therefore interlocking and interdependent.

Recognizing this fact, the EPA has developed a national strategy for the management of solid waste, called the *Integrated Solid Waste Management* (ISWM). The intent of this plan is to assist local communities in their decision making by encouraging those strategies that are the most environmentally acceptable. The EPA ISWM strategy suggests that the list of the most to least desirable solid waste management strategies should be

- Reducing the quantity of waste generated
- Reusing the materials
- Recycling and recovering materials
- Combusting for energy recovery
- Landfilling

That is, when an integrated solid waste management plan is implemented for a community, the first means of attacking the problem should be reducing the waste at the source. This action minimizes the impact of natural resource and energy reserves.

Reuse is the next most desirable activity, but this also has a minimal impact on natural resources and energy. Recycling is the third option, and should be undertaken when most of the waste reduction and reuse options have been implemented. Unfortunately, the EPA confuses recycling with recovery, and groups them together as meaning any technique that results in the diversion of waste. As previously defined, recycling is the collection and processing of the separated waste, ending up as new consumer product. Recovery is the separation of mixed waste, also with the end result of producing new raw materials for industry.

Reprinted by permission of United Media.

The decade of the 1990s has seen dramatic growth in recycling/recovery in the United States. For example, the State of North Carolina was recycling about 10% of residential and commercial solid waste in 1988, and ten years later was running at about 26% recycling rate. In 1988 there were only three curbside collection programs in North Carolina, and ten years later there were 260 such programs serving 3.4 million people, or 45% of the population of the state. In addition, 120 yard trimmings composting facilities process over 700,000 tons of material a year.[18]

The fourth level of the ISWM plan is solid waste combustion, which really should include all methods of treatment. The idea is to take the solid waste stream and to transform it into a nonpolluting product. This conversion may be by combustion, but other thermal and chemical treatment methods may eventually prove just as effective.

Finally, if all of the above techniques have been implemented and/or considered, and there is still waste left over (which there will be), the final solution is landfilling. At this time there really is no alternative to landfilling (except disposal in deep water—which is now illegal), and therefore every community must develop some landfilling alternative.

While this ISWM strategy is useful, it can lead to problems if taken literally. Communities must balance the above strategies to fit their local needs. This is where engineering judgment comes into play, and where the solid waste engineer really earns his/her salary. All the options have to be juggled and the special conditions integrated into the decision. The economics, history, politics, and aspirations of the community are important in developing the recommendations.

SPECIAL WASTES

In addition to the usual residential and commercial solid waste, municipal engineers have to deal with special or unusual waste. These wastes may include items such as inert material (materials that do not degrade, such as rocks), agriculture waste, sludge from both water and wastewater treatment facilities, tires, household hazardous waste, and medical waste. Each of these special wastes must be managed in a way that protects human health and welfare and complies with applicable regulations. While these issues are beyond the scope of this book, solid waste engineers must be prepared to manage these wastes.

FINAL THOUGHTS

As Kermit the Frog laments, "It's not easy being green." A solid waste engineer might modify this to "It's not easy being a green engineer." Often the engineer is placed in a role of the bad guy who, by doing the right thing, receives severe criticism from some segments of society. Sometimes the engineer is accused of appeasing the establishment and destroying our environment.

One alternative used by some engineers confronted with such harsh criticism is to withdraw into their professional shells. If they are only "hired guns," then what do they care what the outcome of any public decision is? They simply do what the client wants and do not voice their own opinions about societal or environmental values. The engineer can argue that the job is value free, and he or she is not making any value decisions. Technology, they argue, is value free.

But this is a classical cop-out. The fact is that engineers *do* decide policy in many ways. On a local and mundane level, engineers make seemingly minor decisions daily that affect all of us. For example, the routing of a local highway across a stream that may be a rare habitat for the endangered Venus fly trap, or prohibiting the use of asphalt containing rubber from old tires are both engineering decisions that will get no press whatever. And on a national level the mere initiation of a major project will result in a momentum that will be difficult to reverse. Thus it is not enough simply to defer engineering decisions to public comment or environmental oversight. The engineers need to introduce environmental concerns into the planning of projects before the plans become public documents. The decisions of the engineers do make a difference.

Private and governmental clients, if they are to get things done, must use engineering know-how. If a community wants to construct a landfill, it *has* to use engineers to do it. Other professionals—such as sociologists, philosophers, epidemiologists, or even planners—are useful, but at the end of the day, the engineers are the ones who actually *design* the landfill. Thus the engineer has the responsibility of viewing his/her role in the broadest sense—to introduce alternatives, concepts, and values that the client may otherwise never consider. For example, if a city hires an engineering firm to design a landfill, it is possible for the engineer to ignore all the aspects of the job except the determination of the best (safest and least expensive) design. What we are suggesting is that the engineer, recognizing that there are many conflicting values involved in such a project, and that it is unlikely that the municipality will seek expert assistance in the resolution of such conflicts, has a responsibility to introduce the value (ethical) questions into the project.

A truly professional engineer will infuse ethics into his/her decision making, and with the increasing pressure on the natural environment, a growing population, and accelerated technological development, environmental ethics will play an ever-increasing role in the engineer's professional responsibilities to society. The engineer *can*, and *should*, make a difference.

REFERENCES

1. Melosi, M. V. 1981. *Garbage in the Cities*. College Station, Tex.: Texas A&M Press.
2. Mumford, L. 1961. *The City in History*. New York: Harcourt, Brace & World.
3. "Trash Timeline: 1,000 Years of Waste." *Waste News*, www.wastenews.com/features.
4. Meadows, D. H., et al. 1972. *The Limits of Growth*. Washington, D.C.: Potomac Associates.
5. Boughey, A. S. 1976. *Strategy for Survival*. Menlo Park, Calif.: W. A. Benjamin, Inc.
6. Kesler, S. E. 1976. *Our Finite Mineral Resources*. New York: McGraw-Hill Book Co.
7. Cross, J. F. 1992. "Pollution Prevention and Sustainable Development." *Renewable Resources Journal* 5: 13–17.
8. Conn, W. D. 1995. "Reducing Municipal Solid Waste Generation: Lessons from the Seventies." *Journal of Resource Management and Technology* 16: 24–27.
9. Sherman, S. 1991. "Local Government Approaches to Source Reduction." *Resource Recycling* 9: 112 and 119.
10. Lober, D. 1996. "Municipal Solid Waste Policy and Public Participation in Household Source Reduction." *Waste Management and Research* 14: 129–145.
11. Goldberg, D. 1990. "The Magic of Volume Reduction." *Waste Age* 21: 98–104.
12. Skumatz, L. A. 1991. "Variable Rates for Solid Waste Can Be Your Most Effective Recycling Program." *Journal of Resource Management and Technology* 19: 1–12.
13. City of Greensboro. 1992. *Waste Stream Characterization Project*. HDR Engineering (quoted in Lober, 1996).
14. *Shop Earth Smart: A Consumer Guide to Buying Products That Help Conserve Money and Reduce Waste*. 1999. The City and County of San Luis Obispo, Calif.
15. Franklin Associates. 1999. *Characterization of Municipal Solid Waste in the United States: 1998 Update*. EPA Report No. EPA 530.
16. Dair, F. R., and R. E. Schwegler. 1974. "Energy Recovery from Landfills." *Waste Age* 5: 6.
17. Brown, H. 1997. "Ash Use on the Rise in the United States." *World Waste*: p. 16.
18. Glenn, J. 1998. "The State of Garbage in America." *BioCycle* (April) 32.
19. Daigger, G. 1992. Quoted in "The New Environmental Age." *Engineering News Record* 228, n. 25 (22 June).

ABBREVIATIONS USED IN THIS CHAPTER

CERCLA = Comprehensive Environmental Response, Compensation, and Liability Act
EIS = environmental impact statement
EPA = Environmental Protection Agency
HDPE = high-density polyethylene
ISWM = Integrated Solid Waste Management
LDPE = low-density polyethylene
MRF = materials recovery facility
MSW = municipal solid waste

NEPA = National Environmental Policy Act
PETE = polyethylene terephthalate
PP = polypropylene
PS = polystyrene
PVC = polyvinyl chloride
RCRA = Resource Conservation and Recovery Act
RDF = refuse-derived fuel
VOC = volatile organic compound

PROBLEMS

1-1. The objective of this assignment is to evaluate and report on the solid waste you personally generate. For one week, selected at random, collect all of the waste you would have normally discarded. This includes food, newspapers, beverage containers, etc. Using a scale, weigh your solid waste and report it as follows:

Component	Weight	Percent of Total Weight
Paper		
Plastics		
Aluminum		
Steel		
Glass		
Food		
Other	_____	_____
		100

Your report should include a data sheet and a discussion. Answer the following questions.

a. How do your percentage and total generation compare to national averages?

b. What in your refuse might have been reusable (as distinct from recoverable), and if you had reused it, how much would this have reduced the refuse?

c. What in your refuse is recoverable? How might this be done?

1-2. One of the least studied aspects of municipal refuse collection is the movement of the refuse from the household to the truck. Suggest a method by which this can be improved. Originality counts heavily here, practicality far less.

1-3. Plastic bags at food stores have become ubiquitous. Often recycling advocates point to the plastic bags as the prototype of wastage and pollution, as stuff that clogs up our landfills. In retaliation, plastic bag manufacturers have begun a public relations campaign to promote their product. On one of the flyers (printed on paper) they say:

> The (plastic) bag does not emit toxic fumes when properly incinerated. When burned in waste-to-energy plants, the resulting by-products from combustion are carbon dioxide and water vapor, the very same by-products that you and I produce when we breathe. The bag is inert in landfills where it does not contribute to leaching bacterial or explosive gas problems. The bag photodegrades in sunlight to the point that normal environmental factors of wind and rain will cause it to break into very small pieces, thereby addressing the unsightly litter problem.

Critique this statement. Is all of it true? If not, what part is not? Is anything misleading? Do you agree with their evaluation? Write a one-page response.

1-4. The siting of landfills is a major problem for many communities. This is often an exasperating job for engineers because the public is so intimately involved. A prominent environmental engineer, Glenn Daigger of CH2M Hill, is quoted as follows:

> Environmental matters are on the front page today because we, the environmental industry, are not meeting people's expectations. They're telling us that accountability and quality are not open questions that ought to be considered. It's sometimes difficult to grasp in the face of all the misinformation out there. Ultimately, we're responsible for accommodating the public's

point of view, not the other way around.[19]

Do you agree with him? Should the engineer "accommodate the public's point of view," or should the engineer impose his/her own point of view on the public since the engineer has a much better understanding of the problem? Write a one-page summary of what you believe should be the engineer's role in the siting of a landfill for a community.

1-5. Suppose you are the engineer employed by a small community, and the town council tells you to design a "recycling" program that will achieve at least 50% diversion from the landfill. What would be your response to the town council? If you agree to try to achieve such a diversion, how would you do it? For this problem you are to prepare a formal response to the town council, including a plan of action and what would be required for its success. Numbers are mandatory. Make up data as needed. This response would include a cover letter addressed to the council and a report of several pages, all bound in a report format with title page and cover.

1-6. Same scenario as Problem 1-5, but now you are asked to develop a zero waste program. That is, the community is to produce no solid waste whatsoever. Respond with a letter to the town council.

1-7. How might you use the principles of waste reduction in doing your food shopping at the food store? Name at least three specific ways in which you might be able to reduce waste in the purchase of your groceries.

1-8. We often see packaging labeled "Made from 100% recycled materials" or "Made from 50% recycled materials." The objective is, of course, to make you (the consumer) believe that the company is environmentally conscious and caring, and thus to make you buy more of their products.

a. Why are such statements as the ones quoted above potentially misleading? What questions would you want to ask the company to determine if they truly are helping with materials recycling?

b. All things considered, if the statements are true, why ought you to buy their products in preference to products with no recycled material? Be specific.

c. If you were working for the company, how would you write the statement ("Made from...") to be more accurate?

1-9. Give three examples of waste reduction that you might be able to implement in your everyday life.

1-10. Consider the meaning and purpose of the various recycling symbols shown in Figure 1-11 on the next page.

a. Which are used solely as marketing tools, and which tell you what you want to know if your intent is to buy recycled materials?

b. Find and cut out a recycling symbol from some product and attach it to your paper. Discuss its purpose and integrity.

1-11. If you were responsible for marketing the paper collected by your city's curbside collection program, what would you tell the purchasing department if they asked you whether the city should buy copy paper that was (1) 75% recycled content (55% pre-consumer and 20% post-consumer) or (2) 50% recycled content paper (10% pre-consumer and 40% post-consumer)? Why?

1-12. What is the content of recycled fiber in the paper used by your school/college/department? What are the goals established by the EPA and your state?

(You can find this for the EPA at www.epa.gov.)

1-13. Of the 6 types of plastic shown in Figure 1-4, how many can be recycled in your community? If a type of plastic is not recyclable, do you think it is misleading (ethical) for the plastic manufacturers to place the chasing arrow design on the container?

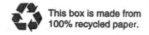

Figure 1-11 *Various recycling symbols.*

2

Municipal Solid Waste Characteristics and Quantities

A great deal of confusion exists in the definition of solid waste, and this leads directly to disagreements on the estimated quantities and composition of solid waste. There are both gross and subtle differences in the types and sources of such material, and indeed a question as to what is and is not solid.

DEFINITIONS

This book is concerned only with the first form of these solid wastes, *municipal solid waste* or *MSW*, which can be further defined as having the following components:

- Mixed household waste
- Recyclables, which may not be limited to the following:
    Newspapers
    Aluminum cans
    Milk cartons
    Plastic soft drink bottles
    Steel cans
    Corrugated cardboard
    Other material collected by the community
- Household hazardous waste
- Commercial waste
- Yard (or green) waste
- Litter and waste from community trash cans
- Bulky items (refrigerators, rugs, etc.)
- Construction and demolition waste

Often these wastes are defined by the way they are collected. Commonly, the mixed household wastes are collected by trucks specially built for that purpose, and the recyclables are collected either with the mixed waste but in a separate compartment or by other vehicles built for that purpose. Yard waste collected may also be mixed in with the household waste, or placed separately in a dedicated vehicle. Commercial wastes use large containers that are emptied into specially built trucks. Construction and demolition wastes are collected in roll-off stationary containers that remain on the job site until full. Bulky items are commonly collected on an "as needed" basis by larger trucks capable of handling large items. Household hazardous wastes are collected periodically by the community, or are taken to hazardous waste collection centers by the homeowners (Figure 2-1).

For many reasons it is convenient to define *refuse* as

- Solid waste generated by households, including mixed nonsorted waste
- Recyclables (whether or not they are collected separately)
- Household hazardous wastes if these are not collected separately
- Yard (or green) waste originating with individual households
- Litter and community trash because the material is produced by individuals
- Commercial waste because it often contains many of the same items as household waste

By our definition, refuse does *not* include

- Construction and demolition debris
- Water and wastewater treatment plant sludges
- Leaves and other green waste collected from community streets and parks
- Bulky items such as large appliances, hulks of old cars, tree limbs, and other large objects that often require special handling

This text is devoted mainly to the collection, disposal, and recovery of materials and energy from *refuse*. The other solid wastes—such as construction and demolition (C & D) wastes—are mentioned only in passing. This slight should not be

Figure 2-1   Household hazardous waste collection facility.

taken as any indication that these wastes are either small in quantities or unimportant in defining their disposal options. C & D wastes, for example, are commonly landfilled in a separate section of a landfill, and these wastes do not require daily cover. Leaves, in some parts of the United States, represent a serious problem to communities during the autumn, and composting facilities are often developed to handle this seasonal load. In the discussions that follow, however, the material of interest is the fraction of MSW previously defined as refuse.

In summary:

(MSW) = (refuse) + (C & D waste) + (sludge) + (leaves) + (bulky items)

Refuse can be defined in terms of *as generated* and *as collected* solid waste. The refuse generated includes all of the wastes produced by a household—whatever is no longer wanted and is to be gotten rid of. Often some part of the refuse, especially organic matter and yard waste, is composted on premises. The fraction of refuse that is generated but not collected is called *diverted* refuse. The *as generated* refuse is always larger than the *as collected* refuse, and the difference is the *diverted* refuse.

In summary:

(as generated refuse) = (as collected refuse) + (diverted refuse)

Sometimes diversion is defined on the basis of MSW instead of refuse. When defined in this way, diverted MSW is that fraction of MSW that is generated but does not find its way to the landfill. The objective is to increase the life of a landfill or to reduce the cost of disposal. One major diversion is the collection of recyclables (aluminum cans, newspaper, etc.) that can be sold on the secondary materials market. The EPA has challenged communities to increase their diversion to 35%, up from the 25% originally established in 1988. California has set an even higher goal of 50% diversion.

The calculation of diversion in terms of either refuse or MSW is important and controversial. Recall that MSW includes refuse plus C & D debris, sludges, leaves, and bulky items. When communities are under state mandates to increase the recycling of consumer products, the calculation often is made using the MSW as the denominator instead of the refuse, thereby achieving large diversion rates. This is demonstrated by the example below.

EXAMPLE 2-1

A community produces the following on an annual basis:

Fraction	Tons per year
Mixed household waste	210
Recyclables	23
Commercial waste	45
Construction and demolition debris	120
Treatment plant sludges	32
Leaves and miscellaneous	4

The recyclables are collected separately and processed at a materials recovery facility. The mixed household waste and the commercial waste go to the landfill, as do the leaves and miscellaneous solid wastes. The sludges are dried and applied on land (not into the landfill), and the C & D wastes are used to fill a large ravine. Calculate the diversion.

If the calculation is on the basis of MSW, the total waste generated is 434 tons per year. If everything not going to the landfill is counted as having been diverted, the diversion is calculated as

$$\frac{23 + 120 + 32 + 4}{434} \times 100 = 41\%$$

This is an impressive diversion. But if the diversion is calculated as that fraction of the *refuse* (mixed household and commercial waste) that has been kept out of the landfill by the *recycling* program, the diversion is

$$\frac{23}{210 + 23 + 45} \times 100 = 8.3\%$$

This is not nearly as impressive, but a great deal more honest.

MUNICIPAL SOLID WASTE GENERATION

Table 2-1 lists some other solid wastes generated in the United States and places the production of MSW into perspective.

Table 2-1   Generation of All Types of Solid Waste in the United States, 1998

Source	Millions of tons annually
Agricultural	250
Industrial	400
Mining	1,200
Municipal solid waste	246
Refuse	208
C & D	30
Sludge	8

Source: (1)

While refuse is not the largest quantity, it is one that will require the greatest ingenuity in transport and disposal. On a national level the quantity of refuse generated has increased markedly over the past 40 years for which data are available. Figure 2-2 shows the change in quantities of various refuse components generated in the United States. Note the dramatic increase in plastics as a major component in solid waste. On a per capita basis waste generation has also increased, as shown in Figure 2-3, although there are indications that the per capita generation seems to be leveling off.

The generation of refuse in a community also varies throughout the year. Figure 2-4 shows the average generation rate of refuse as measured at five different landfills in Wisconsin[2] and waste generation in New Orleans.[3] In Wisconsin the very cold months of the winter result in a low generation rate during January and February, while the New Orleans data show little seasonal variation, as can be expected.

In addition to seasonal variations, refuse generation varies with the day of the week, with Mondays being typically the heaviest days and Fridays the lightest, but this depends on the collection program for the community. Some periods of the year can have as much as 130% of the average, while others are as low as 80% of the average refuse generated. Figure 2-5 is a summary of the variability of refuse generation based on U.S. national averages. For nearly 95% of the year (49 of 52 weeks) the refuse generation does not exceed 15% of the annual mean.

Collection frequency also affects the production of refuse.[4] Generally, the more frequent the collection, the more MSW is produced. Apparently, if the frequency of service is not sufficient, citizens will find other, perhaps less desirable, means of solid waste disposal. On the other hand, this might indicate that more waste diversion activities occur when collection is less frequent.

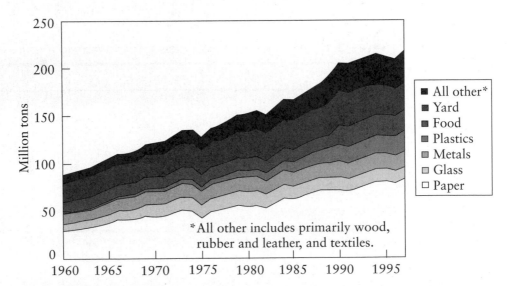

Figure 2-2 Historical trends in municipal solid waste generation and composition in the United States. Source: (1)

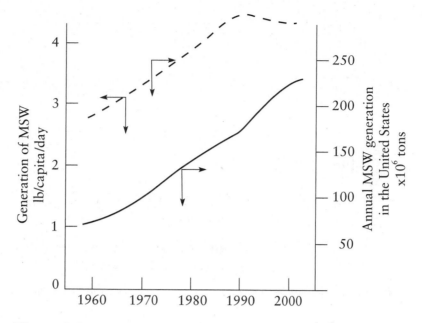

Figure 2-3 Historical trends in municipal solid waste as per capita generation. Source: (1)

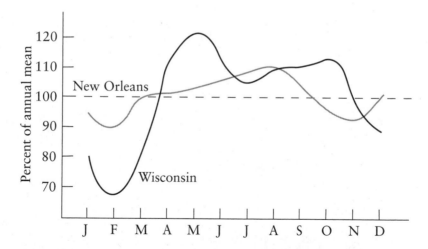

Figure 2-4 Monthly variation in the generation of municipal refuse in Wisconsin and New Orleans. Source: (2, 3)

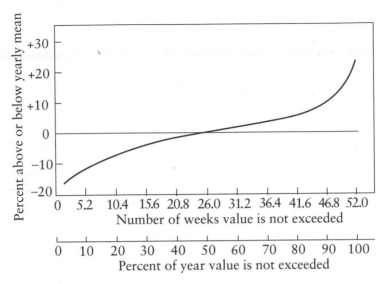

Figure 2-5 Weekly variation in refuse generation. Source: (3)

Income and affluence tend to have a positive effect on refuse generation with the logic that the more expendable income a household has, the more they tend to throw away. Wealthier people tend to read more and have greater amounts of paper wastes. If income can be correlated with the number of people per dwelling, then there seems to be a positive correlation with more refuse generated by people who live in single-family residences than those who live in apartment houses, as shown in Figure 2-6.

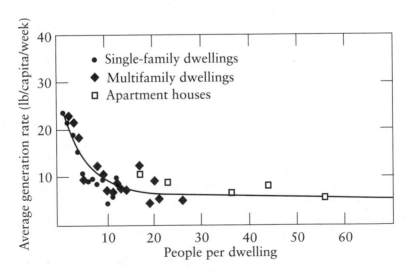

Figure 2-6 Solid waste generation as a function of number of people per dwelling in low-income areas. Source: (5)

On the other hand, more affluent people eat less canned foods and purchase less wasteful packaging. Higher incomes also suggest that all adults work, and this results in the use of restaurants for meals. Thus it is possible to argue that the higher the income level, the lower will be waste generation.[5, 6] Perhaps these variables cancel each other out, and there is little if any effect of affluence on solid waste generation, as found by several researchers.[7, 8, 9]

The effect of population density on solid waste generation is still uncertain. One study (Figure 2-7) found that higher population densities (in terms of people per square mile) can be correlated with higher refuse generation. On the other hand, refuse production from North Carolina counties does not seem to vary with population[10] as shown in Figure 2-8. The disagreement between the two figures may be resolved by noting that in Figure 2-7 the solid waste production is fairly constant for all communities with populations less than about 2000 people per square mile. Figure 2-8 represents North Carolina counties, many of which have low population densities, and thus the refuse production would not be expected to vary greatly.

Variables that do seem significantly to affect the rate of solid waste production are the cost of disposal and the retail sales. The model that had the best fit for North Carolina counties [modified from (11)] is

$$W = 3.725 - 0.34T + 0.323R + 0.059V + 0.227C$$

where W = refuse collected, lb/capita/day
T = disposal fee (or commonly called *tipping fee*), \$/ton
R = per capita retail sales, \$1000/capita
V = value added by manufacturing, \$1000/capita
C = construction in regions, \$1000/capita

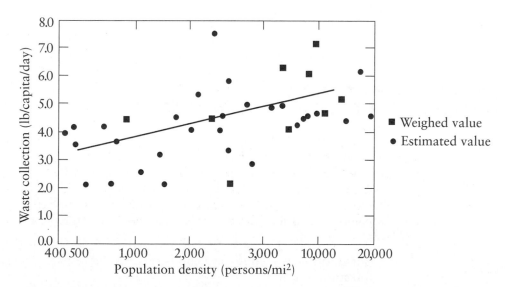

Figure 2-7   Solid waste generation as a function of population density.
Source: (9)

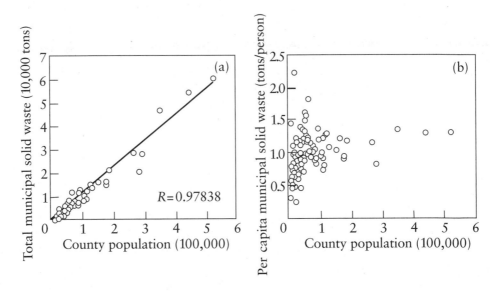

Figure 2-8 *Solid waste generation as a function of population for North Carolina counties. Source: (11)*

The conclusion is therefore that waste generation is governed by the tipping fee (negatively) and retail sales (positively), the amount of manufacturing (positively), and the amount of ongoing construction (positively). This equation should be applied with care to other locations. The data are from mostly rural counties and may not be useful in urban areas.

Nevertheless, an intriguing question is: What would happen in a county that had a lot of retail sales, manufacturing, and construction, and at the same time raised its tipping fees? Would the solid waste then be disposed of surreptitiously? Obviously, individual dumps or illicit backyard burning are not counted in this model because the model is for refuse *collected* not refuse *generated*.

MUNICIPAL SOLID WASTE CHARACTERISTICS

As long as the MSW is to be disposed of by landfill, there is little need to analyze the waste much further than to establish the tons of waste generated and perhaps consider the problems of special (hazardous) materials. If, however, the intent is to collect gas from a landfill and put it to some beneficial use, the amount of organic material is important. When recycling is planned, or if materials or energy recovery by combustion is the objective, it becomes necessary to have a better picture of the solid waste. Some of the characteristics of interest are

Composition by identifiable items (steel cans, office paper, etc.)
Moisture content
Particle size
Chemical composition (carbon, hydrogen, etc.)
Heat value

Density
Mechanical properties
Biodegradability

Composition by Identifiable Items

On a national level, data from published industry production statistics can be used for estimating waste composition. This method is called the *input method* of estimating solid waste production. For example, the annual production of glass is about 12,830,000 tons (11,640,000 tonnes) annually, and we can safely assume that all of this will end up (sooner or later) in waste that is either disposed of or processed for materials recovery. The 1998 EPA estimates of national waste composition based on such data are shown in Table 2-2. The last line in the table indicates an annual national solid waste production of 217 million tons (196 million tonnes)—or on a personal level, a contribution of about 4.3 lb/capita/day (1.6 kg/capita/day).

Table 2-2   Generation of Municipal Solid Waste Components in the United States, 1998

Item	Weight generated (millions of tons)	Percent
Paper and paperboard	81.5	37.6
Glass	12.0	5.5
Ferrous metals	12.3	5.7
Aluminum	3.0	1.3
Other nonferrous metals	1.3	0.6
Plastics	21.5	9.9
Rubber and leather	6.6	3.0
Textiles	8.2	3.8
Wood	11.6	5.3
Other materials	3.8	1.8
Food waste	21.9	10.1
Yard trimmings	27.7	12.8
Miscellaneous inorganic	3.3	1.5
Total	217	100

Source: (1)

The input method of estimating solid waste generation is applicable where the input figures can be obtained from specialized agencies which routinely collect and publish industry-wide data. This system also allows for regular updates of waste generation estimates because gathering the data is expensive. Further, since the data collected by the same institutions include future projections, it is possible to estimate future solid waste generation. The numbers in Table 2-2 were generated using this procedure. National averages are of limited value, however, because they can rarely be used with any degree of precision for local or regional purposes. On the local level the only reliable method of estimating refuse composition and production is to use an *output method* of analysis and perform sampling studies.

Sampling studies for characterizing refuse must be designed so as to produce the most useful and accurate data for the least cost and effort. The two variables of importance in designing such a study are sample size and method of characterizing the refuse. Although manual sampling is still the only truly reliable way of estimating composition, other techniques—such as photogrammetry—hold promise for the future.

Measuring the composition of a totally heterogeneous material, such as mixed municipal refuse, is not a simple task, but some determination of its components is necessary if the various fractions are to be separated and recovered. Some authorities, however, suggest that since a major effort is required to establish the composition with reasonable accuracy, it is often not worth the trouble and expense, and a national average can be used. Composition studies should be used where accurate data are absolutely required for estimating the economics of future solid waste management alternatives.

Measuring Composition by Manual Sampling

The sampling plan drives the waste composition study. Even if the sampling procedure is performed adequately, it means nothing unless the plan can produce valid results. First, the waste has to be accurately represented through proper load selection so as not to bias the final analysis. The truckload to be analyzed has to represent as closely as possible the average production of refuse in the community.

Once the load has been selected, a methodology for producing a sample small enough to be analyzed but big enough to be statistically representative of the MSW must be established. The most frequently used methodology for determining the number of samples required in order to achieve statistical validity is the American Society for Testing and Materials (ASTM) *Standard Test for Determination of the Composition of Unprocessed Municipal Solid Waste* (ASTM designation D 5231-92).[12] This method provides a script to follow when conducting a waste composition study, including a statistically based method for determining the number of samples required to characterize the waste. The number of samples required to achieve the desired level of measurement precision is a function of the component(s) under consideration and the desired confidence level. The calculations are an interactive process, beginning with a suggested sample mean and standard deviation for waste components. Typically, a 90% confidence level is adequate for most studies.[13] As a crude first estimate, sorting and analyzing more than 200 lb (90 kg) in each sample would have little statistical advantage.[14] The question is how many of the 200-lb samples are necessary for the testers to feel statistically confident in the results.

To obtain representative 200-lb (90-kg) samples, ASTM recommends *quartering* and *coning*. Quartering is the separation of a truckload of waste into successive quarters, after thoroughly mixing the contents with a front-end loader. The samples

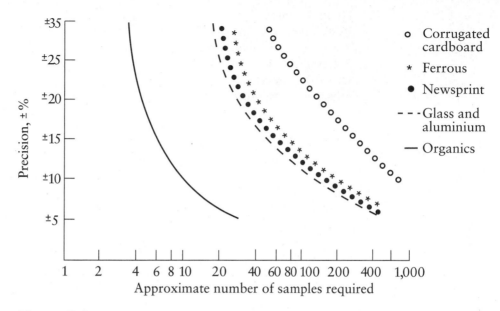

Figure 2-9 *Approximate number of 200-lb samples required to achieve desired precision. Source: (15)*

are then coned again and quartered again until they are about 200 lb (90 kg). The greater the desired precision, the greater will be the number of 200-lb samples analyzed. Figure 2-9, based on the work of Klee and Carruth,[15] represents a good first estimate of the sample sizes required for acceptable precision. For example, if 50 samples of 200 lb each are processed, then the expected precision for the organic fraction would be better than ±5%, the precision for newsprint, aluminum, and ferrous would be about ±15%, and the precision for corrugated cardboard would be only about ±25%. As would be expected, the larger the articles, the more is required to achieve acceptable precision.

It is important to decide early in the sampling program what is to be measured, that is, how many categories of waste are to be used. This decision, of course, depends on what the data are to be used for. In one study that was conducted to estimate the possibility of diverting waste from the landfill through prevention and recycling, 43 different categories were established.[16]

    Paper
        Newsprint
        Magazines
        Corrugated cardboard
        Telephone books
        Office/computer paper
        Other mixed paper
    Plastics
        PETE bottles

> HDPE bottles
> PVC containers
> Polypropylene containers
> Polystyrene
> Other containers
> Films/bags and other rigid plastics

Organics
> Food waste
> Textiles/rubber/leather
> Fines (unidentifiable small organic particles)
> Other organics

Ferrous materials
> Ferrous/bimetal cans
> Empty aerosols
> Other ferrous metals

Nonferrous metals
> Aluminum cans
> Other nonferrous metals

Electronic components
> Parts and materials from computers
> Printers
> Copy machines

Glass
> Clear
> Green
> Brown
> Other glass

Wood
> Lumber
> Pallets
> Other wood

Inerts
> Asphalt roofing materials
> Concrete/brick/rock
> Sheet rock
> Ceiling tiles
> Dirt/dust/ash and other inerts

Yard waste
> Grass clippings
> Leaves
> Trimmings

Hazardous materials
> Lead acid batteries
> Other batteries
> Other hazardous wastes

Such a sampling study obviously would be very cumbersome, extended, and expensive, and it is questionable if such detail is really necessary in most cases. Based on the intended use of the study (e.g., construction of a materials recovery facility), a more pragmatic listing of components for a sampling study might be more useful and certainly less expensive, might be the following:

Paper
    Newsprint
    Corrugated cardboard
    Magazines
    Other paper
Metal
    Aluminum cans
    Steel cans
    Other aluminum
    Other ferrous
    Other nonferrous
Glass
    Clear (flint)
    Green
    Brown
Plastic
    HDPE
    PETE
    Other plastics
Yard wastes
    Wood (branches and lumber)
    Leaves and clippings
Food waste
Other
    Materials that either have little recovery potential or are of low fraction in refuse, such as rubber, ceramics, other glass, bricks, rocks, etc.

Even careful sampling studies yield imprecise information because of the nature of refuse. Not all items can be readily categorized into the desired components. For example, a tin can with an aluminum top and paper wrapper has four components: steel, tin, aluminum, and paper. Regardless of the final classification of this item, inaccuracies are introduced into the final values.

Common contaminants of waste items include moisture, food, and dirt. Although these materials are normal components of the waste stream, they are of concern when they add significantly to the weight of paper, plastic film, yard waste, and containers. During placement in collection vehicles, the contamination increases when waste is squeezed, causing materials to smear or stick together and forcing moisture from food and other wet wastes into other absorbent items. In addition, contamination can occur during sampling and storing as a result of mixing and/or inclement weather.

After sorting, the samples should be taken to a laboratory where they can be weighed, cleaned of contamination, and air dried. Durable items, such as glass and plastic containers, can be washed prior to air drying, and filled containers can be emptied of their contents. If the contaminant category can be identified (for example, food), each category should be properly adjusted for the weight of contamination.

Waste composition studies are essential tools for municipal solid waste management. However, because of a lack of consistent procedure and underfunding of studies, the data provided are frequently inaccurate and imprecise. Too often an insufficient number of samples is obtained, sampling events are not representative of seasonal and economic changes, contamination is not accounted for, and the study is not repeated in response to changes in the community. A poor study may be worse than no study at all if the numbers obtained are to be used for design purposes.

Measuring Composition by Photogrammetry

The difficulty, not to say the less-than-appealing nature of the work, of manual sampling has for some time driven the need for a better and safer way of obtaining solid waste composition data. One of the candidate procedures as a substitute for manual sampling is a photogrammetric technique that involves photographing a representative portion of refuse and analyzing the photograph.[17]

The photograph should be taken directly at the refuse (90° angle) with a wide-angle lens, using 35-mm color slide film, with electronic flash fill to eliminate shadows. The 2×2 slide is then projected onto a screen that has been divided into about 10×10 grid blocks. This grid can either be drawn on a piece of poster board or be a negative transparency projected from an overhead projector. The components in each grid intersection are then identified and tabulated. Using predetermined bulk densities (which include interior space, e.g., in a beverage can), the fraction by weight is then calculated. Some of the bulk densities are listed in Table 2-3.

The photogrammetric technique suffers from two disadvantages. First, its accuracy is dependent on bulk density figures that must be fine tuned. However, some practitioners found this technique to be quite accurate. The second disadvantage is that the time required to analyze one picture is substantial, and it may be faster to simply classify manually and weigh the samples. The technique does have one important advantage—the refuse need not be touched or smelled, and thus there are no problems with disease transmission.

Photogrammetry might also be useful for wastes other than refuse. Construction and demolition wastes, for example, have a limited number of components, and taking pictures of large piles of C & D wastes may be far more effective (not to say safer) than sorting by hand.

Table 2-3   Bulk Densities of Some Refuse Components

Components	Condition	Bulk density (lb/yd^3)*
Aluminum cans	Loose	50–74
	Flattened	250
Corrugated cardboard	Loose	350
Fines (dirt, etc.)	Loose	540–1600
Food waste	Loose	220–810
	Baled	1000–1200
Glass bottles	Whole bottles	500–700
	Crushed	1,800–2,700
Magazines	Loose	800
Newsprint	Loose	20–55
	Baled	720–1000
Office paper	Loose	400
	Baled	700–750
Plastics	Mixed	70–220
	PETE, whole	30–40
	Baled	400–500
	HDPE, loose	24
	Flattened	65
Plastic film and bags	Baled	500–800
	Granulated	700–750
Steel cans	Unflattened	150
	Baled	850
Textiles	Loose	70–170
Yard waste	Mixed, loose	250–500
	Leaves, loose	50–250
	Grass, loose	350–500

*To obtain kg/m^3, multiply by 0.59.

Sources: (18, 19, 20, 21)

A more complete list of bulk densities, useful in calculating recycling and waste reduction, is provided in the Appendix.

Moisture Content

A transfer of moisture takes place in the garbage can and truck, and thus the moisture content of various components changes with time. Newsprint has about 7% moisture by weight as it is deposited into the receptacle, but the average moisture content of newsprint coming from a refuse truck often exceeds 20%. The moisture content becomes important when the refuse is processed into fuel or when it is fired directly. The usual expression for calculating moisture content is

$$M = \frac{w - d}{w} \times 100$$

where M = moisture content, wet basis, %
w = initial (wet) weight of sample
d = final (dry) weight of sample

Some engineers define moisture content on a dry weight basis, or

$$M_d = \frac{w - d}{d} \times 100$$

where M_d = moisture content, dry basis, %

This relationship seems at first irrational because the moisture content can exceed 100%, but in some fields, such as geotechnical engineering, moisture content on a dry basis is useful. In this text moisture is always expressed on a wet basis unless otherwise indicated.

Drying is usually done in an oven at 77°C (170°F) for 24 h to ensure complete dehydration and yet avoid undue vaporization of volatile material. Temperatures above this will melt some plastics and cause one unholy mess.

The moisture content of various refuse components varies widely, as shown in Table 2-4.

The moisture content of any waste can be estimated by knowing the fraction of various components and using either measured values of moisture content or typical values from a list such as Table 2-4. This calculation is illustrated in Example 2-2.

EXAMPLE
2-2

A residential waste has the following components:

Paper	50%
Glass	20%
Food	20%
Yard waste	10%

Estimate its moisture concentration using the typical values in Table 2-4.

Assume a wet sample weighing 100 lb. Set up the tabulation:

Component	Percent	Moisture	Dry weight (based on 100 lb)
Paper	50	6	47
Glass	20	2	19
Food	20	70	6
Yard waste	10	60	4
			Total: 76 lb dry

The moisture content (wet basis) would then be

$$M = \frac{w-d}{w}(100) = \frac{100-76}{100}(100) = 24\%$$

Table 2-4   Moisture Content of Uncompacted Refuse Components

Component	Moisture content	
	Range	Typical
Residential		
Aluminum cans	2–4	3
Cardboard	4–8	5
Fines (dirt, etc.)	6–12	8
Food waste	50–80	70
Glass	1–4	2
Grass	40–80	60
Leather	8–12	10
Leaves	20–40	30
Paper	4–10	6
Plastics	1–4	2
Rubber	1–4	2
Steel cans	2–4	3
Textiles	6–15	10
Wood	15–40	20
Yard waste	30–80	60
Commercial		
Food waste	50–80	70
Mixed commercial	10–25	15
Wood crates and pallets	10–30	20
Construction (mixed)	2–15	8

Source: (20 based on 21)

Typically, the moisture content of loose refuse is about 20% if there have not been rainstorms before collection. During rainy weather, the moisture content can go as high as 40%.

In a refuse truck *moisture transfer* takes place, and the moisture of various components of refuse changes. Paper sops up much of the liquid waste, and its moisture increases substantially. The moisture content of refuse that has been compacted by a collection truck is therefore quite different from the moisture of various components as they are in the can ready for collection.

Particle Size

Any mixture of particles of various sizes is difficult to describe analytically. If these particles are irregularly shaped, the problem is compounded. Municipal refuse is possibly the worst imaginable material for particle size analysis, and yet much of the MSW processing technology depends on an accurate description of particle size.

No single value can adequately hope to describe a mixture of particles. Probably the best effort in that direction is to describe the mixture by means of a curve showing percent of particles (by either number or weight) versus the particle size. The curve can be plotted by unit intervals, as shown in Figure 2-10. The two mixtures shown in these curves have very different particle size distributions. Mixture **A** has mainly uniformly sized particles, while Mixture **B** has a wide variability in particle size, and yet the *average particle size*, defined as that diameter where 50% of the particles (by weight) are smaller than—and 50% are larger than—this diameter, is identical for both mixtures.

Particle diameter of nonspherical particles can be defined in any number of ways, and no one of these definitions of diameter is the "correct" one. Only for spherical particles does the term "diameter" have a strict geometrical meaning. Expressions for defining "diameter" for nonspherical particles is discussed further in the appendix to this chapter.

Although the most accurate expression of particle size distribution is graphical, several mathematical expressions have been suggested. For example, in water engineering, the particle size of filter sand is expressed using the *uniformity coefficient*, defined as

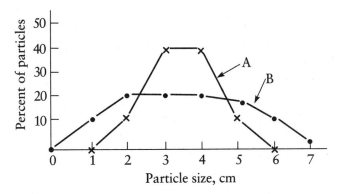

Figure 2-10   Particle-size distribution curves for two mixtures of particles.

$$UC = \frac{D_{60}}{D_{10}}$$

where UC = uniformity coefficient

D_{60} = particle (sieve) size where 60% of the particles are smaller than that size

D_{10} = particle (sieve) size where 10% of the particles are smaller than that size

Figure 2-11 shows a particle size distribution based on a sieve analysis for a raw (as collected) refuse for a study conducted at Pompano Beach, Florida.[22] The results of a study performed in two New England towns, illustrated in Figure 2-12, show a similar particle-size distribution by the ability to pass a sieve.[23]

Chemical Composition

The economic recovery of materials and/or energy often depends on the chemical composition of the refuse—the individual chemicals as well as the heat value. Two common means of defining the chemical composition of refuse are the *proximate analysis* and the *ultimate analysis*. Both descriptions were originally developed for solid fuels, especially coal. The proximate analysis is an attempt to define the fraction of volatile organics and fixed carbon in the fuel, while the ultimate analysis is based on elemental compositions. Some data for both proximate and ultimate analysis published by EPA are tabulated in Table 2-5. These data are of limited value for design purposes because once again the heterogeneous nature of refuse and its variability with geography and with time result in wide ranges. Accurate information for a specific refuse can be attained only by concerted sampling and analysis.

Heat Value

The heat values of refuse are of some importance in resource recovery. Some published values for several fuels are shown in Table 2-6 to illustrate the variability of the fuels according to how they are derived. The production of RDF (refuse-derived fuel) is described further in Chapter 7.

In common American engineering language, heat value is expressed as Btu/lb of refuse, while the proper SI designation is kJ/kg. Commonly the heat values of refuse and other heterogeneous materials are measured with a *calorimeter*, a device in which a sample is combusted and the temperature rise is recorded (see Chapter 7). Knowing the mass of the sample and the heat generated by the combustion, the Btu/lb is calculated (recognizing, of course, that 1 Btu is the heat necessary to raise the temperature of 1 lb of water 1°F).

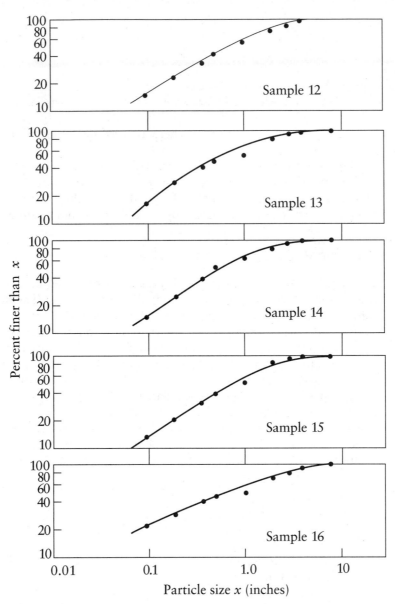

Figure 2-11 Particle-size distribution curve for unprocessed refuse.
Source: (22)

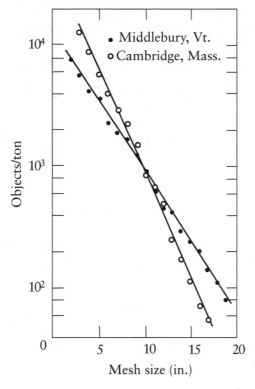

Figure 2-12   Particle-size distribution of municipal solid waste. Source: (23)

Table 2-5   Proximate and Ultimate Chemical Analyses of Refuse

Proximate analysis (percent by weight)	
Moisture	15–35
Volatile matter	50–60
Fixed carbon	3–9
Noncombustibles	15–25
Higher heat value (HHV)	3000–6000
Ultimate analysis (percent by weight)	
Moisture	15–35
Carbon	15–30
Hydrogen	2–5
Oxygen	12–24
Nitrogen	0.2–1.0
Sulfur	0.02–0.1
Total noncombustibles	15–25

Source: (24)

Table 2-6   Heat Value of Fuels

Fuel	Heat value		Composition (wt%)					
	(kJ/kg)	(Btu/lb)	S	H	C	N	O	Ash
Natural gas	54,750	23,170	nil	23.5	75.2	1.22	–	nil
Heating oil (no. 2)	45,000	19,400	0.3	12.5	87.2	0.02	nil	nil
Coal, anthracite	29,500	12,700	0.77	3.7	79.4	0.9	3.0	11.2
Coal, bituminous	26,200	11,340	3.22	4.6	40.0	1.0	6.5	9.0
Coal, lignite	19,200	8300	0.4	2.5	32.3	0.4	10.5	4.2
Wood, hardwood	7180[*]	3090[*]						
Wood, softwood	7950[*]	18,400[*]						
Shredded refuse[a]	10,846	4675	0.1	–	–	–		20.0
RDF[b]	15,962	6880	0.2	–	37.1	0.8		22.6
RDF[c]	18,223	7855	0.1	–	45.4	0.3		6.0
Unprocessed refuse	10,300	4450	0.1	2.65	25.6	0.64	21.2	20.8
Unprocessed refuse			0.13	4.80	35.6	0.9	29.5	28.9
Paper	24,900	7500	0.1	2.7	20.7	0.13	19.1	2.74

[*] Lower Heat Value (LHV); all other heat values are Higher Heat Value (HHV)
[a] Shredded, nonair-classified, ferrous removed, not dried; St. Louis RDF facility
[b] Shredded, air-classified, not dried
[c] Same as above, but oversize from a 3/16-in. screen
Source: (18)

Refuse can be characterized as being made up of organic materials, inorganic materials, and water. Usually, the heat value is expressed in terms of all three components, the Btu/lb, where the sample weight includes the inorganics and water. But sometimes the heat value is expressed as *moisture-free*, and the water component is subtracted from the denominator. A third means of defining heat value is to also subtract the inorganics, so the Btu is *moisture- and ash-free*, the ash being defined as the inorganics upon combustion.

EXAMPLE 2-3

A sample of refuse is analyzed and found to contain 10% water (measured as weight loss on evaporation). The Btu of the entire mixture is measured in a calorimeter and is found to be 4000 Btu/lb. A 1.0-g sample is placed in the calorimeter, and 0.2 g ash remains in the sample cup after combustion. What is the comparable moisture-free Btu and the moisture- and ash-free heat value?

The one pound of refuse would have 10% moisture, so the moisture-free heat value would be calculated as

$$4000 \text{ Btu/lb} \times \frac{1 \text{ g}}{1 \text{ g} - 0.1 \text{ g water}} = 4444 \text{ Btu/lb}$$

Similarly, the moisture- and ash-free heat value would be

$$4000 \text{ Btu/lb} \times \frac{1 \text{ g}}{1 \text{ g} - 0.1 \text{ g water} - 0.2 \text{ g ash}} = 5714 \text{ Btu/lb}$$

The heat value of various components of refuse is quite different, as shown in Table 2-7. As is the case with moisture content, the heat value of a refuse where the fraction of components is known can be estimated by using such estimated heat values for the various components.

Table 2-7   Heat Values of Some Refuse Components

Component	As collected	Btu/lb Moisture-free	Moisture- and ash-free
Cardboard	7,040	7,400	7,840
Food waste	1,800	6,000	7,180
Magazines	5,250	5,480	7,160
Newspapers	7,980	8,480	8,610
Paper (mixed)	6,800	7,570	8,050
Plastics (mixed)	14,100	14,390	16,020
HDPE	18,700	18,700	18,900
PS	16,400	16,400	16,400
PVC	9,750	9,770	9,980
Steel cans	0	0	0
Yard waste	2,600	6,500	6,580

Source: (adapted from 20 from 21)

An important aspect of calorimetric heat values is the distinction between *higher heat value* and *lower heat value*. The higher heat value, or HHV, is also called the *gross calorific energy*, while the lower heat value (LHV) is also known as the *net calorific energy*. The distinction is important in design of combustion units. More on this in Chapter 7.

A comparison of the heat values of coal and refuse-derived fuel (RDF) in Table 2-6 suggests implicitly that RDF can be substituted directly for coal, by using ratios of the heat values. If the heat value of coal is 10,000 Btu/lb and that of RDF is about 7000 Btu/lb, one might conclude (erroneously) that 10 tons of RDF represents the same energy value as 7 tons of coal. The problems in substituting RDF for coal or oil are immense and often impractical, such as running a car on RDF.

Bulk and Material Density

Municipal solid waste has a highly variable bulk density, depending on the pressure exerted, as shown in Figure 2-13.[25] Loose, as it might be placed into a garbage can by the homeowner, the bulk density of MSW might be between 150 and 250 lb/yd^3 (90 and 150 kg/m^3); pushed into the can it might be at 300 lb/yd^3 (180 kg/m^3). In a collection truck that compacts the refuse, the bulk density is normally between 600 and 700 lb/yd^3 (350 and 420 kg/m^3). Once deposited in a landfill and compacted with machinery, it can achieve bulk densities of about 1200 lb/yd^3 (700 kg/m^3). If the covering soil in landfills is included, the total landfilled density can range from about 700 lb/yd^3 for a poorly compacted landfill to as high as 1700 lb/yd^3 (1000 kg/m^3) for a landfill where thin layers of refuse are compacted. Such landfills are

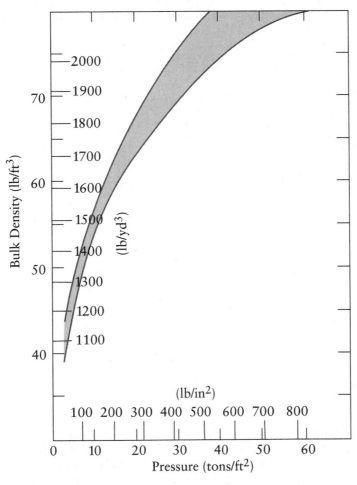

Figure 2-13   Compression of municipal solid waste. Source: (25)

quite dense. For comparison, the density of water is 1690 lb/yd³ (1000 kg/m³). The bulk densities of refuse in various stages of compaction are listed in Table 2-8.

Table 2-8   Refuse Bulk Densities

Condition	Density (lb/yd³)
Loose refuse, no processing or compaction	150–250
In compaction truck	600–900
Baled refuse	1200–1400
Refuse in a compacted landfill (without cover)	750–1250

Tables 2-3 and 2-8 show the densities of refuse and its components as *bulk* densities. This is different from *materials* densities, or densities of materials without any void spaces. Table 2-9 shows such material densities for a number of refuse components. For example, for a steel can, the material density is that of steel, or about 7.7 g/cm³. For an uncrushed empty steel can, about 95% of the volume is air, and the bulk density is then only about 0.4 g/cm³.

Table 2-9   Material Densities Commonly Found in Refuse

Material	Specific gravity	lb/yd³
Aluminum	2.70	4536
Steel	7.70	12,960
Glass	2.50	4212
Paper	0.70–1.15	1190–1940
Cardboard	0.69	1161
Wood	0.60	1000
Plastics		
HDPE	0.96	1590
Polypropylene	0.90	1510
Polystyrene	1.05	1755
PVC	1.25	2106

Because of the highly variable density, MSW quantities are seldom expressed in volumes and are almost always expressed in mass terms as either pounds or tons in the American standard system, or kilograms or tonnes in the SI system. Note that in this text ton = 2000 lb and tonne = 1000 kg.

Mechanical Properties

The compressive strength of some typical MSW constituents is shown in Figure 2-14. A wide variation exists in the amount of energy necessary to obtain volume reduction. The curves tend to be mostly linear, indicating that substantial volume reduction can be achieved by expending greater energy in compaction—a fact understood in the use of solid waste balers.

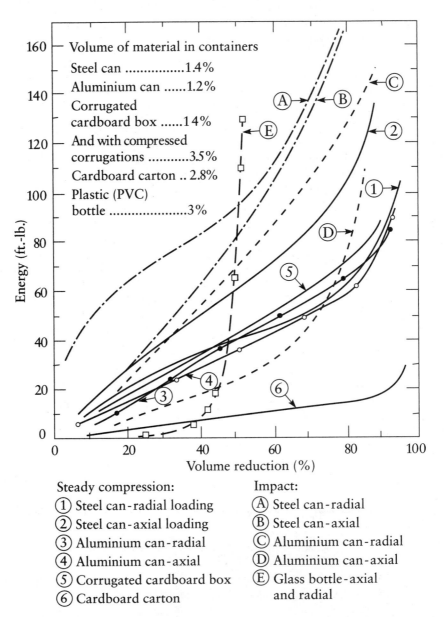

Steady compression:
- ① Steel can-radial loading
- ② Steel can-axial loading
- ③ Aluminium can-radial
- ④ Aluminium can-axial
- ⑤ Corrugated cardboard box
- ⑥ Cardboard carton

Impact:
- Ⓐ Steel can-radial
- Ⓑ Steel can-axial
- Ⓒ Aluminium can-radial
- Ⓓ Aluminium can-axial
- Ⓔ Glass bottle-axial and radial

Figure 2-14 Compressive characteristics of some components of solid waste.
Source: (19)

The tensile stress–strain curves for several refuse components are shown in Figure 2-15. As expected, the steel has the greatest ultimate strength, yielding a computed modulus of elasticity (E) of 28.5×10^6 lb/in^2 (83.8×10^6 kg/cm^2), which compares favorably with the usual value of E of 29 to 30×10^6 lb/in^2 (85 to 88×10^6 kg/cm^2) for carbon and low-alloy steels. The E for aluminum was computed as 10×10^6 lb/in^2 (30×10^6 kg/cm^2), while PVC had an E of only 0.2×10^6 lb/in^2 (0.6×10^6 kg/cm^2), as expected.[26]

Biodegradability

From Table 2-2 the fraction of municipal solid waste that is organic can be listed as in Table 2-10. Using calculated[18] and estimated percentages of degradation, Table 2-10 shows that only about 45% of MSW is potentially biodegradable. Treatment techniques such as composting must take into account that a large fraction of MSW is not biodegradable and that this material must be disposed of by means other than producing useful products using biodegradation.

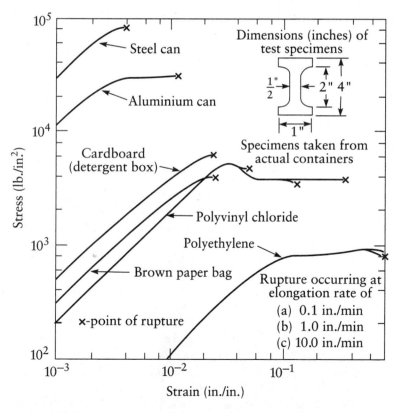

Figure 2-15 Tensile strength of some municipal solid waste components. Source: (26)

Table 2-10   Calculation of Biodegradable Fraction of MSW

Component	Percent of MSW	Percent of each component that is biodegradable
Paper and paperboard	37.6	0.50
Glass	5.5	0
Ferrous metals	5.7	0
Aluminum	1.3	0
Other nonferrous metals	0.6	0
Plastics	9.9	0
Rubber and leather	3.0	0.5
Textiles	3.8	0.5
Wood	5.3	0.7
Other materials	1.8	0.5
Food waste	10.1	0.82
Yard trimmings	12.8	0.72
Miscellaneous inorganic	1.5	0.8
Total	100	

Source: (1, 20)

FINAL THOUGHTS

Continued implementation of reduction, reuse, recycling, and resource recovery systems will reduce both the extraction of raw material and the quantities of waste disposed of into our environment. Increased recovery would mean less landfilling (hence less air and water pollution and other detrimental effects of landfills), less incineration (hence less air pollution), and less blatant open dumping (hence reduction in visual affronts and public health problems).

The rate of the increased use of natural resources for both energy and materials production will at least be slowed by a wider use of secondary materials. It is now commonly accepted that the supply of natural raw materials will be restricted worldwide, and that the United States will feel disproportionately greater pressure as a result of our large consumption and limited domestic supplies. We thus recognize that resource recovery is reasonable, logical, prudent, economical, and feasible. Even having stated this, we find that others still ask, "Is it right?", which is possibly the most difficult question to answer.

Over the course of the development of our basic Western philosophical and ethical framework, environmental concerns have played a minor role. With the exception of Thoreau, St. Francis of Assisi, and a few others, we have not wrestled very long or hard with the problem of environmental ethics. The question posed above is thus one that does not yet have a simple answer, nor are there many philosophers even willing to tackle it. Much work remains to be done in this area, as is the case with economics, engineering, and the other sciences. The absence of an answer should not, however, deter us from thinking out the "good" in this endeavor. We should still seek understanding, perhaps from a spiritual source.

We respond to nature spontaneously in ways that are not easily accounted for by any other means, and there is evidence that we need nature for psychological health. Of course, we can't just make up a new environmental ethic, but we do have spiritual resources in our cultural traditions, and a study of other traditions can both illuminate our own and provide insights into possibly more common elements of human spiritual needs and what satisfies them.

When Aldo Leopold was a young man working for the Forest Service, one of his first assignments was to help in the eradication of the wolf in New Mexico. One day he and his crew spotted a wolf crossing a creek and instantly opened fire, mortally wounding the animal. Leopold scrambled down to where it lay and arrived in time to "watch the fierce green fire dying in her eyes."[27] This experience was significant in Leopold's recognition of the value of predators in an ecosystem and his eventual acceptance of the intrinsic value of all nature. Did Leopold's spirituality see the fire in the wolf's eyes, or was that spirituality transferred from the wolf to Leopold?

It is unlikely that most of us will ever share such a life-changing experience. But it is also true that most of us already understand, somewhere deep down, the significance and meaning of the spiritual dimension of the environmental ethic.

APPENDIX
MEASURING PARTICLE SIZE

For nonspherical particles the diameter of a particle may be defined as any of the following:

$$D = l$$

$$D = \frac{h + w + l}{3}$$

$$D = \sqrt[3]{hwl}$$

$$D = \sqrt{lw}$$

$$D = \frac{w + l}{2}$$

where D = particle diameter
l = length
w = width
h = height

E X A M P L E
2 - 4

Consider nonspherical particles that are uniformly sized as length, $l = 2$, width, $w = 0.5$ and height, $h = 0.5$. Calculate the particle diameter by the various definitions above.

$$D = l = 2; \quad D = \frac{w + l}{2} = 1.25; \quad D = \frac{h + w + l}{3} = 1.0$$

$$D = \sqrt{lw} = 1; \quad D = \sqrt[3]{hwl} = 2.12$$

Note that the "diameter" varies from 1.0 to 2.12, depending on the definition.

When particle size is determined by sieving, the most reasonable definition is

$$D = \sqrt{lw}$$

since only two dimensions must be less than the sieve opening for the particle to fall through. As stated before, the term *diameter* has significance only if the shape is circular, and any other geometrical shape is therefore not really describable by diameter.

When the mixture of particles is nonuniform, the particle size is often expressed in terms of the *mean particle diameter*. Given an analysis of the various diameters of individual particles (such as by sieving), the mean particle diameter can be expressed in a number of ways, including all of the following:

the *arithmetic mean*:

$$D_A = \frac{D_1 + D_2 + D_3 + \dots D_n}{n}$$

the *geometric mean*:

$$D_G = \sqrt[n]{D_1 \times D_2 \times D_3 \times \dots D_n}$$

the *weighted mean*:

$$D_W = \frac{W_1 D_1 + W_2 D_2 + \dots W_n D_n}{W_1 + W_2 + \dots W_n}$$

the *number mean*:

$$D_N = \frac{M_1 D_1 + M_2 D_2 + \ldots M_n D_n}{M_1 + M_2 + \ldots M_n}$$

the *surface area mean*:

$$D_S = \frac{M_1 D_1^3}{M_1 D_1^2 + M_2 D_2^2 + \ldots M_n D_n^2} + \frac{M_2 D_2^3}{M_1 D_1^2 + M_2 D_2^2 + \ldots M_n D_n^2} + \frac{M_n D_n^3}{M_1 D_1^2 + M_2 D_2^2 + \ldots M_n D_n^2}$$

or the *volume mean*:

$$D_V = \frac{M_1 D_1^4}{M_1 D_1^3 + M_2 D_2^3 + \ldots M_n D_n^3} + \frac{M_2 D_2^4}{M_1 D_1^3 + M_2 D_2^3 + \ldots M_n D_n^3} + \frac{M_n D_n^4}{M_1 D_1^3 + M_2 D_2^3 + \ldots M_n D_n^3}$$

where n = number of discrete classifications (sieves)
W = weight in each classification
M = number of particles in each classification

EXAMPLE
2-5

Given the following analysis,

Particle diameter, mm, (D)	60	40	20	5
Weight of each fraction, kg (W)	2	10	5	4
Number of particles, (M)	140	300	1000	2000

calculate the arithmetic mean, geometric mean, weighted mean, number mean, surface area mean, and volume mean.

$$D_A = \frac{(60 + 40 + 20 + 5)}{4} = 31.2 \text{ mm}$$

$$D_G = \sqrt[4]{60 \times 40 \times 20 \times 5} = 21.1 \text{ mm}$$

$$D_W = \frac{(2 \times 60) + (10 \times 40) + (5 \times 20) + (4 \times 5)}{2 + 10 + 5 + 4} = 30.3 \text{ mm}$$

$$D_N = \frac{(140 \times 60) + (300 \times 40) + (1000 \times 20) + (2000 \times 5)}{140 + 300 + 1000 + 2000} = 14.7 \text{ mm}$$

$$D_S = \frac{140 \times 60^3}{140 \times 60^2 + 300 \times 40^2 + 1000 \times 20^2 + 2000 \times 5^2} + \ldots = 40.0 \text{ mm}$$

$$D_V = \frac{140 \times 60^4}{140 \times 60^3 + 300 \times 40^3 + 1000 \times 20^3 + 2000 \times 5^3} + \ldots = 47.4 \text{ mm}$$

REFERENCES

1. Franklin Associates. 1999. *Characterization of Municipal Solid Waste in the United States: 1998 Update.* EPA 530-R-98-007, Washington, D.C. (updated versions of this are available from www.epa.gov).

2. Rhyner, C. R. 1992. "Monthly Variations in Solid Waste Generation." *Waste Management and Research* 10: 67–71.

3. Boyd, G., and M. Hawkins. 1971. *Methods of Predicting Solid Waste Characteristics.* EPA OSWMP, SW–23c, Washington, D.C.

4. Dayal, G., A. Yadav, R. P. Singh, and R. Upadhyay. 1993. "Impact of Climatic Conditions and Socioeconomic Status on Solid Waste Characteristics: A Case Study." *The Science of Total Environment* 136: 143–153.

5. Davidson, G. R. 1972. *Residential Solid Waste Generation in Low Income Areas.* EPA OSWMP.

6. Grossman, D. J., F. Hudson, and D. H. Marks. 1974. "Waste Generation Models for Solid Waste Collection." *Journal of the Environmental Engineering Division* ASCE 100, EE6: 1219–1230.

7. Rathje, W., and C. Murphy. 1992. *Rubbish: The Archeology of Garbage.* New York: Harper-Collins.

8. Ali Khan, M., and F. Buney. 1989. "Forecasting Solid Waste Compositions." *Resources, Conservation and Recycling* 3: 1–17.

9. Cailas, M. D., R. Kerzee, R. Swager, and R. Anderson. 1993. *Development and Application of a Comprehensive Approach for Estimating Solid Waste Generation in Illinois.* Urbana-Champaign, Ill.: The Center for Solid Waste Management and Research, University of Illinois.

10. Henricks, S. L. 1994. *Socio-economic Determinants of Solid Waste Generation and Composition in Florida.* MS thesis, Duke University School of the Environment. Durham, N. C.

11. Hockett, D., D. J. Lober, and K. Pilgrim. 1995. "Determinants of Per Capita Municipal Solid Waste Generation in the Southeastern United States." *Journal of Environmental Management* 45: 205–217.

12. American Society of Testing and Materials. 1992. *A Standard Test Method for Determination of the Composition of Unprocessed Municipal Solid Waste.* D 5231-92. Philadelphia.

13. Reinhart, D., P. M. Bell, B. Ryan, and H. Sfeir. "An Evaluation of Municipal Solid Waste Composition Studies." Orlando, Fla.: University of Central Florida.

14. Sfeir, H., D. R. Reinhart, and P. R. McCauley-Bell. 1999. "An Evaluation of Municipal Solid Waste Composition Bias Sources." *Air and Waste Management Association* 49: 1096–1102.

15. Klee, A. J., and D. Carruth, 1970. "Sample Weights in Solid Waste Composition Studies." *J. Sanit. Eng. Div.* ASCE 96, n. SA4: 945.

16. Miller, C., various articles in *Waste Age.*

17. O'Brien, J. 1977. Unpublished data. Durham, N. C.: Duke Environmental Center, Duke University.

18. National Center for Resource Recovery. 1978. *Incineration.* Toronto: Lexington Books.

19. *Measuring Recycling: A Guide for State and Local Governments.* 1997. EPA 530-R-97-011. Washington, D.C.

20. Tchobanaglous, G., H. Theisen, and S. Vigil. 1993. *Integrated Solid Waste Management.* New York: McGraw-Hill.

21. Neissen, W. 1977. "Properties of Waste Materials." In D. G. Wilson, ed., *Handbook of Solid Waste Management.* New York: Van Nostrand Reinhold.

22. Vesilind, P. A., A. E. Rimer, and W. A. Worrell. 1980. "Performance Characteristics of a Vertical Hammermill Shredder." *Proceedings* 1980 National Waste Processing Conference. Washington, D.C.: ASME.

23. Rhyner, C. R. 1976. "Domestic Solid Waste and Household Characteristics." *Waste Age* (April): 29–39, 50.
24. *Incinerator Guidelines*. 1969. Washington, D.C.: U.S. Dept. of Health, Education, and Welfare.
25. Ruf, J. A. 1974. *Particle Size Spectrum and Compressibility of Raw and Shredded Municipal Solid Waste*. Ph. D. Diss., University of Florida, Gainesville.
26. Trezek, G., D. Howard, and G. Savage. 1972. "Mechanical Properties of Some Refuse Components." *Compost Science* 13, n. 6.
27. Leopold, A. 1966. *A Sand County Almanac*. New York: Ballantine.

ABBREVIATIONS USED IN THIS CHAPTER

ASTM = American Society for Testing and Materials
Btu = British thermal unit
C = chemical symbol for carbon
C & D = construction and demolition
E = modulus of elasticity
EPA = Environmental Protection Agency
H = chemical symbol for hydrogen
HDPE = high-density polyethylene
HHV = higher heat value

LHV = lower heat value
MSW = municipal solid waste
N = chemical symbol for nitrogen
O = chemical symbol for oxygen
PETE = polyethylene terephthalate
RDF = refuse-derived fuel
S = chemical symbol for sulfur
SI = System International, the metric system of units
UC = uniformity coefficient

PROBLEMS

2-1.   A large furniture manufacturer produces 100 tons of industrial solid waste per day. This firm has decided to locate in a small community of 1500 residents. If the wastes from this plant go to the town landfill, which has an estimated life of 4 years, and if the cost of operating the landfill to the town is $10 per ton, how much more will this cost the town, and how much sooner will the town need a new landfill?

2-2.   Estimate, using input analysis, the quantity of paper of different categories that might be obtained from the building where you work or study.

2-3.   A landfill operation uses a tractor to compact the waste. The tractor weighs 8 tons and has two tracks, each 2 ft × 10 ft. Estimate the maximum compaction attainable in the landfill.

2-4.   Headline from the *Chapel Hill News*: "There is enough energy in our garbage to replace all the oil we presently get from Iraq and Iran." Using your knowledge as a solid waste engineer, write a helpful letter to the editor.

2-5.   What is meant by *moisture transfer* in refuse management, and why is this important in studies on refuse composition and materials recovery or energy conversion?

2-6.   What do you believe is the approximate composition of solid waste produced at your university? Indicate your estimates by checking the appropriate boxes in the table on the following page. Make sure your answers sum to 100 percent. Submit as your homework a table showing the fraction of each component.

Component	Percent by weight					
	<1	5	10	20	40	60
Paper	☐	☐	☐	☐	☐	☐
Garbage	☐	☐	☐	☐	☐	☐
Glass	☐	☐	☐	☐	☐	☐
Aluminum	☐	☐	☐	☐	☐	☐
Steel	☐	☐	☐	☐	☐	☐
Plastics	☐	☐	☐	☐	☐	☐
Other	☐	☐	☐	☐	☐	☐

2-7. Use examples and definitions to describe what is meant by *code* and *switch* in material recovery operations.

2-8. How has the plastics industry responded to the threat of restrictive legislation with regard to plastics?

2-9. The town of Chapel Hill, North Carolina, has about 100,000 people (including the students). The Smith Center has a volume of about 180,000 cubic yards. If the town of Chapel Hill used the Smith Center as a refuse disposal site, how long would it take to fill up? Make all necessary assumptions.

2-10. Describe what is meant by *as received*, *moisture-free*, and *moisture- and ash-free* heat value, and how these are calculated.

2-11. The state of California requires its communities to divert 50% of municipal solid waste from landfills. Discuss both the objective and the calculations used to attain such high diversion rates. How would you revise the mandate to achieve what the original intent of the program was to do, to recycle more refuse? Show your calculations.

2-12. Is polystyrene recycled in your community? If it is not recycled, call the community solid waste/recycling office and

ask why it is not. How would you respond to those reasons? If it is, call and ask what happens to it.

2-13. Compare the volume occupied by 100 lb of refuse following baling with the same waste as it would be in a garbage can. Assume the composition shown below. For miscellaneous wastes, assume a loose density of 300 lb/yd^3 and a baled density of 1000 lb/yd^3. What is the density of the combined waste, loose and baled? (Use the representative values for bulk densities found in Table 2-6.)

Component	Percent by weight
Newsprint	21
Office paper	15
Cardboard	8
Glass	12
HDPE	3
PETE	3
Steel cans	5
Yard waste	18
Aluminum cans	4
Miscellaneous	11

2-14. Using the moisture contents (wet basis) of the waste components given in the table below, calculate the overall moisture content of the waste having the composition also shown below. This calculation can be simplified by assuming 100 lb of waste.

Component	Moisture content, percent (wet basis)	Percent by weight
Food	70	10
Paper	6	33
Cardboard	5	8
Plastics	2	5
Textile	10	4
Rubber	2	3
Yard waste	60	18
Metals	3	10
Miscellaneous	6	9

2-15. Calculate the moisture content of the waste in Problem 2-14 on both a wet and dry basis. Why the difference? Which number makes more sense to you, and why?

2-16. Determine the composition of the waste in Problem 2-14 if yard waste is separately collected. How would this affect the moisture content?

2-17. Using the energy content (wet basis) of the waste components given in the table below, calculate the overall energy content of the waste having the composition shown below. This calculation can be simplified by assuming 100 lb of waste.

Component	Energy, Btu/lb	Percent by weight
Food	2,000	10
Paper	7,200	33
Cardboard	7,000	8
Plastics	14,000	5
Textile	7,500	3
Yard waste	2,800	18
Metals	300	10
Miscellaneous	3,000	9

2-18. Determine the composition of the waste in Problem 2-14 if 10% of the wood waste, 50% of the paper, 30% of the metals, and 25% of the plastics is recycled. How would this affect the energy content?

Collection

Solid waste collection is an exercise in reducing entropy. The pieces that make up the solid waste are scattered far and wide, and the role of the collector is to gather this material together into one container. In the United States, as in most other parts of the developed world, solid waste collection systems are invariably person/truck systems. With only a few minor exceptions, the collection of MSW is done by men and women who traverse a town in trucks and then ride with the truck to a site at which the truck is emptied. This may be a disposal site (landfill) or an intermediate stopover where the refuse is transferred from the small truck into trailers, larger vans, barges, or railway cars for long-distance transport.

REFUSE COLLECTION SYSTEMS

The process of refuse collection should be thought of as a multiphase process, and it is possible to define at least five separate phases, as shown in Figure 3-1. First, the individual homeowner must transfer whatever is considered waste (defined as material having no further value to the occupant) to the refuse can, which may be inside or outside the home. The second phase is the movement of the refuse can to the truck, which is usually done by the collection crew, called *backyard collection*. If the can is moved to the street by the waste generator or the home occupant, the system is called *curbside collection*.

More and more separated materials (commonly called *recyclables*) and yard wastes are collected separately, either in the same trucks as the mixed refuse, or in separate vehicles. The analysis below applies to any or all of these materials, although most of the discussion is about mixed (nonseparated) waste. This material is still of greatest concern (and cost) to communities.

The truck must collect the refuse from many homes in the most efficient way possible, and when it is full (or at the end of the day), it must travel to the materials recovery facility, the point of disposal, or the transfer site. The fifth phase of the collection system involves the location of the final destination (materials recovery facility, disposal site, or transfer station). This is a planning problem, often involving more than one community.

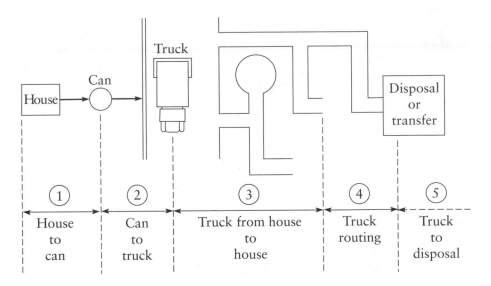

Figure 3-1   Five phases of municipal solid waste collection.

Phase 1: House to Can

The house-to-can phase has received almost no attention or concern by researchers or government because the efficiencies and conveniences gained here are personal and not communal. As discussed later under source separation, one major drawback of collecting separated material is the inconvenience suffered by the individual.

Most communities in the United States use tax funds to operate the solid waste collection and disposal system, or they charge for the service just as they charge for water consumption and wastewater disposal. Such a system gives the generators of waste *carte blanche* to generate as much as they please because the cost is the same regardless of how much they contribute. The need to manage and control the amount of waste generated has led communities to try novel ways of funding solid waste programs. Some communities have adopted a *volume-based fee system* to pay for solid waste collection and specify the containers that must be used. In a volume-based fee system, residents are offered cans in three sizes—such as 30-, 60-, or 90-gallon (110-, 230-, and 340-liter) cans. The fee for refuse service is based on the size of can used. Over 4000 communities have adopted volume-based rates for solid waste collection. The EPA reports typical reduction of 25 to 35% for communities that have gone to volume-based rates.[1] Other communities are taking this approach one step further by weighing every can and charging by actual weight, the so-called *weight-based fee system*.

Volume-based fee systems have generated renewed interest in the home compactor. This device, originally introduced in the 1950s without much success, sits under the kitchen counter and compresses about 20 lb (9 kg) of refuse into a convenient block within a special bag. The bulk density of the compacted refuse is about 1400 lb/yd^3 (830 kg/m^3), achieving a compaction ratio of about 1:5.

EXAMPLE
3-1

A family of four people generates solid waste at a rate of 2 lb/cap/day and the bulk density of refuse in a typical garbage can is about 200 lb/yd^3. If collection is once a week, how many 30-gallon garbage cans will they need, or the alternative, how many compacted 20-lb blocks would the family produce if they had a home compactor? How many cans would they need in that case?

2 lb/cap/day $\times$ 4 persons $\times$ 7 days/week = 56 lb refuse

56 lb/200 lb/yd^3 = 0.28 yd^3

0.28 yd^3 $\times$ 202 gal/yd^3 = 57 gal. They will require two 30-gallon cans.

If the refuse is compacted into 20-lb blocks, they would need to produce 3 such compacted blocks to take care of the week's refuse. If each block of compacted refuse is 1400 lb/yd^3, the necessary volume is

$$\frac{56 \text{ lb}}{1400 \text{ lb/yd}^3} \times 202 \text{ gal/yd}^3 = 8.1 \text{ gal}$$

They would need only one 30-gal can.

Phase 2: Can to Truck

At one time the most common system of getting the solid waste into the truck was the collectors going to the backyard, emptying the garbage cans into large tote containers, and carrying these to the waiting truck. This system was not only expensive in dollar cost to the community, but it was expensive in terms of the extremely high injury rate to the collectors. At one time solid waste collectors had the highest injury rate of any vocation—three times higher than the injury rate for coal miners, for example. Even now, with all of the improvements in collection technology, solid waste collection is still one of the most hazardous jobs in America. A survey by the U.S. Department of Labor's Occupational Safety and Health Administration (OSHA) Statistics Department found that fully 40% of the solid waste workers had missed time during the preceding year due to various injuries, including strains, bruises, and fractures.[2]

The traditional trucks used for residential and commercial refuse collection are rear-loaded and covered compactors called *packers*, and vary in size and design, with 16- and 20-yd^3 (12- and 15-m^3) loads being common (Figure 3-2). The truck size is often limited not by its ability to store refuse but by its wheel weight. Residential streets are not designed to carry large wheel loads, and refuse trucks can easily exceed these limits. Commonly, the refuse is emptied from garbage cans into the back of the packers where it is scooped up by hydraulically operated compaction mechanisms that compress the refuse from a loose density of about 100 to 200 lb/yd^3 (60 to 120 kg/m^3) to about to 600 to 700 lb/yd^3 (360 to 420 kg/m^3). The compaction (packing) mechanism of one manufacturer is shown in Figure 3-3. To reduce injuries and to speed up collection, some solid waste collection companies are changing from rear loaders to side loaders (Figure 3-4).

Figure 3-2   A rear-loading packer truck for collecting residential solid waste.

Figure 3-3   Compacting mechanism for a packer truck.

Two revolutionary changes during the 1990s had a great impact on both the cost of collection as well as the injury rate of the collectors. The first is wide acceptance of the *green-can-on-wheels* idea, where the resident fills a large plastic container on wheels and then pushes it to the curb for collection. These containers can be used for mixed refuse, recyclables, and/or yard waste (Figure 3-5). The collection vehicles are equipped with hydraulic hoists that are used to empty the contents into the truck, as shown in Figure 3-6. The collectors do not come into contact with the refuse, avoiding dangerous materials that can cut or bruise. This system, referred to as *semi-automated collection*, typically requires a driver and one or more collectors. A further development in solid waste collection technology is the *can snatcher*, trucks equipped with long arms that reach out, grab a can, and lift it into the back of the truck (Figure 3-7). Such systems, called *fully automated collection*, are especially useful where the street layout includes alleys behind the houses. Communities

Figure 3-4 *Side-loading packer truck.*

that have converted from the manual system to the fully automated system have saved at least 50% in collection costs, much of it in reduced medical costs.[3]

The second revolutionary development in solid waste collection is the widespread use of plastic bags. Many communities now insist that all refuse be packaged in the plastic bags and that these be taken to the curb for collection. While the collectors still have to lift the bags to the truck, the bags do not weigh very much, and injuries due to strains are almost eliminated. In some communities the cans are no longer used, and the bags are left on the curb for collection.

Plastic bags, of course, have some serious disadvantages. Bags can rip while they are transported to the curb, and can be torn apart by small animals seeking food, resulting in the garbage being spread all over the sidewalk or alley.

Figure 3-5 *Recyclables, yard waste, and mixed refuse at the curb.*

Figure 3-6   Green plastic containers used for solid waste collection.

Phase 3: Truck from House to House

Once the refuse is in the truck, it is compacted as the truck moves from house to house. The higher the compaction ratio, the more refuse the truck can carry before it has to make a trip to the landfill.

Figure 3-7 *Collection with vehicles equipped with "can snatchers." (Courtesy of Volvo.)*

EXAMPLE
3-2

Assume each household produces 56 lb of refuse per week (as in Example 3-1). How many customers can a 20-yd³ truck that compacts the refuse to 500 lb/yd³ collect before it has to make a trip to the landfill?

$$20 \text{ yd}^3 \times 500 \text{ lb/yd}^3 = 10,000 \text{ lb}$$

$$10,000 \text{ lb} / 56 \text{ lb/customer} = 178 \text{ customers}$$

(Note that the refuse weighs 5 tons, and if the truck itself weighs 2 tons, the common 6-ton residential load limit can be exceeded before the truck is full.)

The size of the truck crew can range from one to over five people. If backyard pickup is offered, a larger crew size is needed because the crew must service cans that might be at some distance from the collection vehicle. Curbside pickup requires a smaller crew, and, of course, fully automated systems require only one person. Studies have shown that the greatest overall efficiencies can be attained with the smallest possible crews. For curbside refuse collection, three-person crews do not collect three times as much refuse as a one-person crew.

As a rough guideline, for most residential curbside collections, a single truck should be able to service between 700 and 1000 customers per day if the truck does not have to travel to the landfill. In one California community an automatic collection system allows a truck to average about two and a half loads per day (10-hour shift) at 10 tons per load.[4] Realistically, most trucks can service only about 200 customers before the truck is full and a trip to the landfill is necessary.

EXAMPLE
3-3

Suppose a crew of 2 people requires 2 minutes per stop, at which they can service 4 customers. If each customer generates 56 lb of refuse per week, how many customers can they service if they did not have to go to the landfill?

A working day is 8 hours, minus breaks and travel from and to the garage—say 6 productive hours, $\times$ 60 = 360 minutes. At 2 minutes per stop, a truck should be able to make 180 stops and service $180 \times 4 = 720$ customers.

(Note, however, from Example 3-2, that the truck has to go to the landfill after only 178 customers, or fewer still if its wheel loading is exceeded for the streets!)

An organized way of estimating the amount of time the crew actually works in collecting refuse is to enumerate all the various ways they spend time. The total time in a workday can be calculated as

$$Y = a + b + c(d) + e + f + g$$

where Y = the total time in a workday

$\quad a$ = time from the garage to the route, including the marshaling time, or that time needed to get ready to get moving

$\quad b$ = actual time collecting a load of refuse

$\quad c$ = number of loads collected during the working day

$\quad d$ = time to drive the fully loaded truck to the disposal facility, deposit the refuse, and return to the collection route

$\quad e$ = time to take the final, not always full, load to the disposal facility and return to the garage

$\quad f$ = official breaks including time to go to the toilet

$\quad g$ = other lost time such as traffic jams, breakdowns, etc.

All variables, of course, have to be in consistent time units such as minutes.

Such an analysis, while it may not be very useful for calculations, strikingly demonstrates that a working day is not the same as the time spent collecting solid waste. If the value of b in the above equation, or the amount of time the crew actually spends collecting refuse, is known, the number of customers served by that truck and crew can be estimated.

If a region is fairly homogeneous, travel times (d in the above equation) can be estimated by driving representative routes and generalizing the data. Once sufficient travel-time data are available, the data can be regressed against the "crow-fly" distance. One such regression[5] for New Jersey resulted in the expression

$$d = 1.5D - 0.65$$

where d = actual one-way travel time, min
D = one-way travel time as the crow flies, assuming different truck speeds along the route, min

EXAMPLE
3 - 4

A truck is found to be able to service customers at a rate of 1.25 customers per minute. If they find that the actual time they spend on collection is 4 hours, how many customers can be served per day?

If the crew can service 1.25 customers in one minute, what can they do in 4×60 minutes?

$$\frac{1.25}{1} = \frac{X}{4 \times 60}$$

X = 300 customers per day

If the number of customers that a single truck can service during the day is known, the number of collection vehicles needed for a community can be estimated by

$$N = \frac{S\,F}{X\,W}$$

where N = number of collection vehicles needed
S = total number of customers serviced
F = collection frequency, number of collections per week
X = number of customers a single truck can service per day
W = number of workdays per week

EXAMPLE
3 - 5

Calculate the number of collection vehicles a community would need if it has a total of 5000 services (customers) that are to be collected once per week.

A single truck can service 300 customers in a single day and still have time to take the full loads to the landfill. The town wants to collect on Mondays, Tuesdays,

Thursdays, and Fridays, leaving Wednesdays for special projects and truck maintenance.

$$N = \frac{S\,F}{X\,W} = \frac{5000 \times 1}{300 \times 4} = 4.2 \text{ trucks}$$

The community will need 5 trucks.

Phase 4: Truck Routing

The routing of a vehicle within its assigned collection zone is often called *micro-
routing* to distinguish it from the larger-scale problems (phase 5) of routing to the
disposal site and the establishment of the individual route boundaries. The latter
problem is commonly known as *macrorouting* or *districting*, and is discussed later.

The present question is how to route a truck through a series of one- or two-
way streets so that the total distance traveled is minimized. Put another way, the
objective is to minimize *deadheading*, traveling without picking up refuse. The
assumption is that if a route can be devised that has the least amount of deadheading possible, it is the most efficient collection route.

The problem of designing a route so as to eliminate all deadheading was actually addressed as early as 1736. The brilliant mathematician, Leonard Euler, was
asked to design a route for a parade across the seven bridges of Königsberg, a city
in eastern Prussia, such that the parade would not cross the same bridge twice but
would end at the starting point. The problem is illustrated in Figure 3-8, together

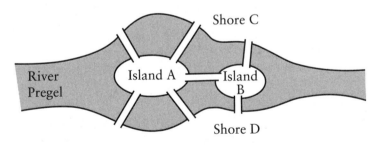

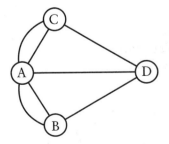

Figure 3-8 The bridges of Königsberg.

with a schematic diagram. The routes are shown by lines called *links*, and the locations are known as *nodes*. The system shown has four nodes and seven links.

Euler not only proved that the assignment was impossible, but he generalized the two conditions that must be fulfilled for any network to make it possible to traverse a route without traveling twice over any road. These two conditions are

1. All points must be connected (one must be able to get from one place to another).
2. The number of links to any node must be of an even number.

The first condition is logical. The second similarly makes sense, in that if one travels to a location such as island **A** or **B** in Figure 3-8, one must be able to get off again—hence two roads. Euler's parade problem had all the nodes with an odd number of links, a clearly impossible situation. The number of links connecting a node designates its degree, and the existence of any odd-degree nodes in a system indicates that a route without deadheading is impossible. A system that has all nodes of even degree is known as a *unicoursal network*, and an *Euler's tour* is theoretically possible.[6]

In the real world, one-way streets, dead ends, and other restrictions can often make a practical application of the theoretical analysis difficult. One-way streets can be considered in Euler's theory by recognizing again that one must be able to get to a node exactly as many times as one leaves it. Thus a node with three one-way streets leading to it and a single one-way street leading away from it immediately makes a network nonunicoursal, even though the number of links at that node is even.

The development of a least-cost route involves making a system unicoursal with the least number of added links. For example, the Königsberg bridge problem would require only two additional (deadhead) links to make the system unicoursal (Figure 3-9). With this system a theoretical Euler's tour exists, and the problem is now one of finding the proper route.

Kwan[7] has provided a means of achieving the most efficient unicoursal network (and also provided the name for this procedure: the *Chinese postman problem*) by observing that networks are really a series of loops where each node appears exactly once. By minimizing the additional connecting links (deadheads) necessary to achieve a unicoursal system, one can in fact achieve an overall optimum system. For example, the unicoursal network of the Königsberg bridge problem shown in Figure 3-9 is clearly a poor choice. (A new bridge is required!) It would make much more sense to trade the two deadheads shown in Figure 3-9 for

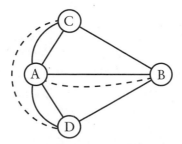

Figure 3-9 One possible route for the king's parade.

the two in Figure 3-10. The latter is an obviously more efficient solution. The skill of the route planner must come into play in such trade-offs, since a shorter street with many traffic problems may, in fact, be a more expensive alternative to a longer but clear street.

Once a unicoursal network has been designed, it remains to route the truck through this network. The method of *heuristic* (commonsensical) *routing* has found wide application.[8] The following set of rules apply to microrouting. Some of these are pure commonsensical judgment, and some are useful guidelines for determining overall strategy when attacking a network.

1.   Routes should not overlap, but should be compact and not be fragmented.
2.   The starting point should be as close to the truck garage as possible.
3.   Heavily traveled streets should be avoided during rush hours.
4.   One-way streets that cannot be traversed in one line should be looped from the upper end of the street.
5.   Dead-end streets should be collected when on the right side of the street.
6.   On hills, collection should proceed downhill so that the truck can coast.
7.   Clockwise turns around blocks should be used whenever possible.
8.   Long, straight paths should be routed before looping clockwise.
9.   For certain block patterns, standard paths, as shown in Figure 3-11, should be used.
10.   U-turns can be avoided by never leaving one two-way street as the only access and exit to the node.

These rules can be used to develop effective routes with minor deadheading. Figure 3-12 is an example of some of these routing rules applied to a large area. Elegant computer programs have been developed by a number of researchers, but in practice it has been found that the tours constructed manually are almost always better than those done by mechanical tour-building codes.[7]

Phase 5: Truck to Disposal

For smaller isolated communities, the macrorouting problem reduces to one of finding the most direct road from the end of the route to the disposal site. For regional systems or large metropolitan areas, however, macrorouting in terms of developing

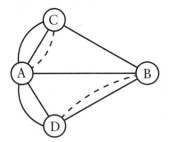

Figure 3-10   An alternative route for the parade.

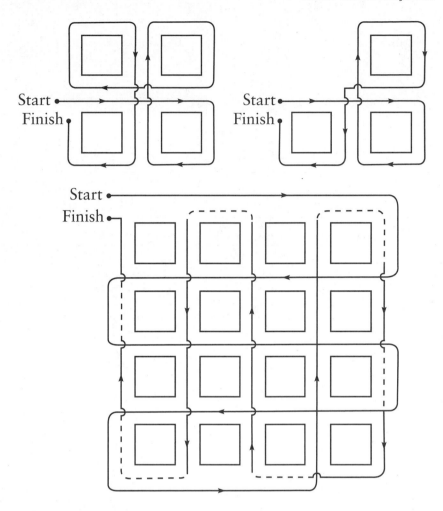

Figure 3-11 *Large loops usually result in efficient collection.*

the optimum disposal and transport scheme can be used to great advantage. The available techniques, called *allocation models*, are all based on the concept of minimizing an objective function subject to constraints, linear programming being the most common technique.

The simplest allocation problem is the assignment of solid waste disposal to more than one disposal site. Often the solution is obvious—the closest sources are allocated first, followed by the next closest, and so on. With more complex systems, however, it becomes necessary to use optimization techniques. The most appropriate one is the transportation algorithm, which is a type of linear programming. This technique is illustrated in the Appendix to Chapter 3.

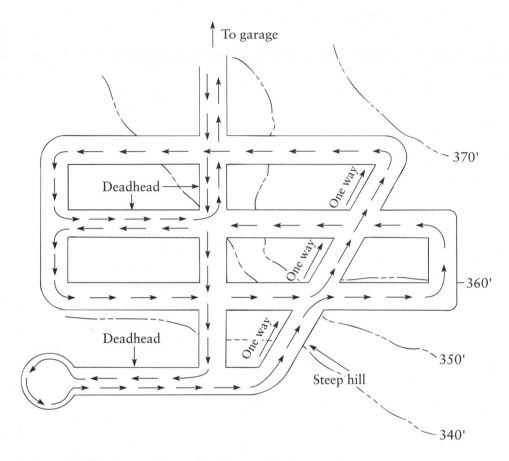

Figure 3-12 A sample routing for a collection truck.

COMMERCIAL WASTES

Commercial solid waste is almost always collected with *dumpsters*, large steel containers that are commonly lifted overhead by the collection truck (Figure 3-13). As with the can snatchers, the driver does not have to get out of the truck, which is both an advantage and a disadvantage. In both cases the driver does not see what has been placed in the container. Hazardous or dangerous materials can be transferred to the truck and could cause dangerous situations in the landfill or combustor. At the landfill the refuse in the full truck is then pushed out (Figure 3-14).

Very large dumpsters are roll-off containers (Figure 3-15) commonly used in construction sites. The roll-off containers typically range in size from 10 to 40 cubic yards. The 10-yd^3 (7.6-m^3) units are used for demolition material, while the 40-yd^3 (30-m^3) units can be used as rural transfer stations for household garbage. The containers are pulled onto trucks that then take the full containers to the landfills. A special truck transports one roll-off container at a time to the landfill.

Figure 3-13 A dumpster used for commercial collection. (Courtesy of Dempster Dumpster.)

Figure 3-14 Dumpster collection truck being emptied at a landfill.

Figure 3-15   Roll-off containers. (Courtesy of Volvo.)

TRANSFER STATIONS

When the waste disposal unit is remote to the collection area, a *transfer station* is employed. At a transfer station waste is transferred from smaller collection vehicles to larger transfer vehicles, such as a tractor and trailer, a barge, or a railroad car.

Transfer stations can be quite simple, or they can be complex facilities. The design of the facility is based on its intended use, with small transfer stations typically relying on a tipping floor where collection vehicles drop their loads. Waste can then be loaded into open-top trailers using a wheeled loader. More complex facilities might employ pits for vehicles to dump into. Transfer vehicles can then be loaded by using a compacting unit. A facility might also have a tunnel for the transfer vehicle to drive into. Chutes or an opening in the floor would allow these units to be loaded by having waste pushed over the edge into them. Some typical transfer stations are shown in Figure 3-16.

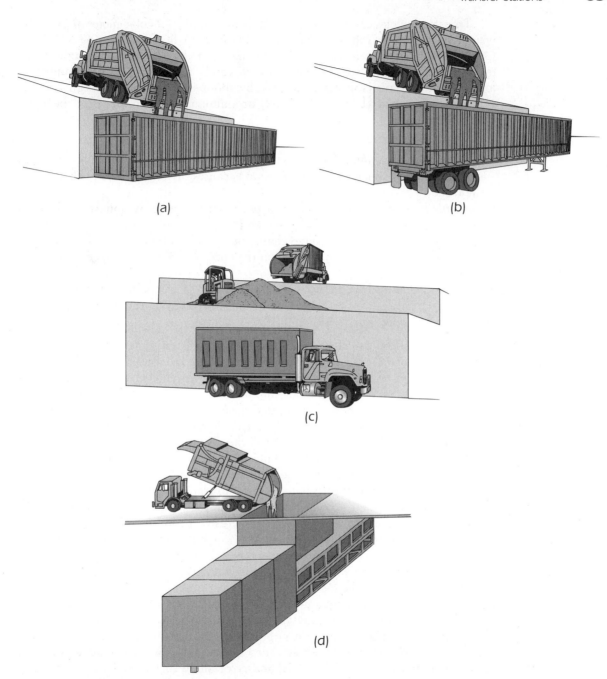

Figure 3-16 Several typical transfer stations. (a) Dump to container. (b) Dump to trailer. (c) Store and dump to truck trailer. (d) Dump to compactor.

Transfer vehicles can be as large as 105 cubic yards (80 cubic meters). The total weight of a transfer vehicle is limited by allowable wheel loads up to 80,000 lb (36,000 kg). In addition many states also have weight limits on each axle. Some transfer stations employ scales so that trailers can be loaded to their maximum payload weight. The available payload is usually about 40,000 lb (18,000 kg). Transfer vehicles are constructed from either steel or aluminum, which can affect the available payload.

One of four methods is used to unload the transfer trailer:

- *Live bottom*, or *walking floors*, on the floors of the vehicles. The back-and-forth movement of the longitudinal floor sections causes the refuse to be pushed out of the trailer.
- *Push blade*, similar to the blade in a packer truck. A telescoping rod pushes a blade from the front of the trailer, which forces the waste out.
- *Drag chains*. Some vehicles have chains on sprockets that go from the front to the back of a trailer, and by pulling on the chain the refuse is dragged out of the vehicle.
- *Tipper*. Some units have no unloading mechanism, and a large tipper at the landfill lifts the entire transfer vehicle up at an angle, causing the door to open and the refuse to slide out.

Transfer stations can also transfer refuse to trains or even barges. In New York City solid waste was transferred onto barges, which then moved the refuse to an island landfill. In Seattle solid waste is transferred to railroad shipping containers that are then placed on a rail car for shipment to eastern Oregon. Durham, North Carolina, sends all of its refuse by railroad to southern Virginia. On a much smaller scale residents in rural areas bring their waste to transfer stations with 40-yd^3 (30-m^3) roll-off bins, which are then hauled to the disposal site.

The decision to build or not to build a transfer station is often an economic decision. If the one-way haul distance from the point of full collection vehicle to the discharge point is short, then it is likely that no transfer station is needed. On the other hand, if the discharge point is far away and the collection vehicle will have to be away from its primary role of collecting refuse for too long, then a transfer station might be warranted. The relationship is illustrated in Figure 3-17. Where the two curves cross is the breakeven point. Longer distance will warrant the construction of a transfer station while shorter hauls will make it uneconomical.

In situations where the sophistication of linear programming models is not warranted, a brute-force technique using a simple grid system can be of value. In this case the region is divided into equal blocks on an X–Y grid, and the solid waste generation is then estimated based on population. The sites for transfer stations and final disposal facilities are initially screened to eliminate obviously inadequate areas (e.g., urban areas for landfills). Trial-and-error siting of facilities is then used to obtain the most reasonable combination of solid waste disposal facilities.

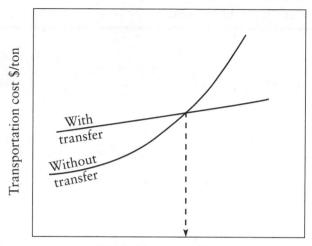

Figure 3-17   Breakeven point of transfer stations.

COLLECTION OF RECYCLABLE MATERIALS

Recycling entered the mainstream of solid waste management in the 1990s. No longer was recycling conducted by the underfunded, idealistic individuals, but rather multinational garbage companies were now involved. Recycling, regardless of the price paid for recycled material, could be profitable. While some long for the days of the idealistic recycler, no one can dispute that more material is being recycled today. At the present time approximately half of the population of the US is being served by curbside recycling.[9]

Two factors have caused recycling to succeed. First, government has provided leadership in the area of waste reduction. Over 40 states have adopted waste reduction goals, and some states have set mandatory goals, with noncompliance resulting in fines of up to $10,000 per day. Second, the public has accepted the concept that recycling is not free. Just as residents pay for garbage service, they also now pay for recycling. Over 7000 curbside recycling programs have been implemented, resulting in 56 million tons—or 27% of all solid waste—being recycled. Table 3-1 highlights the progress of recycling in the United States.

Table 3-1   Collection of Recyclables, 1997

Material	Recovered (millions of tons)	Recovered (percent of generation)
Paper and paperboard	34.9	41.7
Glass	2.9	24.3
Ferrous metals	4.7	38.4
Aluminum	0.9	31.2
Other nonferrous metals	0.8	63.4
Plastics	1.1	5.2
Rubber and leather	0.7	11.7
Textiles	1.1	12.9
Wood	0.6	5.1
Other materials	0.8	20.1
Food waste	0.6	2.6
Yard trimmings	11.5	41.4
Total recycled MSW	60.7	28.0

Source: (10)

Some groups, such as the Grassroots Recycling Network, see this as only the beginning and are calling for a zero waste goal.[11] A major barrier to the increasing expansion of recycling is finding markets for the material. Recycling cannot occur without markets for the recycled material, and markets are created by using post-consumer material in the manufacture of a product.

Producers are being urged to take responsibility for the products they produce to ensure that the material is either recyclable or that it uses post-consumer recycled material. For example, the soft drink industry sells its products in aluminum cans that can easily be recycled, and these cans already contain a high percentage of recycled aluminum. On the other hand, the industry also sells product in PETE plastic containers that are much more difficult to recycle and are not used in the manufacture of new bottles (unlike in Europe and Australia where the soft drink industry uses plastic containers with recycled plastic).

Because the recycling industry is still so young, there has not been time to settle into a standard system that has a high probability of succeeding in every application. Each community has almost an infinite number of options in how to design the recycling program.

Some communities have their citizens place all the recyclable material into one container, and a separate truck takes the recyclable to a materials recovery facility (*clean MRF*) for processing. Because the recyclables are not separated into specific items but placed all into one container, this is called *comingled* collection of recyclable materials. The MRF for processing comingled materials is much smaller than a mixed waste (*dirty MRF*) because it has to process only the separated recyclable material. However, the trade-off is that another truck must travel the same route to pick up the mixed nonrecyclable waste. In addition the residents must separate the recyclable materials and place them in a container for this option to work.

An innovation that allows one truck to pick up both solid waste and comingled recyclables is the development of a dual-compartment packer truck and split container (Figure 3-18). The split 90-gal (340-liter) container has one side for recyclable materials and one side for solid waste. This container is collected by an automated truck fitted with dual compartments. The contents of the container remain split during the dumping process, and the mixed refuse can then be compacted to achieve a full load. The advantage of this system is that only one truck and driver is needed to collect both the solid waste and recyclable materials. However, the amount of material collected must be balanced so that both compartments fill at the same rate. If only one compartment is full and the other is not, the system loses its advantage. Because the truck now has to discharge at two locations, the landfill and MRF must be near each other to maintain the efficiency of this option.

Another option for collecting comingled recyclable material is the *blue bag system* in which the recyclables are placed in a special blue bag provided to each customer. The same truck that picks up the solid waste also picks up the blue bags in a separate compartment. The bags are removed manually at the transfer station and the contents processed separately from the solid waste.

Figure 3-18 Dual-compartment packer truck.

One disadvantage of any system that collects comingled recyclables is the resulting contamination of the paper products by residual liquids in the glass, aluminum, or plastic containers. In addition the sorting costs for comingled recyclables can be high. As a result, many communities have initiated collection programs where the residents separate the materials and place them into two or three containers. For example, bottles and cans go in one container, newspaper in a second, and mixed paper in the third. This eliminates the need to sort newspaper, which is the most common recyclable (by weight) and shifts the cost of sorting to the waste generator. This system requires trucks with multiple compartments (Figure 3-19). While this method has a low initial cost, the participation rate may be low. When Los Angeles switched from this method to a comingled container, the recycling rate increased by 40%.[12]

Yard waste is also typically placed in a separate container. A third truck is needed to collect the yard waste and take this material to a mulching or composting operation. In addition to the above items, some communities have implemented curbside used motor oil collection. Residents are allowed to place their used motor oil at the curb where it is collected and recycled into new oil. Other communities have tried curbside collection of batteries.

Both experience and mathematical modeling suggest that collection of comingled recyclables on a per-ton basis is more expensive than the collection of mixed refuse. As would be expected, the cost gain on a per-ton basis decreases with both participation rate and the total amount of recyclables collected. If the income from the sale of recycled materials is taken into account, however, the cost of collecting recycled items is not significantly different from the cost of collecting mixed, unsorted refuse.

A curbside recycling program typically diverts between 5 and 10% of the residential waste and costs between $1.00 and $4.00 per month per customer, depending on various options. A yard waste program can divert between 15 and 30% of the residential waste and also costs between $1.00 and $4.00 per month. One study concluded that the cost of a typical suburban recycling program is between $114 and $120 per ton of material collected, based on a 50% set-out rate.[13]

Everett et al. have developed a model that estimates the cost of collecting partially separated (comingled) recyclables at curbside.[14] The model estimates the time to collect such waste and to sort it at the truck into various individual components. Three variables in the model are travel time, sorting time, and waiting time (at stop lights, for example).

Travel time is estimated as

$$T = \frac{D}{L(1 - e^{-kD})\,C}$$

where T = travel time between two consecutive stops, seconds
$\quad\quad D$ = travel distance between two consecutive stops, m (ft)
$\quad\quad L$ = calculated coefficient, km h^{-1} (mile h^{-1})
$\quad\quad k$ = calculated coefficient, m^{-1} (ft^{-1})
$\quad\quad C$ = conversion factor, 1.467 for American standard units, 0.278 for SI units

Figure 3-19 Multicompartment truck for collecting separated recyclable materials.

The factors L and k vary with crew size, the maximum speed attained by the vehicle, the acceleration, driver ability, and road conditions. The values of L and k, as estimated by Everett et al., are shown in Table 3-2.

Table 3-2 Travel Time Coefficients

Crew size	L (km m^{-1})	k (m^{-1})
One person	18.8	2.2×10^{-2}
Two persons	24.3	1.7×10^{-2}

Source: (14)

The time to walk to the containers and bring them back to the truck can be estimated as

$$W = 0.86A$$

where W = walk time, seconds
A = average walk distance at a single stop, m

The sorting time is estimated as

$$S = 21.3 + 2.4B$$

for a one-person crew, and

$$S = 23.2 + 1.8B$$

for a two-person crew,
where S = average sorting time of a single set-out, seconds
B = average set-out amount, kg

On average, the set-out amount for a typical curbside program varies between 0.7 and 12 kg.

Finally, the truck has to navigate the streets, and if the traffic is heavy and/or if there are many stop signs and traffic lights, there could be considerable wait time. This can be estimated as

$$E = M_S N_S + M_L N_L$$

where E = average wait time at traffic lights and stop signs, s
M_S = mean time spent at stop signs, s
N_S = number of stop signs
M_L = mean time spent at traffic lights, s
N_L = number of traffic lights

The model, with these variables, has been found to predict the time for collection within 10% of the actual value.[14]

Estimating Success of a Curbside Recycling Program

Whenever a curbside program is initiated in a community, the leaders want to know how successful it has been. Success can be measured in many ways, including these:

- Fraction of refuse components diverted from the landfill, tons/week
- Fraction of households participating in the program; participation can be defined in many ways, such as having set out recyclables at least once a month

- Fraction of households participating on any given week
- Profit from the sale of recyclables, $/week

All of these methods are useful, but they have to be well defined and used consistently.

LITTER AND STREET CLEANLINESS

Litter is a special type of MSW. It is distinct from other types of MSW in that it is a solid waste that is not deposited into proper receptacles. We usually think of litter as existing in public places, but litter could be on private premises as well. Although litter is usually considered to be a visual affront only, it may also be a health hazard. Broken glass and food for rats are but two examples. It is also a drain on our economic resources, because the public must pay to have it collected and removed when it is on public property.

The collection of litter is of secondary importance to a community because it does not represent a critical public service, as do police and fire protection, water treatment, and collection of refuse from residences and commercial establishments. Litter removal is expensive, costing U.S. municipalities millions annually.

The composition of roadside litter can vary considerably from place to place, as can the method of data collection. One major problem with any litter data analysis is that the reports fail to specify the guidelines used in the collection and identification of litter, and seldom specify the way in which the percentages of the various components were calculated. For example, a broken bottle can be counted as either one item or many items, depending on the guidelines. Similarly, the results can be calculated as a percent of the total of each of the following items:

- Items by actual count
- Total weight of litter
- Volume of litter
- Visible items by actual count

Because of the problem of not having a standard counting technique, the following guidelines for conducting litter studies are suggested:[15]

1. Count as one item all pieces larger than 2.5 cm (1 in.). This count includes removable tabs from beverage cans.
2. Do not count rocks, dirt, or animal droppings.
3. Count as one item all pieces of any item clearly belonging together, such as a broken bottle. Otherwise count each piece of glass, newspaper, and so on, singly.
4. Do not count small, readily decomposable material, such as apple cores.
5. For roadside litter surveys, measure all items within the officially designated right of way.
6. Empty liquids out of all bottles and cans before collecting them.

The litter survey, if conducted along a road, should be started by driving along the road at a slow speed and having a passenger record the visible items into a tape

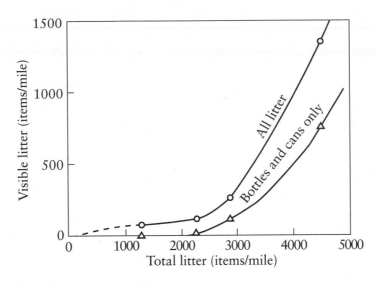

Figure 3-20 *Results of a litter survey on a rural road. Source: (15)*

recorder for future transcription. Next, the litter is identified, recorded, and manually collected. The items should be separated during collection into as many components as feasible. The collected items are then weighed and the volume measured. The relationship between visible items and total items along a roadside is shown in Figure 3-20. It is interesting that along fairly clean roads, the visible fraction is only about 6% of the total litter count! These data also confirm that a large fraction of our visible litter is bottles and cans.

For community litter surveys, the photometric technique developed for Keep America Beautiful, Inc. (KAB), has found wide acceptance.[16] The blockfaces of a community are first numbered, and a preliminary sample size is established. About 5% of the blockfaces are usually adequate. Using the random number technique, the blockfaces and the locations on those blockfaces to be measured are selected. As shown in Figure 3-21, a marker is located in the front center of the survey area, and a chalkboard is used to identify the location and date. To facilitate the counting of litter from the developed photographs, a picture is taken of a clean pavement laid out with white marking tape in a l-ft grid, 6 ft wide, 16 ft long, parallel to a street curb. A transparency of the grid is prepared, and the resulting 96-square grid is placed on top of each litter photograph. The litter is then counted and classified using a magnifying glass. The first photographs are used for establishing the baseline litter conditions. The litter rating (L) is calculated for each picture (location) as the squares containing some litter compared to the total number of squares (%). After initial baseline photographs are analyzed and the L is calculated, the number of sampling sites necessary can be calculated as

$$N = (22.8) S^2$$

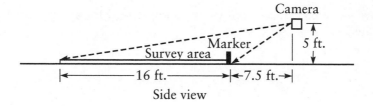

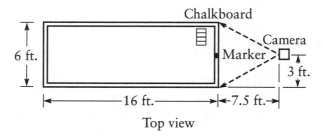

Figure 3-21 Keep America Beautiful litter measurement technique.
Source: (16)

where N = sample size needed to make a 0.5-point difference between two average
litter ratings (Ls) in an area, significant at the 90% confidence level
S^2 = variance of the litter ratings (L) of the initial photographs

The variance is calculated as

$$S^2 = \frac{\displaystyle\sum_{i=1}^{n} F(L_i)^2}{n-1} - \frac{\left(\displaystyle\sum_{i=1}^{n} F L_i\right)^2}{n(n-1)}$$

where L_i = litter rating of the ith photograph
n = total number of photographs
F = frequency, or the number of photographs with any one L

EXAMPLE
3-6

Suppose that a town has 600 blockfaces, and that a 5% sample, or 30 blockfaces,
is photographed as explained above. The litter ratings (L) are as shown:

L	Number of photographs, F	FL	FL^2
1	6	6	6
2	3	6	12
3	1	3	9
4	4	16	64
5	1	5	25
		36	116

How many blockfaces need to be photographed for a litter survey? The S^2 is calculated as

$$S^2 = \frac{116}{30-1} - \frac{(36)^2}{30(29)} = 2.5$$

If $S^2 = 2.5$, the number of sampling sites necessary is

$$N = 22.8\ (2.5) = 57$$

In other words, $57 - 30 = 27$ more sites are needed in order to have a statistically satisfactory baseline.

Litter can theoretically be controlled by cognitive, social, and technological means. A cognitive solution would be convincing people not to litter; a social solution would be depriving the public of items that might become litter, or fining them heavily if they are caught; and a technical solution would be simply cleaning up after littering has occurred.

The first option demands an explanation of why people litter, a question requiring studies on the psychology of litterers. In one study[17] the actions of 272 persons were observed when they bought a hot dog wrapped in paper. Of interest was the final deposition of the wrapper. Ninety-one people chose to dispose of the wrapper improperly (they littered). The probability of any one person littering, based on this sample, could be calculated as

$$E = 0.019 + 0.414(A) + 0.1654(C) + 0.1532(D)$$

where E is the probability that a person would litter; and A, C, and $D = 0$, except that $A = 1$ if the person is 18 years old or younger, $C = 1$ if there are no trash cans conveniently located, and $D = 1$ if the area is already dirty with litter.

From the study it is clear that age is quite important, with younger people being much more likely to litter than are older persons. There was no statistical difference between 19- to 26-year-olds and persons older than 26 years. Gender was found to be statistically insignificant. Because the study was conducted in 1973, its validity to today's urban populations may be questionable. Intuitively, however, the role of younger persons as the major contributors to urban litter remains valid.

EXAMPLE
3-7

Calculate the probability of a 40-year-old person littering a dirty street that has no convenient trash cans.

$$E = 0.019 + 0 + 0.1654 + 0.1532 = 0.33$$

That is, of 100 people answering that description, 33 would probably litter the street.

Such studies yield clues as to how persons might be induced not to litter (e.g., put out more trash cans and clean up the street) and who the target population is (e.g., young people).

Other litter psychology studies have been directed at finding out what motivates people not to litter. In one study[18] movie theater patrons during Saturday matinees were asked by several means not to litter the theater. The total quantity of litter was then measured and used as an indicator of the success of that control approach. The results showed that measures such as personal exhortation for cooperation and anti-litter cartoons had no effect on litter, but that payment of money for pieces of litter at the end of the showing resulted in about a 95% reduction in litter. The clear indication is that self-interest, such as placing a substantial deposit on beverage containers, is an effective force in convincing people not to litter.

The second method of litter control is to prevent items that might become litter from ever reaching the consumer. For example, in the earlier example of the hot dog wrapper, it would seem reasonable to suggest that 100% litter-free results could be obtained by not giving customers a paper wrapper around their hot dog. The banning of tearaway metal tabs on beer and soft drink cans is a practical means of controlling this type of litter.

The third method of litter control is to clean up the mess once it has occurred. This system is commonly used in sports stadiums and other public areas where no effort is made to ask people to properly dispose of their waste. For roadside litter it seems that the most economical litter control alternative is actually frequent cleanup.

An effective means of litter control is to enlist the help of the community by having organizations "adopt" a section of roadway. Organizations as varied as church groups, Rotary Clubs, sports teams, and even private businesses have agreed to keep sections of roadways clean by conducting periodic litter pick-ups. This method of litter control not only keeps roadways and streets cleaner, but it also brings the litter problem down to the personal level. Anyone who has contributed a Saturday to the hot and dirty job of collecting litter along the roadway will not throw trash out the car window, and will be critical of those who do.

Attempts have also been made at designing mechanical litter collection machines. One towed device has proven both inexpensive and effective. It works by having a series of rotating plastic teeth that fling the litter into a collection basket, much like a leaf collector connected to a lawn mower.[19] A more sophisticated and ambitious unit, developed by a major manufacturer of beverage containers, uses a vacuum arm on a truck to suck up the roadside litter while cruising at highway speed.[20]

Finally, street cleanliness can be negatively affected by the very people who collect the household refuse. In a study at the University of Florida, litter was measured using the KAB method on typical residential streets before and after garbage collection.[21] An almost 300% increase in litter was noted. The collection supervisor believed the reason was twofold. First, he believed that his automated collection vehicles were poorly designed and that they were not always able to get everything into the truck. The second reason was that the city had initiated a volume-based refuse collection charge. He suggested that the residents had been practicing the

"Seattle Stomp," the tricky two-step that originated in Seattle when that city switched to volume-based collection. Residents figured they could get more refuse into a can if they stomped on the refuse and packed it into the can. This led to the garbage being stuck in the can and to the potential for spillage during the transfer to the truck.

FINAL THOUGHTS

In "the good old days" the garbage man came twice a week, went to your backyard and collected all your garbage. Today some communities have three different-sized garbage cans, three recycling containers, a yard waste container, and a used oil container. The instructions on how to put your garbage out are almost as complicated as those for operating your VCR. Yet, with all this complexity, it is working. Why? Because Americans have an environmental ethic and recognize the need for integrated waste management. More people recycle than vote.

The limits to recycling are still unknown. Learned people are debating whether it is possible to even think about recycling all the materials we consider waste. Most likely, fundamental principles of physics[22] and thermodynamics[23] and mass transfer[24] make the goal of 100% collection of materials impossible. It is quite likely that when the environmental effects of recycling are compared to the impact of disposal options, there will be some point at which more recycling will actually have increasingly detrimental effects on energy use and materials conservation. At the present time, however, one of the major objectives in solid waste engineering is to effectively and efficiently collect all the materials people no longer want. Because the collection of municipal solid waste accounts for between 50% and 75% of the total cost of refuse management, it is important that solid waste engineers properly design collection systems.

Finally, it is important to recognize that garbage collection is still a labor-intensive job. Any system developed is going to be only as good as the men and women who operate it. As one solid waste director of a major city said, "There is nothing more beautiful than watching my fleet of one hundred garbage trucks leaving the yard silhouetted against the rising sun."

APPENDIX
DESIGN OF COLLECTION SYSTEMS

Systems analyses can be used to design collection systems so as to minimize the cost. Consider the simple system pictured in Figure 3-22. The waste generated at four sources (denoted by centroids of the collection area, a poor assumption, especially if the disposal sites are close to the collection routes) is to be allocated to two disposal sites. The objective is to achieve this in a minimum-cost manner.

At the same time, several requirements must be met (constraints in an optimization model).

1. The capacity of each disposal site (e.g., a landfill) is limited.
2. The amount of refuse disposed of must equal the amount generated.
3. The collection route centroids cannot act as disposal sites; or the total amount of refuse hauled from each collection area must be greater than or equal to zero.

The following notation is adopted:

x_{ik} = quantity of waste hauled from source i to disposal site k, per unit time
c_{ik} = cost per quantity of hauling the waste from source i to disposal site k
F_k = disposal cost per waste quantity at disposal site k (capital plus operating)
B_k = capacity of disposal site k, in waste quantity per unit time
W_i = total quantity of waste generated at source i, per unit time
N = number of sources i
K = number of disposal sites k

The problem then boils down to minimizing the following objective function:

$$\sum_{i=1}^{N}\sum_{k=1}^{K} x_{ik}\, c_{ik} + \sum_{k=1}^{K}\left(F_k \sum_{i=1}^{N} x_{ik} \right)$$

subject to the following constraints:

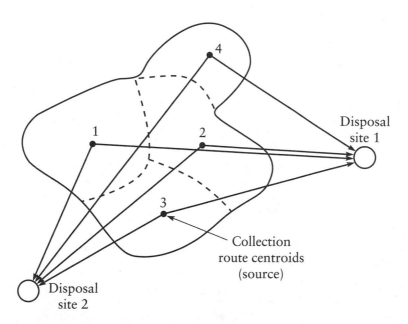

Figure 3-22 *Elements of a simple solid waste management system.*

Constraint 1. The sum of all the solid waste hauled out of each section of the community must be equal to or less than the capability of the disposal sites to receive that waste, or

$$\sum_{i=1}^{N} x_{ik} \leq B_k \text{ for all } k$$

Constraint 2. The sum of all the waste hauled from a section of the community (to any disposal site) has to equal the amount generated in that section, or

$$\sum_{k=1}^{K} x_{ik} = W_i \text{ for all } i$$

Constraint 3. The waste hauled out has to be positive (not negative), or

$$x_{ik} \geq 0 \quad \text{for all } i, k$$

The first term in the objective function is transportation costs, and the second term is disposal costs. For the case shown in Figure 3-13, the objective function is:

Minimize $[x_{11}c_{11} + x_{21}c_{21} + x_{31}c_{31} + x_{41}c_{41} + x_{12}c_{12} + x_{22}c_{22} + x_{32}c_{32} + x_{42}c_{42} + F_1(x_{11} + x_{21} + x_{31} + x_{41}) + F_2(x_{12} + x_{22} + x_{32} + x_{42})]$

subject to the following constraints:

$$x_{11} + x_{21} + x_{31} + x_{41} \leq B_1$$

$$x_{21} + x_{22} + x_{32} + x_{42} \leq B_2$$

$$x_{11} + x_{21} + x_{31} + x_{41} = W_1$$

$$x_{12} + x_{22} + x_{32} + x_{42} = W_2$$

$$x_{11} \geq 0, x_{12} \geq 0, \ldots, x_{23} \geq 0, x_{24} \geq 0$$

This problem can be solved using any linear programming algorithm. The *transportation algorithm* is particularly useful for such applications.

EXAMPLE
3-8

Assume the solid waste generation and disposal figure for the system pictured in Figure 3-22 is as follows:

| | | Cost of transport, c_{ik} | |
| | Generation, W_i | To site 1 | To site 2 |
Source i	(tonnes/week)	($/tonne)	($/tonne)
1	100	5	12
2	130	7	5
3	125	4	8
4	85	13	6

Disposal site, k	Capacity, B_k	Cost, F_k ($/tonne)
1	450	4
2	200	6

We can hand-calculate the cost for any option. Suppose all of the waste was to be sent to disposal in site 1 ($k = 1$). The cost for delivering the refuse from the first source ($i = 1$) is

100 tonnes/wk x $5/tonne = $500/wk

Similarly for the other sections, the cost would be $910, $500, and $1105 per week, or a total cost of $3015/wk.

The processing cost is

440 tons/wk x 4 $/ton = $1760/wk

The total cost is therefore $4775/wk.

This is only one solution, and might not be the least-cost solution. To find the least-cost solution (minimize cost), we can use the transportation algorithm and find the following:

From source i	To disposal site k	Waste hauled (tonnes/week)	Transport cost ($/week)	Processing cost ($/week)
1	1	100	500	400
2	1	25	175	100
2	2	105	525	630
3	1	125	500	500
4	2	85	510	510

Therefore, the total system minimum cost is $4350/week, substantially less than if all the refuse was shipped to disposal site 1. Note that considerable capacity remains unused in landfill 2.

Systems in which transfer stations are used can also be optimized by systems analysis, using the scheme introduced previously. Figure 3-23 shows the same community with four sources of waste and two disposal sites, but now a transfer station is placed in the town. The trucks now have K disposal points and J intermediate facilities. As before, these facilities have processing costs, annualized capital plus operating costs. F_j is the annual cost for the transfer stations, and F_k is the annual capital and operating costs for the disposal sites. The other variables are

c_{ij} = cost per quantity of hauling the waste from source i to intermediate facility j

c_{jk} = cost per quantity of hauling the waste from intermediate facility j to final disposal facility k

x_y = quantity of waste hauled from source i to intermediate facility j, per unit time

x_k = quantity of waste hauled from intermediate facility j to final disposal facility k, per unit time

B_j = capacity of intermediate facility j, in waste quantity, per unit time

P_j = proportion of waste at intermediate facility j that, after processing, remains for disposal; P_j = 1.0 if the facility is a transfer station, but P_j = 0.2 if it is an incinerator

J = number of intermediate facilities j

The problem now is to minimize the following objective function:

$$\sum_{i=1}^{N}\sum_{j=1}^{J}c_{ij}x_{ij}\sum_{i=1}^{N}\sum_{k=1}^{K}c_{ik}x_{ik} + \sum_{j=1}^{J}\sum_{k=1}^{K}c_{jk}x_{jk}$$

$$+\sum_{j=1}^{J}F_{j}\sum_{i=1}^{N}x_{ij} + \sum_{k=1}^{K}F_{k}\sum_{i=1}^{N}x_{ik}$$

This object function is subject to the following constraints:

1. The quantity of waste generated at source i, W_i, must equal the sum of all the waste hauled from that source to the J intermediate sites and K disposal points.

$$\sum_{j=1}^{J}x_{ij} + \sum_{k=1}^{K}x_{ik} = W_i \text{ for all } i$$

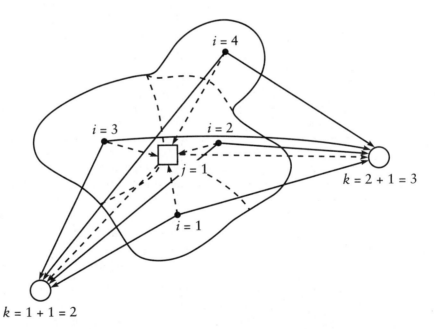

Figure 3-23 Adding transfer stations to the collection system plan.

2. The capacity of the jth intermediate site, B_j, must be more than or equal to the total waste brought to it. If this constraint is omitted, the model can be used to determine the required capacity.

$$\sum_{i=1}^{N} x_{ij} \leq B_j \text{ for all } j$$

3. B_k, the capacity of the final disposal site k, which might be influenced by the number of trucks a tipping floor or landfill site can handle, the compaction capacity at a landfill, and so on, should not be exceeded by the waste brought in directly from the collection sites or from the intermediate facilities.

$$\sum_{i=1}^{N} x_{ik} + \sum_{i=1}^{J} x_{jk} \leq B_k \text{ for all } k$$

4. Whatever waste is shipped in to an intermediate site must be shipped out to a disposal site. The proportion of waste that remains for disposal after any processing is denoted by P_j.

$$P_j \sum_{i=1}^{N} x_{ij} - \sum_{k=1}^{K} x_{jk} = 0 \text{ for all } j$$

5. The nonnegativity constraints are

$$x_{ij} \geq 0; x_{ik} \geq 0; x_{jk} \geq 0 \text{ for all } i, j, k$$

REFERENCES

1. *Unit-Based Pricing in the United States: A Tally of Communities*. 1999. Washington, D.C.: EPA.
2. U.S. Department of Labor, Occupational Safety and Health Statistics Department. 1998. Washington, D.C. As reported by *Waste Age* (July): p. 20.
3. *Getting More for Less; Municipal Solid Waste and Recyclables Collection Workbook*. 1996. Washington, D.C.: Solid Waste Association of North America.
4. DeWeese, A. 2000. "Mandates Motivate to Automate." *Waste Age* (February).
5. Greenberg, M. R., et al. 1976. *Solid Waste Planning in Metropolitan Regions*. New Brunswick, N.J.: The Center for Urban Policy Research, Rutgers University.
6. Liebman, J. C., J. W. Male, and M. Wathne. 1975. "Minimum Cost in Residential Refuse Vehicle Routes." *Journal of the Environmental Engineering Division*, ASCE, v. 101, n. EE 3: 339–412.
7. Kwan, K. 1962. "Graphic Programming Using Odd or Even Points." *Chinese Math.* 1: 207–218.
8. Shuster, K. A., and D. A. Schur. 1974. *Heuristic Routing for Solid Waste Collection Vehicles*. EPA OSWMP SW-113. Washington, D.C.
9. Goldstein, N. 1997. "The State of Garbage in America." *BioCycle* 38, n. 4: 60–67.
10. Franklin Associates. 1998. *Characterization of Municipal Solid Waste in the United States*. EPA 530-R-980-007. Washington, D.C.
11. *Zero Waste*. The Grassroots Recycling Network. www.grrm.org/zw5.html.
12. "Single Stream Recycling: The Future Is Now." 2000. *Recycling Today* 38, n. 1: 79.
13. Miller, C. 1993. *The Cost of Recycling at the Curb*. Washington, D.C.: National Solid Waste Management Association.
14. Everett, S. Maratha, R. Dorairaj, and P. Riley. 1998. "Curbside Collection of Recyclables I: Route Time Estimation Model." *Resources, Conservation and Recycling* 22: 177–192.
15. Vesilind, P. A. 1976. *Measurement of Roadside Litter*. Durham, N. C.: Duke Environmental Center, Duke University.
16. *The Photometric Index*. n.d. Stanford, Conn.: Keep America Beautiful.
17. Finnie, W. C. 1973. "Field Experiments in Litter Control." *Environment and Behavior* 5 n. 2.
18. Burgess, R. L., et al. 1971. "An Experimental Analysis of Anti-litter Procedures." *Journal of Applied Behavioral Analysis* 4 n. 2.
19. Hart, F. D., et al. 1973. *Design and Development of a Machine to Remove Litter from the Roadside*. Raleigh, N. C.: School of Engineering, North Carolina State University.
20. National Center for Resource Recovery. 1973. *Municipal Solid Waste Collection*. Lexington, Mass.: Lexington Books.
21. Schert, J. 2000. Center for Solid and Hazardous Waste Management, University of Florida.
22. Georgescu-Roegen, N. 1976. *Energy and Economic Myths*. New York: Pergamon.
23. Biaciardi, C., E. Tiezzi, and S. Ulgiati. 1993. "Complete Recycling of Matter in the Frameworks of Physics, Biology and Ecological Economics." *Ecological Economics* 8: 1–5.
24. Converse, A. O. 1996. "Letter to the Editor on Complete Recycling." *Ecological Economics* 19: 193–194.

ABBREVIATIONS USED IN THIS CHAPTER

EPA = Environmental Protection Agency
KAB = Keep America Beautiful, Inc.
MRF = materials recovery facility

MSW = municipal solid waste
OSHA = Occupational Safety and Health Administration

PROBLEMS

3-1. Determine the time required to complete a filling and emptying cycle for a refuse collection vehicle serving a residential area if the following conditions pertain:
the truck volume = 20 yd^3,
each location has on average two containers, 80 gallons each, 75% full,
refuse is picked up twice per week,
the truck has a compaction ratio of 1:2,
the pickup time is 1.58 min per service,
the truck spends 20 min at the disposal site,
it takes 35 min to drive to the disposal site.

3-2. Design an innovative system for transporting refuse from the kitchen to the curb. Cost should not be a major consideration. Imaginative ideas that just might work are required.

3-3. On a map of your campus (or any other convenient map) develop an efficient route for refuse collection, assuming that each blockface must be collected.

3-4. Visit City Hall and obtain the accident records for city employees. Report on the relative accident rate of solid waste workers.

3-5. The haul distance is 15 miles as the crow flies, and the anticipated average truck speed is 35 mph. Estimate the one-way haul time.

3-6. Using a study hall or social lounge as a laboratory, study the prevalence of litter by counting the items in the receptacles versus the items improperly disposed of. Each day vary the conditions as follows:

Day 1: Normal (baseline).
Day 2: Remove all receptacles except one.
Day 3: Add additional receptacles; more than normal.

Plot the percent properly disposed of versus the number of receptacles. Discuss the implications of your results.

3-7. Estimate the time to collect refuse in green roll-out containers from a 1000-foot-long street if the homes are 50 feet apart, on both sides of the road, and a two-person crew is doing the collecting. Estimate that it takes 30 seconds to empty the contents of each green container into the truck.

3-8. Using the principles of heuristic routing, develop a collection route for the streets shown in Figure 3-24. Each blockface must be collected (i.e., one side of street collection). Eliminate all blind blockfaces (it is possible!) and minimize left-hand turns.

3-9. Design a refuse collection route for the suburban development shown in Figure 3-25. Note the large and busy 4-lane highway and other features. What criteria should you use to judge the suitability of your route? Justify your route on the basis of these criteria. Assume the following: Collection is required on all streets; the trucks come from town and return to town; there is no median strip on the 4-lane highway (left-hand turns are possible); all streets other than the highway are small residential streets; the

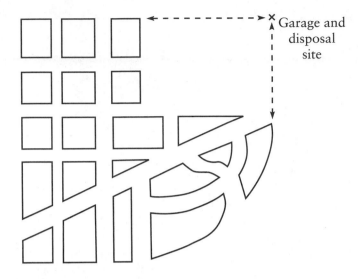

Figure 3-24   See Problem 3-8.

small bridge has a weight limit less than the weight of the empty collection vehicle.

3-10. How would you discourage littering in front of your university engineering building?

3-11. Suppose you are in charge of evaluating an anti-litter campaign for your campus. How would you obtain quantitative information on the success or failure of the campaign?

3-12. Two small communities, Alpha and Beta, each have a small landfill. They are considering building a joint landfill to serve both communities. (See map, Figure 3-26.)

   a. Write an objective function that can be used to minimize the total cost of disposal, where the towns would have the option of taking the refuse to their own landfill or to the common facility. They will not, however, be able to take the refuse to each other's landfill. Use the following variables, and state all the constraints:

F_i = fixed-cost landfill i ($100,000 per year, regardless of size)

C_{ki} = transportation cost from town k to landfill i ($2.00 per km per ton)

x_k = refuse generated in town k (2000 metric tons per year for Alpha, and 3500 metric tons per year for Beta.)

y_i = 1 if the landfill is operated, 0 if it is not

c_i = operating cost of landfill i ($10.00 per metric ton)

   b. Solve the objective function for two conditions: (1) when the towns use their own landfills exclusively, and (2) when only the central shared facility is used by both towns.

   c. Use an optimization software package to find the optimum (least-cost) solution.

3-13. Your client is a small community of 5000 people. They want to begin a municipally operated refuse collection and disposal program and ask your advice on the purchase of a collection vehicle. They want

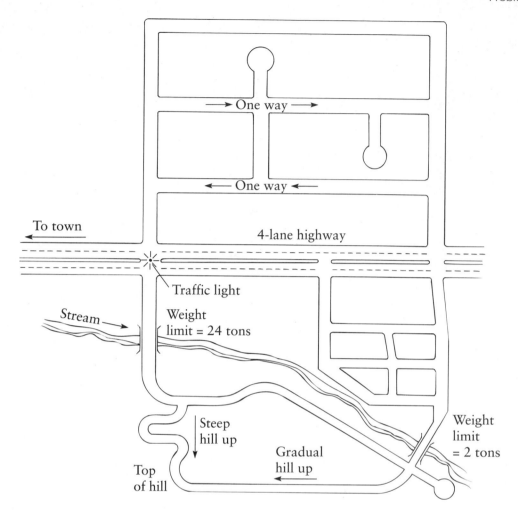

One way →

← One way ←

To town

4-lane highway

Traffic light

Stream →

Weight
limit = 24 tons

Steep
hill up

Gradual
hill up

Weight
limit
= 2 tons

Top
of hill

Figure 3-25 *See Problem 3-9.*

to collect refuse once a week. What would you recommend?

3-14. Given the street configuration shown in Figure 3-25, and if all blockfaces are to be collected, is a unicoursal (Euler's) tour possible? Show your work.

3-15. A community has two landfills. The cost per ton of disposing of refuse in the landfills is exactly the same. There are three collection routes in town. Write an objective function for minimizing the transportation cost (only) from the three collection routes to the two disposal sites.

3-16. Write an objective function for minimizing the transportation cost (only) for the community shown in Figure 3-27. Assume that the cost of disposal is the same at both disposal sites and therefore the disposal cost can be ignored.

3-17. On the street map shown in Figure 3-28, determine first if an Euler's Tour is possible. Show why or why not. Then show how you would route a truck so as to achieve the least number of deadheads. You are to collect on both sides of the street at once.

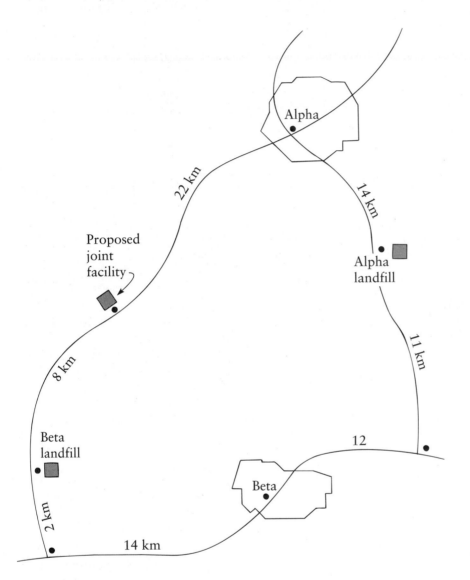

Figure 3-26 See Problem 3-12.

3-18. A community of 25,000 people wants to initiate public garbage collection. If the collection is to be once a week, what size truck would they need? (Note: Available truck sizes are 10, 14, 16, and 20 cubic yards.)

3-19. The Mayor of Recycleville has asked you to develop an integrated waste management program. Residents currently pay $10 per month and can put out up to 5 32-gallon cans per week. You offer the Mayor an automated system with two price options:

Three collection routes
in the community

Figure 3-27 See Problem 3-16.

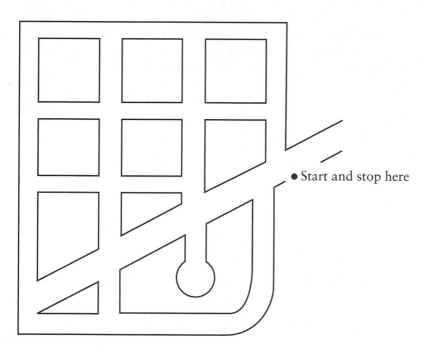

• Start and stop here

Figure 3-28 See Problem 3-17.

Container size, gallons	Option 1	Option 2
32	$9.00	$12.00
64	$18.00	$15.00
96	$27.00	$18.00

In addition you will provide the residents a 64-gallon commingled recycling container and a 96-gallon green waste container.

a. What are the advantages of using the program developed by you?

b. Are residents opting for the 96-gallon container paying more for this service? (*Hint*: Think of the cost per gallon of material.)

c. Which price option would likely have the higher recycling rate, and why? Look at the EPA web site for some case studies to support your reasoning.

d. What impact would collecting recyclables one week and collecting green waste the next week have on the recycling rate?

3-20. Estimate the number of commercial locations that can be serviced in one day by the collecting vehicle described below:

truck volume = 20 yd^3
compaction ratio = 2
container volume = 6 yd^3
on average, containers are 80% full
drive between locations = 8 minutes
at site time (loading dumpster into truck) = 15 min
haul time (to transfer or disposal) = 20 min (one way)
to and from garage (at beginning and end of day) = 15 min
workday = 8 h
off-route factor = 0.25 (breaks, lunch, other necessities)
Assume all vehicles making the trip to transfer or disposal are full.

Landfills

Regardless of how much reuse, recycling, and energy recovery is achieved, some fraction of the MSW must be returned to the environment. This chapter provides a discussion of how residues that have no value can best be managed and disposed of. Although the majority of the text is devoted to the disposal of raw, untreated MSW, it is equally applicable to any fraction of MSW that remains after the recoverable materials have been removed.

Discounting outer space (although the EPA did a feasibility study of sending waste into space in the 1970s—it was not economical) and air (from whence it eventually comes back to the earth's surface), the only two locations for the ultimate disposal of wastes are (1) in the oceans and other large bodies of water, or (2) on or in land. With rare exceptions, most solid waste may no longer be legally dumped into oceans. Most developed countries have enacted strong ocean dumping legislation, and the once ubiquitous refuse- or sludge-loaded barges have all but disappeared. The last barge has taken New York's refuse to the Staten Island landfill, and the Fresh Kills landfill is being closed. The remaining ocean disposal problems appear to result from the discharge of refuse from ships. Long-standing maritime tradition holds that all manner of nonfloating refuse can be discharged from the stern, a fact well known to sea gulls, who follow ships for long distances. The total amount of refuse deposited on the sea bottom in some of the more frequently traveled shipping lanes (such as the North Atlantic) must be impressive as well as depressing. Philosophically, therefore, since ocean disposal is merely a storage process and not a method of treatment, it makes little sense to use the oceans as dumping sites. Little is said in this chapter on ocean disposal, and only the option of land disposal (as our only remaining alternative) is discussed below.

A landfill is an engineered method for land disposal of solid or hazardous wastes in a manner that protects the environment. Within the landfill biological, chemical, and physical processes occur that promote the degradation of wastes and result in the production of *leachate* (polluted water emanating from the base of the landfill) and gases. In the US, MSW landfills are regulated under Subtitle D of the Resource Conservation and Recovery Act (Public Law 94-580) passed in 1976. Specific landfill design and operational criteria were issued under 40 CFR Part 258 in 1991.

The European Union (EU) Council Directive on Landfilling of Waste has also identified the need to optimize final waste disposal methods and seeks to ensure uniform high standards of landfill operation and regulation throughout the

European Union. These standards require a strategy that, by 2014, limits the quantity of biodegradable wastes entering the landfill to 35% of the total amount by weight of biodegradable waste produced in 1995. Consequently, most waste in Europe will be incinerated or composted prior to landfilling. This policy is in direct contrast to a recent trend in the US and many other parts of the world to operate the landfill as a bioreactor. The bioreactor landfill provides control and process optimization, primarily through the addition of leachate or other liquid amendments, if necessary. Beyond that, bioreactor landfill operation may involve the addition of wastewater sludge and other amendments, temperature control, and nutrient supplementation. The bioreactor landfill attempts to control, monitor, and optimize the waste stabilization process rather than contain the wastes as prescribed by most regulations.

PLANNING, SITING, AND PERMITTING OF LANDFILLS

A solid waste engineer, when asked to take on the job of managing the solid waste system for a major city, had only one question: "Do the existing landfills have enough remaining capacity to last until I retire?"

No essential public facility, with the possible exception of an airport, is more difficult to plan, site, and permit than a landfill. In the last 15 years the number of operating landfills has decreased from about 8000 to 2800, primarily due to stringent requirements under the RCRA (40 CFR Part 258). However, landfill capacity is greater now than it was a decade ago due to the expansion of many sites to take advantage of economies of scale achieved in operating the large modern landfill.

Commonly, it now takes over ten years to go through the process of opening a new landfill. During the ten-year process many of the rules will change, including regulations, permits, and approval requirements. In addition, public opposition and even lawsuits during the process are more than likely. The difficulty of siting landfills makes disposal in outer space appear to be a more attractive alternative! Clearly, one quality that is needed when siting a landfill is perseverance.

Planning

In planning for solid waste disposal, communities must look many years into the future. A ten-year time frame is considered short-term planning. Thirty years seems to be an appropriate time frame because after thirty years it becomes difficult to anticipate solid waste generation and new disposal technology.

The first step in planning for a new landfill is to establish the requirements for the landfill site. The site must provide sufficient landfill capacity for the selected design period and support any ancillary solid waste functions, such as leachate treatment, landfill gas management, and special waste services (i.e., tires, bulky items, household hazardous wastes). Some sites also house facilities for handling recyclable materials (material recovery facilities). To determine landfill capacity, the disposal requirements for the community or communities must be estimated. A landfill that is too small will not have an adequate service life and will not justify the expense of building it. On the other hand, a landfill that is too large may eliminate many potential sites and will result in high up-front capital costs that would preclude the community from constructing other needed public facilities.

Records from the last several years provide a historical guide of disposal quantities. In some cases this information can be very accurate if scales were used at existing landfills. In other cases the information is suspect because the quantities of solid waste delivered to the landfills were only estimated. Finally, by using the historical population, a per capita disposal rate can be calculated. Typical US per capita solid waste production rates are in the range between three to seven pounds per person per day. Of course, the potential for importation or exportation of solid waste must be considered.

If possible, in-place density (the density once the refuse has been compacted in the ground) should be estimated. If an existing landfill is available, density can be easily determined by routinely conducting aerial surveys of the landfill and then calculating the volumes. This method includes the volume of cover material in the calculation. If dirt is used as daily and final cover, 20 to 50% of the volume of the landfill may be cover material. An in-place density of 1200 lb/yd^3 (700 kg/m^3) is typical.

When some materials are recovered from solid waste, the compaction characteristics may change markedly, and it is then necessary to estimate the compaction of the waste by individual refuse components. Bulk densities of the components can be used in such calculations, even though the actual compaction may be quite different. Table 4-1 lists some bulk densities that can be used for this purpose, and Example 4-1 illustrates the procedure. A more complete listing of bulk densities is found in the Appendix.

Table 4-1   Bulk Densities of Some Uncompacted Refuse Components

Material	g/cm^3	lb/ft^3
Light ferrous (cans)	0.100	6.36
Aluminum	0.038	2.36
Glass	0.295	18.45
Miscellaneous paper	0.061	3.81
Newspaper	0.099	6.19
Plastics	0.037	2.37
Corrugated cardboard	0.030	1.87
Food waste	0.368	23.04
Yard waste	0.071	4.45
Rubber	0.238	14.9

Source: (1)

Volume, mass, and density calculations for mixed materials can be simplified by considering a container that holds a mixture of materials, each of which has its own bulk density. Knowing the volume of each material, the mass is calculated for each contributing material, added, and then divided by the total volume. In equation form,

$$\frac{(\rho_A \times V_A) + (\rho_B \times V_B)}{V_A + V_B} = \rho_{(A + B)}$$

where ρ_A = bulk density of material A
ρ_B = bulk density of material B
V_A = volume of material A
V_B = volume of material B

When there are more than two different materials, the above equation is extended.

If the two materials at different densities are expressed in terms of their weight fraction, then the equation for calculating the overall bulk density is

$$\frac{M_A + M_B}{\left[\dfrac{M_A}{\rho_A}\right] + \left[\dfrac{M_B}{\rho_B}\right]} = \rho_{(A + B)}$$

where M_A = mass of material A
M_B = mass of material B
ρ_A = bulk density of material A
ρ_B = bulk density of material B

If more than two materials are involved, the above equation is extended.

The volume reduction achieved in refuse baling or landfill compaction is an important design and operational variable. If the original volume of a sample of solid waste is denoted by V_o, and the final volume, after compaction, is V_c, then the calculation of the volume reduction is

$$\frac{V_c}{V_o} = F$$

where F = fraction remaining of initial volume as a result of compaction
V_o = initial volume
V_c = compacted volume

Because the mass is constant (the same sample is compacted so there is no gain or loss of mass), and volume = mass/density, volume reduction can also be calculated as

$$\frac{\rho_o}{\rho_c} = F$$

where ρ_o = initial bulk density
ρ_c = compacted bulk density
F = fraction remaining of initial volume as a result of compaction

EXAMPLE
4-1

For illustrative purposes only, assume that refuse has the following components and bulk densities:

Component	Percentage (by weight)	Uncompacted bulk density (lb/ft^3)
Miscellaneous paper	50	3.81
Garden waste	25	4.45
Glass	25	18.45

Assume that the compaction in the landfill is 1200 lb/yd^3 (44.4 lb/ft^3). Estimate the percent volume reduction achieved during compaction of the waste. Estimate the overall uncompacted bulk density if the miscellaneous paper is removed.

The overall bulk density prior to compaction is

$$\frac{50 + 25 + 25}{\dfrac{50}{3.81} + \dfrac{25}{4.45} + \dfrac{25}{18.45}} = 4.98 \ \text{lb/ft}^3$$

The volume reduction achieved during compaction is

$$\frac{4.98}{44.4} = 0.11$$

or the required landfill volume is approximately 11 percent of the volume required without compaction. If the mixed paper is removed, the uncompacted density is

$$\frac{25 + 25}{\dfrac{25}{4.45} + \dfrac{25}{18.45}} = 7.18 \ \text{lb/ft}^3$$

Over the life of a landfill the density may change. The following factors could affect the volume requirement of the landfill:

- New regulations—many states have adopted waste diversion/recycling goals. If fully implemented, diversion rates as high as 70% may be achieved.
- Competing facilities—other landfills may exist or be planned that would receive some of the planned waste, or other existing landfills may close, resulting in the importation of solid waste.
- Different cover options—dirt used as daily and final cover at a landfill may consume 20 to 50% of the available landfill volume, depending on the size of the

landfill. The new landfill may rely on foam, tarps, or mulched green waste for cover and thus significantly increase the volume available for the solid waste.

- Nonresidential waste changes—a per capita generation rate takes into account all nonresidential waste currently going into the landfill. In some cases the nonresidential waste can account for more than 50% of the landfill volume needed. Some examples are large military facilities, agriculture operation, and manufacturing facilities. Either the closing or opening of these facilities can significantly affect future waste projections.

EXAMPLE
4-2

Calculate the required 20-yr landfill capacity for a community with the population projection, per capita waste generation rate, and diversion rate shown in the table. Note that the waste generation is expected to increase at approximately 3% per year through 2005 and then remain constant. Note also that the community is expected to increase its rate of waste diversion to 35% in 2004 through an aggressive recycling and yard waste composting program. Assume a soil daily cover is used that accounts for 25% of the landfill volume.

Year	Population (000)	Per Capita Generation Rate, lb/cap/day	Diversion, fraction	Waste to Landfill, tons	Waste to Landfill, cu yd
2001	105.4	5.6	0.25	8.08E+04	1.35E+05
2002	108.6	5.8	0.28	8.28E+04	1.38E+05
2003	109.8	6	0.3	8.42E+04	1.40E+05
2004	112.2	6.2	0.35	8.25E+04	1.38E+05
2005	115.2	6.4	0.35	8.75E+04	1.46E+05
2006	117.7	6.4	0.35	8.94E+04	1.49E+05
2007	121.1	6.4	0.35	9.19E+04	1.53E+05
2008	124.7	6.4	0.35	9.47E+04	1.58E+05
2009	128.4	6.4	0.35	9.75E+04	1.62E+05
2010	133.4	6.4	0.35	1.01E+05	1.69E+05
2011	139.1	6.4	0.35	1.06E+05	1.76E+05
2012	144.5	6.4	0.35	1.10E+05	1.83E+05
2013	150.7	6.4	0.35	1.14E+05	1.91E+05
2014	155.6	6.4	0.35	1.18E+05	1.97E+05
2015	163.1	6.4	0.35	1.24E+05	2.06E+05
2016	169.4	6.4	0.35	1.29E+05	2.14E+05
2017	175.3	6.4	0.35	1.33E+05	2.22E+05
2018	181.4	6.4	0.35	1.38E+05	2.30E+05
2019	187.7	6.4	0.35	1.43E+05	2.38E+05
2020	194.3	6.4	0.35	1.48E+05	2.46E+05
Total				2.15E+06	3.59E+06

A typical calculation for one year's volume is

$$\frac{(\text{Population}) \times (\text{Per Capita Gen Rate}) \times (1 - \text{Diversion}) \times (365 \text{ days/yr})}{1200 \text{ lb/yd}^3}$$

Total landfill waste volume is 3.59×10^6 yd^3. To account for the volume requirement for the cover soil:

$$0.25 \, (T) + 3.59 \times 10^6 = T$$

$$T = 4.79 \times 10^6 \text{ yd}^3$$

A landfill with an average depth of 125 ft would have a footprint of 23.8 acres.

Siting

Once the size of the landfill has been determined, it is then necessary to find an appropriate site. While the concept is simple, the execution is far from easy. A button seen at one landfill siting public hearing, NOPE—Not On Planet Earth, can best describe the challenges faced when siting a landfill. Politicians are more likely to embrace the principles of NIMTO—Not In My Term of Office.

For a local community or regional agency, an appropriate landfill site is typically within the geographic boundary of the agency. The relationships between municipalities are such that few would allow a neighboring community to site a landfill in its jurisdiction if the host community was not a participant. For a private company the geographic location is not as critical. Recently, private companies have been siting large remote landfills in rural areas. Host fees are then paid to the local community. For example, both Seattle, Washington, and Portland, Oregon, ship waste to a private regional landfill in eastern Oregon.

Once the geographical boundary of the potential site has been determined, unsuitable locations should be identified. This process is a pass/fail test referred to as a *fatal flaw analysis*. Some fatal flaws are established in regulations such as Subtitle D of the Resource Conservation and Recovery Act. For example, no landfill can be sited near an active seismic fault or an airport. Other criteria are subjective and may be established by the local community, such as no landfill will be sited within one mile of a school. Because of local regulations and criteria it is not possible to provide a complete list of potential fatal flaws, but the following should be considered as fatal flaws:

- The site is too small.
- The site is on a flood plain.
- The site includes wetlands.
- A seismic zone is within 200 ft of the site.
- An endangered species habitat is on the site.

- The site is too close to an airport (not within 5000 ft for propeller aircraft, 10,000 ft if turbine engine aircraft).
- The site is in an area with high population density.
- The site includes sacred lands.
- The site includes a groundwater recharge area.
- Unsuitable soil conditions (e.g., peat bogs) exist on the site.

All of the areas exhibiting one or more of the above fatal flaws should be highlighted on the search map. If no unhighlighted areas remain within the geographic boundary, then it will not be possible to site a new landfill. However, if areas do remain, the siting process goes to the next step that involves developing a ranking system. In addition to the fatal flaws there are subjective ranking criteria, which are also used. In many cases the public is asked to participate in developing the ranking criteria. For example, a site that has good access might receive five points, while a site with poor access might receive only one point. Other qualitative features might include relative population density, land use designation, groundwater quality, visual and noise impacts, site topography, site ownership, soil conditions, and proximity to the centroid of solid waste generation.

Once the criteria are developed, they are applied to any remaining areas that do not have fatal flaws. Potential sites are developed and ranked. The top several sites are then designated for more detailed investigation that includes on-site analysis of habitat, groundwater, and soil conditions, plus physical surveys and initial environmental assessment. Once this information is available, additional public hearings are held, and the elected officials select a site or sites for the preliminary engineering design, permitting application and detailed environmental impact report. Typically, the public is opposed to landfills, even though they may be far distances from their point of interest, such as their home; but their opposition is most intense when the landfill is to be constructed close to their home. Much of this opposition can be attributed to such undesirable developments as additional traffic, noise, odor, and litter. Most important is the belief that the siting of a landfill will reduce property values and decrease their quality of life.

There is little question that siting a landfill near residential areas will reduce property values. In one study a typical town was configured with various neighborhoods having different property values and a hypothetical landfill was placed in this town. Real estate assessors were asked to estimate the effect of this new landfill on the property values of existing homes. As expected, as the value of the property increased, so did the percentage drop in the property value.[2] Most dramatic was the far-reaching effect of the landfill. For the more expensive homes the effect on property values exceeded a three-mile radius from the proposed landfill site. These findings were confirmed in a study using actual home values.[3] Figure 4-1 shows that the largest impact on private property values are on the most expensive homes that are closest to the landfills.[4]

Because there is no compensation for the loss in the value of homes, is it any wonder that there is such strong opposition to the siting of landfills? Should society pay property owners some fraction of their loss as a fair compensation to the siting of undesirable public facilities close to their property? However, if these owners are compensated, what about the owners near other locally undesirable land uses (LULUs), such as an airport or a wastewater treatment plant?

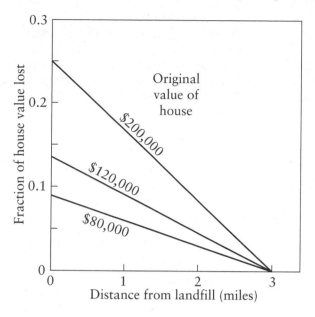

Figure 4-1 Effect on property values of siting a new landfill at a distance of three miles. Source: (4)

These questions relate directly to the issues of *environmental justice*. Undesirable facilities, such as landfills, are often placed in less economically advantaged neighborhoods because the residents do not have the resources to fight City Hall and because the land values in those areas are already low. In one classic case of taking advantage of less-advantaged people, the town council of Chapel Hill, North Carolina, made a solemn promise to a minority neighborhood that as soon as the new landfill to be sited near their homes was full, the town would find another location in another part of the county and turn the old landfill into a community park. The landfill was finally closed 20 years later (at least ten years after it was originally supposed to have been capped). The new town administration claimed that these promises to the neighborhood were null and void because they had no legal standing. The members of the present council also argued that they personally did not make these promises and therefore they should not be bound by promises made by previous town councils![5]

Permitting

On 9 October, 1991, the EPA issued 40 CFR Part 258, regulations pertaining to landfills under Subtitle D of the Resource Conservation and Recovery Act, with implementation two years later. States were required to incorporate the federal standards into their regulations. In addition the states had the flexibility of adding more stringent requirements. Subtitle D addressed such issues as location, design requirements, operating conditions, groundwater monitoring, landfill closure and post-closure, and financial assurance.

After extensive hearings across the country, the EPA decided to use a combination of *performance standards* and *design standards*. A design standard specifies a specific design, for example, a requirement for a composite liner. On the other hand, the requirement for no off-site migration of gas is a performance standard. How this standard is met is up to the applicant. In addition to Subtitle D requirements, landfills are usually subject to permitting for land use conformance, air emissions, groundwater and surface water discharge, operations, extraction for cover material, and closure.

LANDFILL PROCESSES

Biological Degradation

Refuse is approximately 75 to 80% organic matter, composed mainly of proteins, lipids, carbohydrates (cellulose and hemicellulose), and lignins. Approximately two-thirds of this material is biodegradable, one-third is recalcitrant (Figure 4-2). The biodegradable portion can be further divided into a readily biodegradable fraction (food and garden wastes) and a moderately biodegradable fraction (paper, textiles, and wood). Figure 4-3 summarizes the predominant biodegradation pathways for the decomposition of major organic classes in solid waste.

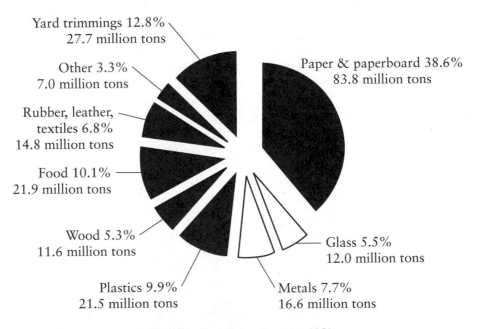

Yard trimmings 12.8%
27.7 million tons

Other 3.3%
7.0 million tons

Rubber, leather,
textiles 6.8%
14.8 million tons

Food 10.1%
21.9 million tons

Wood 5.3%
11.6 million tons

Plastics 9.9%
21.5 million tons

Paper & paperboard 38.6%
83.8 million tons

Glass 5.5%
12.0 million tons

Metals 7.7%
16.6 million tons

Figure 4-2   Organic fraction in refuse. Source: (48)

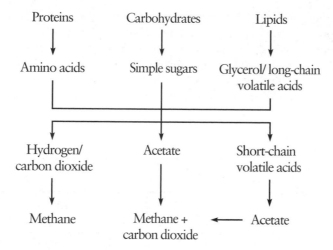

Figure 4-3 *Predominant decomposition pathways for common organic waste constituents.*

The landfill ecosystem is quite diverse due to the heterogeneous nature of waste and landfill operating characteristics. The diversity of the ecosystem promotes stability; however, the system is strongly influenced by environmental conditions, such as temperature, pH, the presence of toxins, moisture content, and the oxidation-reduction potential. The landfill environment tends to be rich in electron donors, primarily organic matter. The dominant electron receptors are carbon dioxide and sulfate. Seven key physiological microbial groups that participate in rate-limiting stabilization steps of fermentation and methanogenesis are listed in Table 4-2.

Table 4-2   Important Microbial Groups Promoting Anaerobic Waste Degradation

Microbial group	Substrate
Amylolytic bacteria	Starches
Proteolytic bacteria	Proteins
Cellulolytic bacteria	Cellulose
Hemicellulolytic bacteria	Hemicellulose
Hydrogen-oxidizing methanogenic bacteria	Hydrogen
Acetoclastic methanogenic bacteria	Acetic acid
Sulfate-reducing bacteria	Sulfate

Source: (6)

A number of landfill investigation studies[7] have suggested that the stabilization of waste proceeds in five sequential and distinct phases. During these phases, the rate and characteristics of leachate produced and gas generated from a landfill are not only dissimilar, but also reflect the microbially mediated processes taking place inside the landfill. The phases experienced by degrading wastes are described below. Leachate characteristics during the waste degradation phases are summarized in Table 4-3.

Table 4-3   Landfill Constituent Concentration Ranges as a Function of the Degree of Landfill Stabilization

Parameter	Transition	Acid formation	Methane fermentation	Maturation
Chemical oxygen demand, mg/l	480–18,000	1500–71,000	580–9760	31–900
Total volatile acids, mg/l as acetic acid	100–3000	3000–18,800	250–4000	0
Ammonia, mg/l-N	120–125	2–1030	6–430	6–430
pH	6.7	4.7–7.7	6.3–8.8	7.1–8.8
Conductivity, μS/cm	2450–3310	1600–17,100	2900–7700	1400–4500

Source (7)

Phase I—Initial Adjustment Phase

This phase is associated with initial placement of solid waste and accumulation of moisture within landfills. An acclimation period (or initial lag time) is observed until sufficient moisture develops and supports an active microbial community. Preliminary changes in environmental components occur in order to create favorable conditions for biochemical decomposition.

Phase II—Transition Phase

In the transition phase the field capacity is often exceeded, and a transformation from an aerobic to an anaerobic environment occurs, as evidenced by the depletion of oxygen trapped within the landfill media. A trend toward reducing conditions is established in accordance with shifting of electron acceptors from oxygen to nitrates and sulfates, and the displacement of oxygen by carbon dioxide. By the end of this phase measurable concentrations of chemical oxygen demand (COD) and volatile organic acids (VOAs) can be detected in the leachate.

Phase III—Acid Formation Phase

The continuous hydrolysis (solubilization) of solid waste, followed by (or concomitant with) the microbial conversion of biodegradable organic content, results in the production of intermediate volatile organic acids at high concentrations throughout this phase. A decrease in pH values is often observed, accompanied by metal species mobilization. Viable biomass growth associated with the acid formers (acidogenic bacteria), and rapid consumption of substrate and nutrients are the predominant features of this phase.

Phase IV—Methane Fermentation Phase

During Phase IV intermediate acids are consumed by methane-forming consortia (methanogenic bacteria) and converted into methane and carbon dioxide. Sulfate and nitrate are reduced to sulfides and ammonia, respectively. The pH value is elevated, being controlled by the bicarbonate buffering system, and consequently supports the growth of methanogenic bacteria. Heavy metals are removed from the leachate by complexation and precipitation.

Phase V—Maturation Phase

During the final state of landfill stabilization, nutrients and available substrate become limiting, and the biological activity shifts to relative dormancy. Gas production dramatically drops, and leachate strength stays steady at much lower concentrations. Reappearance of oxygen and oxidized species may slowly be observed. However, the slow degradation of resistant organic fractions may continue with the production of humic-like substances.

The progress toward final stabilization of landfill solid waste is subject to the physical, chemical, and biological factors within the landfill environment, the age and characteristics of landfilled waste, the operational and management controls applied, as well as the site-specific external conditions.

Leachate Production

Leachate Quantity

The total quantity produced can be estimated either by using empirical data or a water balance technique that sets up a mass balance among precipitation, evapotranspiration, surface runoff, and soil moisture storage.[8, 9, 10] Landfills operated in the Northeast have been designed using leachate generation rates of 1200–1500 gallons per acre per day (gpad) (11,200–14,000 liters/hectare/day) during active phases, 500 gpad (4,700 L/ha/day) following temporary closure, and 100 gpad (930 L/ha/day) following final closure. A 1993 study found landfills in arid regions generate only 1 to 7 gpad (9 to 60 L/ha/day).[11] Clearly, climatic conditions significantly affect leachate generation rates.

A water balance uses site-specific data to track water volumes, as shown schematically in Figure 4-4. Prior to closure, which generally involves use of an impermeable cap, the water balance can be described. Some fraction of precipitation (dependent on runoff characteristics and soil type and conditions) will percolate through the cover soil, and a fraction of this water is returned to the atmosphere through evapotranspiration. If the percolation exceeds evapotranspiration for a sufficiently long time, the amount of water the soil is able to hold (its *field capacity*) will be exceeded. The field capacity is the maximum moisture the soil (or any other material such as refuse) can retain without a continuous downward percolation due to gravity.

Without plants, the soil on top of a landfill eventually reaches its field capacity so that any excess moisture will displace the moisture in the soil. With plant cover, the plants may extract water from the soil and release it by evapotranspiration, thus drying the soil to below field capacity.

All layers of a mixture of soil and refuse in the compacted landfill also have a field capacity, or the ability to retain moisture. If the field capacity of this mixture is exceeded, the liquid (leachate) will drop to the next lowest soil/refuse layer. A soil/refuse mixture that does not attain field capacity discharges essentially no water to the deeper layers.

In the finished landfill, if the field capacity of the soil covering waste is exceeded, the water percolates through the soil and into the buried solid waste. If, in turn, the field capacity of the waste is exceeded, leachate flows into the leachate

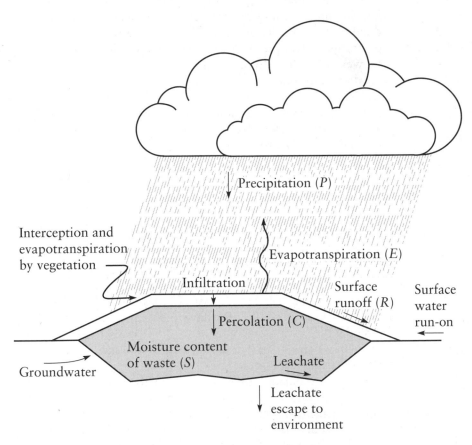

Figure 4-4 Schematic of components of water balance within a landfill.

collection system. The water balance method is a means of calculating if and by how much field capacities are exceeded, and thus is a way of calculating the production of leachate by landfills.

Some rough estimates can be used to develop the necessary calculations. Surface runoff coefficients, for example, can be estimated for different soils and slopes, as shown in Table 4-4. Precipitation data are available through the Weather Bureau. Evapotranspiration rates are also available, or can be calculated using the method of Thornthwaite,[12] which takes into account the fact that evapotranspiration is reduced as the soil moisture drops. In other words, plants use less water in dry weather.

Table 4-4   Runoff Coefficients for Grass-Covered Soils

Surface	Runoff coefficient
Sandy soil, flat to 2% slope	0.05–0.10
Sandy soil, 2% to 7% slope	0.01–0.15
Sandy soil, over 7% slope	0.15–0.20
Heavy soil, flat to 2% slope	0.13–0.17
Heavy soil, 2% to 7% slope	0.18–0.22
Heavy soil, over 7% slope	0.25–0.35

Source: (13)

The *lysimeter* can also be used to estimate evapotranspiration.[14] In most cases, however, yearly average figures are sufficiently accurate for design. Figure 4-5 shows average yearly evapotranspiration in the United States.

The field capacities of various soils and wastes are listed in Table 4-5. The field capacity of compacted waste has been estimated as 20 to 35% by volume, or about 30%, which translates to 300 mm of water/m of MSW. This value must be corrected for moisture already contained in the refuse, which is about 15%. A net field capacity of 150 mm/m of refuse (1.8 in./foot) is therefore not unreasonable.[8]

Table 4-5   Field Capacities of Various Porous Materials

Material	Field capacity, as mm water/m of soil
Fine sand	120
Sandy loam	200
Silty loam	300
Clay loam	375
Clay	450
Solid waste	200–350

Source: (8)

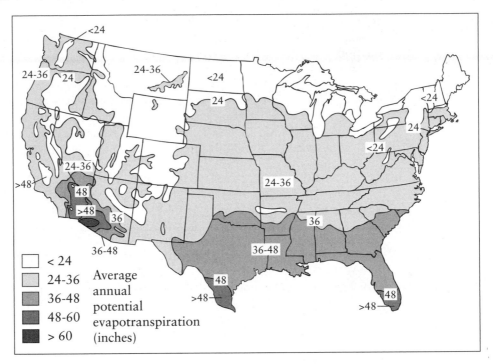

Figure 4-5 *Average annual potential evapotranspiration in the continental United States. The potential evapotranspiration will occur if the soil is completely saturated, hence these figures are highest probable values, in inches of water. The actual evapotranspiration will always be lower. Source: (13)*

The actual calculations of leachate production involve a one-dimensional analysis of water movement through soil and the compacted refuse, as shown in Figure 4-6. The figure also defines the symbols used in the calculation, which are based on the following mass balance equation:

$$C = P(1 - R) - S - E$$

where C = total percolation into the top soil layer, mm/yr
    P = precipitation, mm/yr
    R = runoff coefficient
    S = storage within the soil or waste, mm/yr
    E = evapotranspiration, mm/yr

The net percolation calculated for three areas of the United States is shown in Table 4-6. Note that the net percolation for a Los Angeles landfill is zero. This explains the absence of leachate in Southern California landfills.

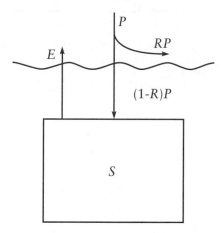

Figure 4-6   Mass balance of moisture in a landfill.

Table 4-6   Percolation in Three Landfills

	Precipitation (mm/yr) P	Runoff coefficient R	Evapotranspiration (mm/yr) E	Percolation (mm/yr) C
Cincinnati	1025	0.15	658	213
Orlando	1342	0.07	1173	70
Los Angeles	378	0.12	334	0

Source: (8)

Using the figures from Table 4-6, it is possible to estimate the number of years before leachate is produced, since the soil/waste mixture will absorb the percolated water until its field capacity is reached. In Los Angeles there will seldom be leachate generated by a landfill because of the lack of percolation. In Orlando, for a landfill 7.5 m deep, the first leachate should appear in 15 years. In Cincinnati, for a landfill depth of 20 m, the first leachate should appear in 11 years. These calculations are based on the assumption that moisture moves as a wetting front. Often there are channels and cracks in the waste that permit preferential flow. In addition some leachate is generated as a result of precipitation running over the landfill face. Therefore leachate may, and frequently is, produced long before these calculations suggest.

EXAMPLE
4-3

Estimate the percolation of water through a landfill 10 m deep, with a 1-m cover of sandy loam soil. Assume that this landfill is in southern Ohio, and that

P = 1025 mm/yr
R = 0.15
E = 660 mm/yr
Soil field capacity, F_s = 200 mm/m
Refuse field capacity, F_r = 300 mm/m, as packed

Assume further that the soil is at field capacity when applied, and that the incoming refuse has a moisture content of 150 mm/m, and therefore has a net absorptive capacity of 150 mm/m. Percolation through the soil cover is

$$C = P(1 - R) - S - E = 1025 \, (1 - 0.15) - 0 - 660 = 211 \text{ mm/y}$$

The moisture front will move

$$\frac{211 \text{ mm/y}}{150 \text{ mm/m}} = 1.4 \text{ m/y}$$

or it will take

$$\frac{10 \text{ m}}{1.4 \text{ m/y}} = 7.1 \text{ y}$$

to produce a leachate that will be collected at a rate of (211 mm × area of landfill) per year.

Once a final cap is placed on the landfill, the water balance must include a consideration of the permeability of the cap and the reduced ability of the water to percolate into the waste. Obviously, leachate volume will be reduced following landfill closure.

The model for leachate generation most frequently used today, the Hydrologic Evaluation of Landfill Performance (HELP), was developed by the U.S. Army Corps of Engineers.[15] This model requires detailed on-site morphology and extensive hydrologic data to perform the water balance. The HELP computer program, which is currently in its third version, is a quasi-two-dimensional hydrologic model of water movement across, into, through, and out of a landfill. Site-specific information is needed for precipitation, evapotranspiration, temperature, wind speed, infiltration rates, and watershed parameters, such as area, imperviousness, slope, and depression storage. The model accepts weather, soil, and design data and uses solution techniques that account for the effects of surface storage, snowmelt, runoff, infiltration, evapotranspiration, vegetative growth, soil moisture storage, lateral subsurface drainage, leachate recirculation, unsaturated vertical drainage, and leakage through soil, geomembrane, or composite liners. A variety of landfill systems can be modeled, including various combinations of vegetation, cover soils, waste cells, lateral drain layers, low-permeability barrier soils, and synthetic geomembrane liners. The HELP model is most useful for long-term prediction of leachate quantity and comparison of various design alternatives; however, it is not suitable for prediction of daily leachate production.

Leachate Quality

Within a landfill, a complex sequence of physically, chemically, and biologically mediated events occurs. As a consequence of these processes, refuse is degraded or transformed. As water percolates through the landfill, contaminants are leached from the solid waste. Mechanisms of contaminant removal include leaching of inherently soluble materials, leaching of soluble biodegradation products of complex organic molecules, leaching of soluble products of chemical reactions, and washout of fines and colloids. The characteristics of the leachate produced are highly variable, depending on the composition of the solid waste, precipitation rates, site hydrology, compaction, cover design, waste age, sampling procedures, interaction of leachate with the environment, and landfill design and operation. Some of the major factors that directly affect leachate composition include the degree of compaction and composition of the solid waste, climate, site hydrology, season, and age of the landfill. Table 4-7 shows the wide variation of leachate quality as determined by various researchers.

The capacity for leachate polluting streams and lakes is expressed by the Biochemical Oxygen Demand (BOD) which estimates the potential for using oxygen. For the sake of comparison, the BOD of raw sewage is about 180 mg/l. Another means for expessing pollutional potential is COD, or Chemical Oxygen Demand. COD is always greater than BOD.

Table 4-7   Ranges of Various Parameters in Leachate

Parameter	Ehrig 1989	Qasim and Chiang 1994	Florida landfills Grosh, 1996 (mean value)	National database (mean value)
BOD (mg/L)	20–40,000	80–28,000	0.3–4660 (149)	0–100,000 (3761)
COD (mg/L)	500–60,000	400–40,000	7–9300 (912)	11–84,000 (3505)
Iron (mg/L)	3–2100	0.6–325	—	4–2200
Ammonia (mg/L)	30–3000	56–482	BDL–5020 (257)	0.01–2900 (276)
Chloride (mg/L)	100–5000	70–1330	BDL–5480 (732)	6.2–67,000 (3691)
Zinc (mg/L)	0.03–120	0.1–30	BDL–3.02 (0.158)	0.005–846 (0.23)
Total P (mg/L)	0.1–30	8–35	—	0.02–7 (3.2)
pH	4.5–9	5.2–6.4	3.93–9.6	6.7–8.2
Lead (mg/L)	0.008–1.020	0.5–1.0	(29.2 ± 114)	0.00–2.55 (0.13)
Cadmium (mg/L)	< 0.05–0.140	< 0.05	(7.52 ± 23.9)	0.0–0.564 (0.0235)

BDL = below detection limit
Source: (16)

Leachate tends to contain a large variety of organic and inorganic compounds at relatively low concentration that can be of concern if groundwater and surface water contamination occurs. These compounds are often constituents of gasoline and fuel oils (aromatic hydrocarbons such as benzene, xylene, and toluene), plant degradation by-products (phenolic compounds), chlorinated solvents (such as used in dry cleaning), and pesticides. Inorganic compounds of concern are lead and cadmium, which come from batteries, plastics, packaging, electronic appliances, and light bulbs.

The quality of leachate directly affects viable leachate treatment alternatives. Because leachate quality varies from site to site and over time, neither biological treatment nor physical/chemical treatment processes separately are able to achieve high treatment efficiencies. A combination of both types of treatment is one of the more effective process trains for the treatment of leachate. Physical/chemical processes are needed for the pretreatment of young leachate to make it amenable to biological treatment, and to hydrolyze some refractory organic compounds found in leachate from older landfills. Biological treatment is primarily used to stabilize degradable organic matter found in young and middle aged leachates.

Gas Production

Gas Quantity

Landfill operators, energy recovery project owners, and energy users need to be able to project the volume of gas produced and recovered over time from a landfill. Recovery and energy equipment sizing, project economics, and potential energy uses depend on the peak and cumulative landfill-gas yield. The composition of the gas (percent methane, moisture content) is also important to energy producers and users. Proper landfill management can enhance both yield and quality of gas.

Mathematical and computer models for predicting gas yields are based on population, per capita generation, waste composition and moisture content, percent actually landfilled, and expected methane or landfill-gas yield per unit dry weight of biodegradable waste.[17] Mathematical models can also be used to model extraction systems, including layout, equipment selection, operation optimization, failure simulation, and problem determination and location within existing extraction systems.[18] Four parameters must be known if gas production is to be estimated with any accuracy: gas yield per unit weight of waste, the lag time prior to gas production, the shape of the lifetime gas production curve, and the duration of gas production.

In theory the biological decomposition of one ton of MSW produces 15,600 ft^3 (442 m^3) of landfill gas containing 55% methane (CH_4) and a heat value of 530 Btu/ft^3 (19,730 kJ/m^3). Since only part of the waste converts to CH_4 due to moisture limitation, inaccessible waste (plastic bags), and nonbiodegradable fractions, the actual average methane yield is closer to 3,900 ft³/ton (100 m³/tonne) of MSW. Theoretical landfill gas production potential for the U.S. is estimated at 1.4 trillion ft³/y (33 × 10¹² m³/y). However, a more conservative estimate would put the figure at roughly 500 billion ft³/y (14 billion m³/y), with an annual oil equivalent energy potential of 5.6 million tons/y.[19]

Gas production data collected at landfills across the United States show that there is significant variation in the data due to differences in environmental conditions and landfill management. The methane generation usually is between 1 to 2 ft³/lb (0.06 to 0.12 m³/kg) of waste on a dry basis, over a period of 10 to 40 years. The recoverable fraction for energy use is between 50 and 90%.[20]

Once the expected yield is determined, a model is selected to describe the pattern of gas production over time. As mentioned above, many models have been proposed, ranging from single valued, to linear increase/linear decline, to exponential decline. The EPA has published a model called LandGEM based on the following equation:

$$Q_T = \sum_{i=1}^{n} 2kL_o M_i e^{-kt_i}$$

where Q_T = total gas emission rate from a landfill, volume/time
$\quad n$ = total time periods of waste placement
$\quad k$ = landfill gas emission constant, time⁻¹
$\quad L_o$ = methane generation potential, volume/mass of waste
$\quad t_i$ = age of the ith section of waste, time
$\quad M_i$ = mass of wet waste, placed at time i

This model can be downloaded at no cost at http://www.epa.gov/ttn/catc. The following example shows how this model is applied for a simple case.

EXAMPLE 4-4

A landfill cell is open for three years, receiving 165,700 tonnes of waste per year (recall that 1 tonne = 1000 kg). Calculate the peak gas production if the landfill-gas emission constant is 0.0307 yr⁻¹ and the methane generation potential is 140 m³/tonne.

For the first year

$$Q_T = 2\ (0.0307)\ (140)\ (165,700)(e^{-0.0307(1)}) = 1,381,000 \text{ m}^3$$

For the second year this waste produces less gas, but the next new layer produces more, and the two are added to yield the total gas production for the second year. These results are plotted as Figure 4-7. The annual peak gas production is found from this figure as about 4,000,000 m³/yr.

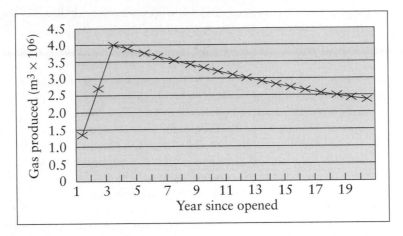

Figure 4-7   Landfill gas generated in a landfill. (See Example 4-4.)

The lag period prior to methane generation may range from a few weeks to a few years, depending on landfill conditions. The duration of gas production is also influenced by environmental conditions within the landfill. For the LandGEM model no lag period is assumed, and the duration is controlled by the magnitude of the landfill-gas emission constant.

A practical guide to forecasting of yield—based on waste generation, the ability of wastes to decompose, and the efficiency of collecting gases from that decomposition—uses the following assumptions:[21]

- 80% of the MSW generated is landfilled,
- 50% of the organic material will actually decompose,
- 50% of the landfill gas generated is recoverable, and
- 50% of the landfills are operating within a favorable pH range.

By using the percentages above, the actual methane yield turns out to be one-tenth of the theoretical yield. Unfortunately, estimates and rules of thumb are of little use for sizing equipment, determining project life, and analyzing project economics for a specific landfill. Gas generation may then be more closely estimated by the use of models, and/or by laboratory tests of waste samples, small- and large-scale pilot tests, and test cells incorporated in the landfill itself. Sources of error include inability to select representative landfill samples or well sites, inability to accurately determine the volume from which a well is actually extracting, and variation in gas generation across the landfill and over time (seasonally). Of course, once a recovery system has been installed and is functioning at steady state, reasonably precise recovery data can be collected to predict future results.

Gas Quality

Because of the prevailing anaerobic conditions within a biologically active landfill, these sites produce large quantities of gas composed of methane, carbon dioxide, water, and various trace components such as ammonia, sulfide, and nonmethane volatile organic carbon compounds (VOCs). Tables 4-8 and 4-9 provide typical composition data for MSW landfill gas. Landfill gas is generally controlled by installing vertical or horizontal wells within the landfill. These wells are either vented to the atmosphere (if gas migration control is the primary intent of the system) or connected to a central blower system that pulls gas to a flare or treatment process.

The gas can pose an environmental threat because methane is a potent greenhouse gas, and many of the VOCs are odorous and/or toxic. However, the gas has a high energy content and can be captured and burned for power, steam, or heat generation. Treatment of landfill gas is required prior to its beneficial use. Treatment may be limited to condensation of water and some of the organic acids, or may include removal of sulfide, particulates, heavy metals, VOCs, and carbon dioxide. Some of the more innovative uses of gas include power generation using fuel cells, vehicle fuel (compressed or liquid natural gas), and methanol production.

Table 4-8   Typical Constituents of MSW Landfill Gas

Component	% by volume (dry)
Methane	45–60
Carbon dioxide	40–60
Nitrogen	2–5
Oxygen	0.1–1.0
Ammonia	0.1–1.0
Hydrogen	0–0.2

Source: (22)

Table 4-9   Trace Organic Compounds Found in Florida Landfill Gas

| Compound | Concentration in ppb | |
	Average	Range
Benzene	240	BDL–470
Chlorobenzene	247	BDL–1978
Chloroethane	161	BDL–796
Cis-1,2-Dichloroethylene	76	BDL–379
Dichlorodifluoromethane	858	BDL–1760
1,1-Dichloroethane	241	BDL–1113
Dichlorotetrafluoroethane	229	BDL–544
Ethylbenzene	5559	2304–7835
M,P-Xylene	8872	3918–12675
Methylene chloride	94	BDL–749

(continued)

Table 4-9 (continued)

Compound	Concentration in ppb	
	Average	Range
o-Xylene	2872	1244–4378
p-Dichlorobenzene	372	BDL–699
Styrene	449	BDL–963
Tetrachloroethylene	131	BDL–561
Toluene	6313	2443–12215
Trichloroethylene	19	BDL–149
Trichlorofluoromethane	20	BDL–160
Vinyl chloride	333	BDL–1448
1,2,4-Trimethylbenzene	1328	244–2239
1,3,5-Trimethylbenzene	573	BDL–1140
1,1,1-Trichloroethane	268	BDL–1137
4-Ethyltoluene	1293	265–2646

BDL = below detection limit
Source: (23)

LANDFILL DESIGN

The landfill design and construction must include elements that permit control of leachate and gas. The major design components of a landfill, as shown in Figure 4-8, include the liner, the leachate-collection and management system, gas management facilities, stormwater management, and the final cap.

Liners

The liner system is required to prevent migration of leachate from the landfill and to facilitate removal of leachate. It generally consists of multiple layers of natural material and/or geomembranes selected for their low permeability. Soil liners usually are constructed of natural clays or clayey soils. If natural clay materials are not readily available, commercial clays (bentonite) can be mixed with sands to produce a suitable liner material. Geomembranes are impermeable (unless perforated) thin sheets made from synthetic resins, such as polyethylene, polyvinyl chloride, or other polymers. High-density polyethylene (HDPE) tends to be used in MSW landfill liners most commonly because it is resistant to most chemicals found in landfill leachates.

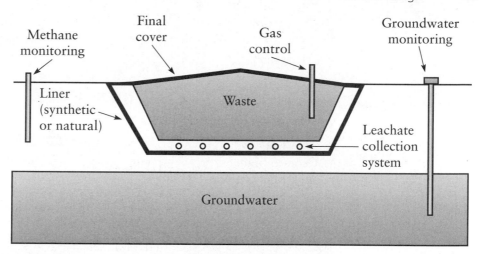

Figure 4-8   *Design components in a Subtitle D landfill.*

Landfills may be designed with single, composite, or double liners, depending on the applicable local, state, or federal regulations (see Figure 4-9). A single liner is constructed of clay or a geomembrane. A composite liner, which is the minimum liner required by RCRA Subtitle D, consists of two layers; the bottom is a clay material and the top layer is a geomembrane. The two layers of a composite liner are in intimate contact to minimize leakage. A double liner may be either two single liners or two composite liners (or even one of each). Figure 4-10 shows a synthetic liner on a side slope, ready for the earth cover. The clay layer has already been installed under this synthetic liner.

Each liner is provided with a leachate collection system. The collection system separating the two liners is a leak detection system—a series of pipes placed between the liners to collect and monitor any leachate that leaks through the top liner. Recently, the geosynthetic clay liner has been introduced for use as the top component in the double liner system. This liner is composed of a thin clay layer (usually sodium bentonite) supported by geotextiles (a geosynthetic filter) or geomembranes. The geosynthetic clay liner is easily placed in the field and uses up less volume, allowing for more volume to be used for waste deposition.

Clearly, the more layers that are included, the more protective the liner system will be. The costs, however, increase dramatically. A composite liner can cost as much as $250,000 per acre. Because the liner is so critical to groundwater protection, an exhaustive quality control/quality assurance program is required during liner installation.

Leachate Collection, Treatment, and Disposal

Leachate is directed to low points at the bottom of the landfill through the use of an efficient drainage layer composed of sand, gravel, or a geosynthetic material. Perforated pipes are placed at low points to collect leachate and are sloped to allow the moisture to move out of the landfill.

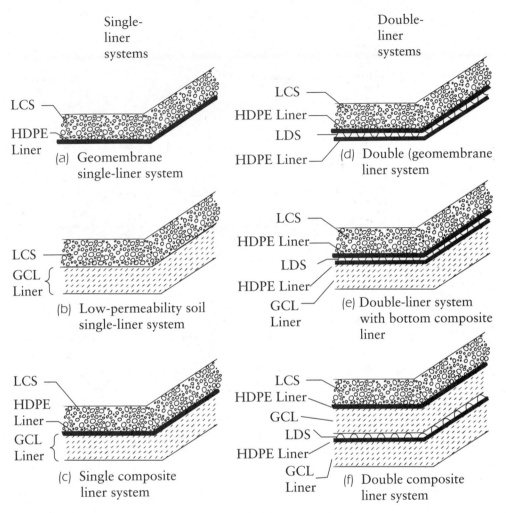

Figure 4-9   *Examples of liner systems in municipal solid waste landfills. LCS = leachate-collection system, GCL = geosynthetic clay liner, LDS = leachate-detection system.*

Leachate Collection and Storage

The primary purpose of lining a landfill cell is to minimize the potential for groundwater contamination. The liner serves as a barrier between the buried waste and the groundwater and forms a catch basin for leachate produced by the landfill. The leachate that is collected within the cell must be removed from above the liner as quickly as possible since the RCRA Subtitle D regulations restrict the head of leachate (free liquid depth) on a liner system to 30 cm. Leachate is typically

Figure 4-10   *Synthetic liner on slope, ready for earth cover.*

removed by two means: gravity flow or pumping. The various components of a leachate collection system for a MSW landfill typically include the following:

- Protective and drainage layers
- Perforated collection lateral and header pipes
- Pump station sump
- Leachate pumps
- Pump controls
- Pump station appurtenances
- Force main or gravity sewer line

Table 4-10 provides general guidelines for leachate collection system components based on a survey of landfill design engineers.[24]

Leachate removed from the landfill cell(s) is temporarily stored on site until it can be treated, recirculated, or transported off site for final treatment and disposal. Storage of leachate is also important for equalization of flow quantities and constituent quality to protect downstream treatment facilities. The typical leachate storage alternatives are surface impoundments and tanks.

Table 4-10 Design Guidance for Leachate-Collection System Components

Parameter	Range	Median
Leachate loading rate (gpd/ac)	600–1000	750
Maximum leachate head (in.)	9–12	11
Pipe spacing (ft)	60–400	180
Collection pipe dia. (in.)	6–8	8
Collection pipe material	PVC or HDPE	HDPE
Pipe slope (%)	0.5–2	1
Drainage slope (%)	0.2–2	1

Source: (24)

Leachate-Collection System Design Equations and Techniques

Because of federal regulations[25] that restrict leachate head to 12 in. (30 cm) on top of the liner, much attention has been devoted to predicting this value. The drainage length, drainage slope, permeability of the drainage materials, and the leachate impingement rate control this depth on the liner.

Darcy's law in conjunction with the law of continuity can be used to develop an equation to predict the leachate depth on the liner based on anticipated infiltration rates, drainage material permeability, distance from the drain pipe, and slope of the collection system.[26, 27]

For $R < 1/4$:

$$Y_{max} = (R - RS + R^2 S^2)^{1/2} \left[\frac{(1 - A - 2R)(1 + A - 2RS)}{(1 + A - 2R)(1 - A - 2RS)} \right]^{1/2.4}$$

For $R = 1/4$:

$$Y_{max} = (R - RS + R^2 S^2)^{1/2} \exp\left[\frac{1}{B} \tan^{-1}\left(\frac{2RS - 1}{B} \right) - \frac{1}{B} \tan^{-1}\left(\frac{2R - 1}{B} \right) \right]$$

For $R > 1/4$:

$$Y_{max} = \frac{R(1 - 2RS)}{1 - 2R} \exp\left[\frac{2R(S - 1)}{(1 - 2RS)(1 - 2R)} \right]$$

where $R = q/(K \sin^2\alpha)$, unitless
$A = (1 - 4R)^2$, unitless
$B = (4R - 1)^2$, unitless
$S = \tan \alpha$, slope of liner, unitless

Y_{max} = maximum head on liner, ft
   L = horizontal drainage distance, ft
   α = inclination of liner from horizontal, degrees
   q = vertical inflow (infiltration) per unit of horizontal area, ft/day
   K = hydraulic conductivity of the drainage layer, ft/day

These equations are obviously cumbersome to use, and a more conservative but far-easier-to-use equation has been proposed:[28]

$$Y_{max} = \frac{P}{2}\left(\frac{q}{K}\right)\left[\frac{K\tan^2\alpha}{q} + 1 - \frac{K\tan\alpha}{q}\left(\tan^2\alpha + \frac{q}{K}\right)^{1/2}\right]$$

where Y_{max} = maximum saturated depth over the liner, ft
   P = distance between collection pipes, ft
   q = vertical inflow (infiltration), defined in this equation as from a 25-year, 24-hour storm, ft/day

This equation can be used to calculate the maximum allowable pipe spacing based on the maximum allowable design head, anticipated leachate impingement rate, slope of the liner, and permeability of the drainage materials. The equation suggests that, holding all other parameters constant, the closer together the pipes are placed (at greater construction cost), the lower the head will be. A reduced head on the liner results in a lower hydraulic driving force through the liner, and the consequence of a puncture in the liner is likewise reduced.

EXAMPLE
4-5

Determine the spacing between pipes in a leachate-collection system using granular drainage material and the following properties. Assume that in the most conservative design all stormwater from a 25-year, 24-hour storm enters the leachate collection system.

Design storm (25 years, 24 hours) = 8.2 in = 0.00024 cm/s
Hydraulic conductivity = 10^{-2} cm/s
Drainage slope = 2%
Maximum design depth on liner = 15.2 cm

$$P = \frac{2\,Y_{max}}{\left(\dfrac{q}{K}\right)\left[\dfrac{K\tan^2\alpha}{q} + 1 - \dfrac{K\tan\alpha}{q}\left(\tan^2\alpha + \dfrac{q}{K}\right)^{1/2}\right]}$$

$$P = \frac{2(15.2)}{\left(\dfrac{0.00024}{0.01}\right)\left[\dfrac{0.01(0.02)^2}{0.00024} + 1 - \dfrac{0.01(0.02)}{0.00024}\left((0.02)^2 + \dfrac{0.00024}{0.01}\right)^{1/2}\right]} = 1428 \text{ cm}$$

Geosynthetic drainage materials, or geonets, have been introduced to increase the efficiency of the leachate collection systems over such natural materials as sand or gravel. A geonet consists of a layer of ribs superimposed over each other, providing highly efficient in-plane flow capacity (or transmissivity). If a geonet is used between the liner and drainage gravel, the spacing between the collection pipes can be estimated from:[29]

$$\theta = \frac{q P^2}{4 Y_{max} + 2 P \sin \alpha}$$

where θ = transmissivity of the geonet, ft^2/day

Leachate Treatment and Disposal

In wet areas leachate treatment, use, and disposal represents one of the major expenses of landfill operations—not only during the active life of the landfill but also for a significant period of time after closure. Due to the cost implications, considerable attention must be given to selecting the most environmentally responsible, cost-effective alternative for leachate treatment and disposal. The optimal treatment solution may change over time as new technologies are developed, new regulations are promulgated, and/or leachate quality varies as a function of landfill age.

Leachate-treatment needs depend upon the final disposition of the leachate. Final disposal of leachate may be accomplished through co-disposal at a wastewater treatment plant or through direct discharge, both of which could be preceded by on-site treatment. Leachate treatment is often difficult because of high organic strength, irregular production rates and composition, variation in biodegradability, and low phosphorous content (if biological treatment is considered). Several authors have discussed leachate treatment options,[30–33] which are briefly summarized in Table 4-11. Generally, where on-site treatment and discharge is selected, several unit processes are required to address the range of contaminants present. For example, a leachate treatment facility at the Al Turi Landfill in Orange County, New York, uses polymer coagulation, flocculation, and sedimentation, followed by anaerobic biological treatment, two-stage aerobic biological treatment, and filtration prior to discharge to the Wallkill River.[34] Pretreatment requirements may address only specific contaminants that may create problems at the wastewater treatment plant. High-lime treatment has been practiced at the Alachua County, Florida, Southwest Landfill to ensure low heavy-metal loading on the receiving treatment facility.

The most economical alternative for leachate treatment is often to transport the wastewater off-site to a Publicly Owned Treatment Works (POTW) or commercial wastewater treatment facility. This option allows landfill owner/operators to focus on their primary solid waste management charge while letting wastewater experts handle the treatment of contaminated liquids. There are also economies of scale for larger treatment plants that cannot be realized for small plants designed to treat only the flow from a single landfill. Off-site treatment of leachate also alleviates some of the permitting, testing, monitoring, and reporting requirements for the landfill owner. Disposal to POTWs sometimes generates opposition from the plant owner, typically a municipal entity or a private operator running the plant on a contract basis for a municipal owner. While the acceptance of leachate certainly merits careful consideration, the waste stream should be evaluated as any other industrial wastewater stream. Leachate is generally more comprehensively characterized than other industrial waste streams. Modern Subtitle D landfills have rigorous waste acceptance criteria to keep hazardous, toxic, radioactive and other noncompatible waste streams out of the landfill. Control of waste input has the effect of improving the quality of the leachate.

Table 4-11 Summary of Leachate Treatment Options

Treatment option	Removal objective	Comments
Biological		Best used on "young" leachate
Activated sludge	BOD/COD	Flexible, shock resistant, proven, minimum SRT increases with increasing organic strength, > 90% BOD removal possible
Aerated lagoons	BOD/COD	Good application to small flows, > 90% BOD removal possible
Anaerobic	BOD/COD	Aerobic polishing necessary to achieve high-quality effluent
Powdered activated carbon/act. sludge	BOD/COD	> 95% COD removal, > 99% BOD removal
Physical/Chemical		Useful as polishing step or for treatment of "old" leachate
Coagulation/ Precipitation	Heavy Metals	High removal of Fe, Zn; moderate removal of Cr, Cu, Mn; little removal of Cd, Pb, Ni
Chemical oxidation	COD	Raw leachate treatment requires high chemical dosages, better used as polishing step
Ion exchange	COD	10–70% COD removal, slight metal removal
Adsorption	BOD/COD	30–70% COD removal after biological or chemical treatment
Reverse osmosis	Total dissolved solids (TDS)	90–96% TDS removal

Source: (16)

The Clean Water Act (CWA) mandates that sources directly discharging into surface waters must comply with effluent limitations listed in the National Pollution Discharge Elimination System (NPDES) permit(s). The CWA also requires the EPA to promulgate nationally applicable pretreatment standards that restrict pollutant discharges for those who discharge wastewater indirectly through sewers flowing to wastewater treatment plants. National pretreatment standards are established for those pollutants in wastewater from indirect discharges that may pass through or interfere with wastewater treatment plant operations. In addition wastewater treatment plants are required to implement local treatment limits applicable to their industrial indirect discharges to satisfy any local requirements (40 CFR 403.5).

Emerging Technologies for the Treatment of Leachate

Because the treatment of leachate is both difficult and expensive, new techniques are being developed for managing this potential pollutant.[35]

Reverse osmosis has nothing to do with osmotic pressure or osmosis, but is the common name for forcing a fluid through a semipermeable membrane, thus separating soluble components on the basis of molecular size and shape. It is one of the treatment systems capable of removing dissolved solids. A high-pressure pump forces the leachate through a membrane, overcoming the natural osmotic pressure, and dividing the leachate into two parts, a water stream (permeate or product) and a concentrated (brine) stream. Molecules of water pass through the membrane while contaminants are flushed along the surface of the membrane and exit as brine. Typical recovery rates (percentage of the feed that becomes permeate or clean water) range from 75 to 90%.

Direct osmosis concentration is a cold-temperature-membrane process that separates waste streams in a low-pressure environment. The system operates by placing a semipermeable membrane between the leachate and an osmotic agent, typically a salt brine with a concentration of approximately 10%. The semipermeable membrane allows the passage of water from the leachate (without outside pressure) into the salt brine, but rejects the contaminants found in the leachate. As the process continues, the brine becomes dilute. This process is called osmosis, and it continues until the water concentrations on both sides of the membrane are equivalent.

Evaporation will actually dispose of the water component of water-based waste streams, such as leachate. This technology can reduce the total volume of leachate to be managed to less than 5% of the original volume. Other technologies (reverse osmosis, ultra/microfiltration, and conventional treatment systems) separate the waste stream into two components, but the water produced by these other treatment technologies must still be disposed of. Landfill-gas-fueled evaporation is a technology that effectively integrates the control of landfill gas and landfill leachate. Evaporative systems typically require an air permit for the flare and usually a modification to the landfill's solid waste permit to address the leachate management practices.

A *vapor compression distillation* process differs from an evaporation process in that a clean effluent is produced. Leachate is introduced into the VCD system through a recirculation loop. This loop constantly pumps leachate and concentrates at a high rate from the bottom of a steam disengagement vessel through a primary heat exchanger and back again. The leachate and concentrate enter the disengagement vessel through a tangential nozzle at a velocity sufficient to create a cyclonic separation of steam from the liquid. As the leachate and concentrate are rapidly recirculated, active boiling occurs on the inside of the primary heat exchanger and within the cyclonic pool formed inside the disengagement vessel.

The *mechanical vapor recompression* process uses the falling film principle in a vacuum. The core of the process is a polymeric evaporation surface (heat transfer element), on which water can boil at a temperature of 50 to 60 °C (120 to 140 °F). The process functions similarly to a heat pump. The raw leachate is pumped through two parallel heat exchangers and into the bottom of the evaporation vessel. From there, a circulation pump transfers a small volume of the leachate into the top of the vessel, where it is evenly distributed on the heat transfer element.

Several different types of *land treatment systems* are available for treating landfill leachates. However, these systems are usually sized for small leachate flows, since a significant amount of land could be required to handle larger flows. The three most common forms of land treatment systems found at MSW landfills are constructed wetlands, windrow composting, and growing poplar trees.

Leachate Recirculation

Landfills can be used as readily available biological reactors for the treatment of leachate. The landfill can be essentially transformed into an engineered reactor system by providing containment using liners and covers, sorting and discrete disposal of waste materials into dedicated cells, collection and recirculation of leachate, and management of gas. Such landfills are capable of accelerated biochemical conversion of wastes and effective treatment of leachate.

Most sanitary landfills are traditionally constructed so the leachate is collected and removed. The rate of stabilization in "dry" landfills may require many years, thereby extending the acid formation and methane fermentation phases of waste stabilization over long periods of time. Under these circumstances, decomposition of biodegradable fractions of solid waste will be impeded and incomplete, often preventing commercial recovery of methane gas and delaying closure and possible future reuse of the landfill site. Furthermore, during single-pass operations, the leachate must be drained, collected, and treated prior to final discharge.

In contrast, leachate recirculation may be used as a management alternative that requires the containment, collection, and recirculation of leachate back through the landfilled waste. This option offers more rapid development of active anaerobic microbial populations and increases reaction rates (and predictability) of these organisms. The time required for stabilization of the readily available organic constituents can be compressed to as little as two to three years rather than the usual 15- to 20-year period. This accelerated stabilization is enhanced by the routine and uniform exposure of microorganisms to constituents in the leachate, thereby providing the necessary contact time, nutrients, and substrates for efficient conversion and degradation. Hence leachate recirculation essentially converts the landfill into a dynamic *anaerobic bioreactor* that accelerates the conversion of organic materials to intermediates and end products.

The advantages of leachate recirculation have been well documented. Application of leachate recirculation to full-scale landfills has occurred with increasing frequency in recent years. Leachate is returned to the landfill using a variety of techniques, including wetting of waste as it is placed, spraying of leachate over the landfill surface, and injection of leachate into vertical columns or horizontal trenches installed within the landfill. It is important to design and operate other landfill components—such as gas management systems, leachate collection, and final and intermediate cover—so that they are compatible with bioreactor operation.

To optimize bioreactor operations, the waste moisture levels must be controlled by the rate of leachate recirculation, which is a function of waste hydraulic conductivity and the efficiency of the leachate introduction technique. To enhance the effect of leachate, the recirculation operations should be moved from one area to another, pumping at a relatively intense rate for a short period of time, then moving to another area. Empirical data provide some guidance for rates of moisture input of approximately 2 to 4 m^3/day/linear m of trench width (22 to 44 ft^3/day/ft of trench width) and 5 to 10 m^3/day (175 to 350 ft^3/day) per well. Field experimentation is required, however, to determine site-specific capacity.

The quantity of liquid supplied is a function of such waste characteristics as moisture content and field capacity. In some cases the infiltration of moisture resulting from rainfall is insufficient to meet the desired waste moisture content for optimal decomposition, and supplemental liquids (i.e., leachate from other areas, water, wastewater, or biosolids) may be required. Sufficient liquid supply must be ensured to support project goals. For example, the goal of moisture distribution might be to bring all waste to field capacity. The most efficient approach to reach field capacity is to increase moisture content through wetting of the waste at the working face and then uniformly to reach field capacity through liquid surface application or injection.

The addition of supplemental liquids increases the base flow of leachate from the landfill. This additional flow must be considered during design, especially following rain events when large amounts of leachate may be generated. Sufficient leachate storage must be provided to ensure that peak leachate generation events can be accommodated. While a properly designed and operated landfill will minimize extreme fluctuations of leachate generation with rainfall events, in wet climates leachate generation will at times exceed the amount needed for recirculation. Other factors—such as construction, maintenance, regulations, etc.—may also dictate that leachate not be recirculated from time to time. Therefore, it is very important to have contingency plans in place for off-site leachate management for times when leachate generation exceeds on-site storage capacity.

Leachate recirculation should be controlled to minimize outbreaks and to optimize the biological processes. Grading the cover to direct leachate movement away from side slopes, providing adequate distance between slopes and leachate injection, eliminating perforations in recirculation piping near slopes, and avoiding cover that has hydraulic conductivity significantly different from the waste can control seeps. In addition it may be desirable to reduce initial compaction of waste to facilitate leachate movement through the waste. A routine monitoring program designed to detect early evidence of outbreaks should accompany the operation of any leachate recirculation system. Alternate design procedures, such as early capping of side slopes and installation of subsurface drains, may also be considered to minimize problems with side seepage.

The depth of leachate on the liner is a primary regulation in the US to protect groundwater and is a major concern for regulators approving bioreactor permits. Control of head on the liner requires the ability to maintain a properly designed leachate collection system, monitor head on the liner, store or dispose of leachate outside of the landfill, and remove leachate at rates two to three times the rate of normal leachate generation. Several techniques are used to measure head on the liner, including sump measurements, piezometers, bubbler tubes, and pressure transducers. Measuring the head with currently available technology provides local information regarding leakage potential; however, for a more realistic evaluation a more complete measurement may be required.

The construction, operation, and monitoring of leachate recirculation systems will affect daily landfill operations. If a leachate recirculation system is to be used, it should be viewed as an integral part of landfill operations. Installation of recirculation systems must be coordinated with waste placement, and should be considered during planning of the fill sequence.

Landfill Gas Collection and Use

Gas generated within a landfill will move by pressure gradient, following paths of least resistance. Uncontrolled migrating gas can collect in sewers, sumps, and basements, leading to tragic consequences if explosions occur. To prevent gas migration, gas vents or wells must be provided.

There are two basic systems for gas emissions control: passive collection and active extraction. Passive systems collect landfill gas using vent collectors and release the gas to the atmosphere without treatment or conveyance to a common point. Passive vents are often provided using natural convective forces within the landfill to direct gas to the atmosphere. Passive vents may reach only a few feet below the cap or may reach up to 75% of the landfill depth, designed in a similar manner to the active extraction well described below. Typical spacing for a passive vent is one per 9000 yd^3 (7500 m^3).

Active extraction systems link collection wells with piping and extract the gas under vacuum created by a central blower. Active extraction wells may be vertical or horizontal wells, although vertical wells are more frequently employed. Vertical wells are installed in landfills using auger or rotary drills. A typical extraction well is shown in Figure 4-11. Wellheads provide a means of controlling the vacuum applied at the well as well as monitoring of gas flow rate, temperature, and gas quality. Spacing of wells is a function of gas flow. Table 4-12 provides general guidance on well construction.

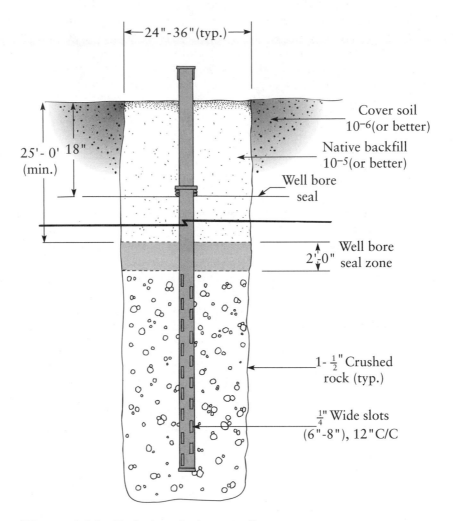

Figure 4-11 Typical vertical gas well.

Landfill gas is extracted by central blowers that create negative pressure in the pipe network. These blowers are sized according to the volume of gas they must move. The greater the required flow, the greater must be the negative pressure that must be created, and therefore the more energy is required. The collection system should be designed to minimize head loss by providing sufficiently large pipes and by minimizing the number of valves and bends in the pipe. However, large pipes can be costly, and the design must balance the cost of the pipes and valves against the energy requirements of the blowers.

Table 4-12 Guidelines for Construction of Vertical Gas Collection Systems

Parameter	Recommendation
Well depth	75% of depth or to water table, whichever comes first
Perforations	Bottom 1/3 to 2/3 Minimum 25 ft below surface
Casing	3–8 in. PVC or HDPE, telescoping well joint
Spacing (on center)	Interior collection system 200–500 ft Perimeter collection system 100–250 ft
Well density	One well per 1/2 to 2 acres
Minimum gas-collection piping slope	3 percent
Well bore diameter	12 to 36 in. is standard (24, 30, 36 in. most common)

Source: (36)

The velocity of the flow through the piping system can be estimated using the continuity equation if compressibility is neglected.

$$Q = vA$$

where Q = landfill-gas flow rate, ft^3/sec
$\quad$ v = landfill-gas velocity, ft/sec
$\quad$ A = cross-sectional interior area of the pipe, ft^2

The head loss through the pipe can be estimated using the Darcy-Weisbach equation, usually stated as

$$\Delta P = \frac{\rho f L v^2}{2gD}$$

where ΔP = pressure drop, ft
$\quad$ f = Darcy-Weisbach friction factor, function of pipe roughness and diameter
$\quad$ L = length of pipe, ft
$\quad$ v = velocity, ft/sec
$\quad$ g = gravitational constant, ft/sec^2
$\quad$ D = diameter, ft
$\quad$ ρ = gas density, lb/ft^3

The Darcy-Weisbach friction factor can be read from a Moody diagram, found in all fluid mechanics textbooks. The diagram is used by first calculating ϵ/D, where ϵ is the pipe roughness, function of the pipe material. The rougher the pipe, the greater is the value of ϵ.

This equation gives the pressure drop in feet of water if all length units are in feet and the fluid is water. If the fluid is a gas, the equation can be modified, taking into account the density of the fluid, as

$$\Delta P = \frac{(0.0096)\rho f L v^2}{Dg}$$

where ΔP = pressure drop when a gas is flowing in the pipe, ft of water
ρ = density of the gas, lb/ft^3

The density of the gas is calculated using the universal gas law,

$$\rho = \frac{MP}{RT}$$

where ρ = density of the gas, lb/ft^3
M = molecular weight of the gas, lb/mole
P = pressure, lb/ft^2
R = universal gas constant, lb-ft/mole-$°R$
T = absolute temperature, $°R$ ($°F$ + 460)

Typically, the pressure inside the collection pipes is close to atmospheric, about 30 inches of water, or 13.6 lb/in^2, or about 1958 lb/ft^2. The molecular weight of the gas can be estimated as 28 lb/mole, with the universal gas constant then having units of lb-ft/$°R$.

EXAMPLE
4-6

Calculate the pressure drop in 800 ft of 4-in.-diameter PVC pipe carrying 500 ft^3/min of landfill gas at 120°F. Assume the pressure in the pipeline is close to atmospheric, about 13.6 $lb/in.^2$, and the molecular weight of the landfill gas is 28 lb/mole. Assume the gas is incompressible.

The area of a 4-in.-diameter PVC Schedule 40 pipe (4.026 in. ID) is 0.0884 ft^2. From the continuity equation:

v = (500/60)/0.0884 = 94 ft/sec

The roughness, ϵ, for PVC pipe is 0.000005 ft. Thus ϵ/D = 0.00005/(4/12) = 0.000015. From the modified Moody diagram, the friction factor is read as f = 0.016.

The gas density is estimated assuming the pressure at 1958 lb/ft^2, the molecular weight at 28 lb/mole, the universal gas constant at 1543 lb-ft/$°R$ and the absolute temperature at 120°F + 460. Thus

$$\rho = \frac{(28)(1958)}{(1543)(120 + 460)} = 0.0613 \text{ lb/ ft}^3$$

The pressure drop is then calculated as

$$\Delta P = \frac{(0.0096)(0.0613)(0.016)(800)(94)^2}{(4.026/12)(32.2)}$$

$$= 6.2 \text{ ft of water}$$

Repeating the calculations for 6- and 8-in. pipes, the pressure drops are 0.8 ft and less than 0.1 ft, respectively, of water column.

Assuming head loss in the valves and fittings at 20% of the total pressure drop and adding an additional 10 in. of water column to maintain negative pressure at the wellhead, the 4-in. pipe requires a negative blower pressure at the suction side of the blower at 84 in. water column. The 6-in. pipe requires about 20 in. water column, and the 8-in. pipe requires less than 12 in.

Technical Issues in Landfill Gas Use

Some of the technical issues regarding the use of landfill gas include gas composition, the effects of corrosives and particulates on equipment, potential energy losses, and gas extraction and cleanup. Energy users are concerned with the problems and solutions associated with the use of landfill gas as a fuel source. Much information can be obtained from equipment manufacturers as to the fundamentals and site-specific applications of gas extraction and cleanup.

The physical, chemical, and combustion characteristics of landfill gas can have significant impact on energy recovery equipment selection and operation. Trace organics, such gases as hydrogen sulfide and others, and particulates can cause corrosion and excessive wear. Carbon dioxide, nitrogen, and water vapor, with various inert materials, may reduce efficiency. Variability of gas composition and production rate over time exacerbates the problems associated with gas cleanup operations and energy applications.

The primary source of trace gases is from discarded volatile materials and their transformation by-products. Although most trace gases, primarily hydrocarbons, are harmless to energy use, halogenated hydrocarbons may cause problems upon combustion. Volatile and nonvolatile acidic hydrocarbons—such as the organic acids found in untreated landfill gas—are also highly corrosive. Trace constituents have been reported to cause corrosion, combustion chamber melting, and deposits on blades of turbine engines, as well as internal combustion engines.[37]

Hydrogen sulfide and water vapor can also have corrosive effects. The use of landfill gas as a vehicle fuel requires their removal due to corrosion problems when they condense during gas compression and cooling. Hydrogen sulfide in concentrations as low as 100 ppm may lead to corrosion in piping, storage tanks, and engines.[38] Landfill gas with hydrogen sulfide and halide concentrations, chiefly chloride and fluoride, as low as 21 ppm and 132 ppm, respectively, require pretreatment prior to application in fuel cells.[39] Stringent cleanup technology is also applied to remove the trace constituents from landfill gas when purifying it to pipeline quality natural gas.[37]

Landfills contain a large amount of soil and other particulate matter. Extraction systems can dislodge and take up these particulates into the gas stream. Deposits in engines and buildup in oil results in increased wear with a simultaneous decrease in lubricant capacity and life. Particulates can be removed by gas filtration or gas refrigeration. Dimethyl siloxane, a gaseous silicon compound, will combust to produce silica deposits in internal combustion engine cylinders and gas turbines, resulting in decreased combustion chamber volume, increased compression ratio, a tendency to detonate, and abrasion of valve stems and guides from hardened deposits.[20, 40]

The purpose of treatment systems is to remove particulates, condensates, and trace compounds, and to upgrade landfill-gas quality for direct use or for energy/synthetic fuel conversion. Filtration and condensate removal are the more common cleanup approaches, whereas refrigeration and desiccation are used less frequently or as a means to enhance cleanup efficiency.[20]

Applications of Landfill Gas Use

Landfill gas can be *flared* (burned) on site, but this is not a beneficial application of this resource. Beneficial energy recovery systems include direct use, electricity generation, and conversion to chemicals or fuels.

Boilers and other direct combustion applications are by far the cheapest and easiest use options. Direct uses of landfill gas to replace or supplement coal, oil, propane, and natural gas have been successfully demonstrated. Applications include boiler firing, space heating, cement and brick kilns, sludge drying, and leachate drying and incineration. In most cases gas cleanup consists of little more than condensate removal. There is little risk for the end user in terms of gas quality, use, and continuity of supply. The payback period can be as little as a few months, but is dependent on the negotiated price paid for the fuel replaced. The typical discounts of 10 to 20 percent are influenced by the pipeline distance and the gas quantity, quality, and variations permitted. The ideal situation is one where a user, located within a two-mile radius of the landfill, could accept all of the gas generated on a continuous basis.[41]

The potential for use as *vehicle fuel* exists if the gas is upgraded to natural gas quality, vehicles are modified to operate on some form of natural gas, and refueling stations are widely available and equipped for dispensing natural gas in the proper form. The technology is well established, and gas-powered stationary internal combustion engines have been available for decades. Anaerobic digester gas is processed into fleet vehicle and engine generator fuels in several locations in Florida, including Plantation and Tampa.[43] In New Zealand many vehicles can already run on upgraded landfill gas, and the number is increasing.[19] Worldwide over 700,000 vehicles, many of them passenger cars, are fueled by natural gas.[40] However, in most countries the use of landfill gas as a vehicle fuel is limited to landfill or other municipal fleets with a limited range of operation. Until a greater number of vehicles, modified for processed landfill gas use, provide the demand necessary to support a more diverse infrastructure, the generation of landfill gas will outpace the fuel requirements of a landfill fleet.[19]

The benefits of using landfill gas as a vehicle fuel include improved air quality benefits that extend from reduced flare emissions to lower emissions from vehicles that burn landfill gas as an alternative to diesel. Results of demonstration projects show substantial vehicle NO_x emission reductions, with minimal CO increases, when using fuel produced from landfill gas. Fuel economy and engine performance do not suffer significantly. The use of landfill gas in vehicles offers greater economic benefits than power generation using treated or even untreated landfill gas.[42]

Conversion of landfill gas to synthetic fuels and chemicals is also possible, if not economically feasible. The technologies include hydrocarbon production by the Fischer-Tropsch process and methanol synthesis by high-pressure chemical catalysis and partial biological oxidation. Gas-based chemical processes for synthesizing acetic acid and other compounds are also available. These technologies were developed for large-scale production of synfuels using coal-gas feedstocks. Production ventures were costly, and their products were expensive. The typical landfill can produce only one to ten percent of the gas required for the size of plant considered for these technologies and processing techniques.[20]

Electrical power generation (for internal combustion engines and gas turbines) is by far the most common landfill gas-to-energy application. A number of proven technologies with a range of economies of scale, easily adapted to landfill gas; a highly developed and distributed transmission infrastructure; and a virtually limitless market make landfill-gas-to-electricity projects some of the easiest and most profitable use alternatives. Eighty-five projects, capable of generating 344 MW of electricity, represent three-fourths of all landfill-gas-to-energy projects in the United States. Of these, internal combustion engines account for 61 of the 85 projects, which are responsible for generating just over half (51%) of the electricity produced at landfill sites.[41]

Projects are set up according to the perceived electrical power generating capacity and the number of generating units. If landfill gas production is insufficient to support at least one MW of power generation, it is generally deemed economically unsuitable. Internal combustion engines are typically used at sites capable of producing less than three MW. Three to five engines are employed per project. Turbines, with higher horsepower and performance when operating at full load, are used at landfills where gas quantity can support greater than three MW, requiring only one or two turbine units for generating electricity.[41]

Electrical power generation (fuel cells) is another well-established technology that, until recently, was subject to unfavorable economics when using landfill gas. Fuel cells are basically electrochemical batteries utilizing molten carbonate or phosphoric acid, fueled by coal, petroleum, natural gas, or other such hydrocarbon feedstocks. Hydrogen from the converted fuel combines with oxygen to produce electricity. The advantages of fuel cells over other use options include higher energy efficiency, availability to smaller landfills, minimal byproduct emissions, minimal labor and maintenance, and minimal noise impact.[41] Economic analysis reveals the benefits of fuel cells comparable to their environmental impacts. Locations with higher population densities generally have greater air emission problems. Likewise, commercial and industrial power rates are greater for these highly industrialized, highly urbanized areas. With a large market of potentially large profits, strict emissions standards, and generous offsets and credits, fuel cells are ideal for commercially viable electricity generation.[39]

Purification to pipeline-quality gas is also an attractive option for the use of landfill gas. There are major differences between landfill gas and pipeline-quality natural gas in composition and energy content. Landfill gas has a lower Btu content, combusts at a lower temperature, is more corrosive, and contains much greater concentrations of undesirable gases (CO_2, O_2, N_2) and harmful halocarbons than pipeline-quality natural gas. Diligent extraction and stringent cleanup are therefore necessary to render landfill gas virtually devoid of all components save methane. The required gas cleanup, an expensive and complex process for other use alternatives, includes nearly complete CO_2 removal. So expensive is the conversion of landfill gas to natural gas that only large landfills can attain the economies of scale necessary to support operations. Although other, less elaborate projects exist, fewer than ten U.S. landfills produce gas for pipeline use. Commercially attractive in the early 1980s, when gas prices were high, projects operating today all have favorable, long-term contracts.[15, 20]

Nontechnical Issues Relating to Landfill Gas Use

The beneficial use of landfill gas has safety, environmental, and energy implications. There also exist economic (as well as regulatory) issues for direct use, energy production, or co-generation. For any landfill gas project these nontechnical issues must be identified, and their effects evaluated, quantified, and ultimately factored into an overall cost–benefit analysis.

Energy recovery for off-site use generally involves investor capital for defraying extraction, cleanup, and use project costs, and generates revenue for landfill owners/operators. Generating revenue from landfill gas is easier said than done. Less than two percent of landfills in the United States have landfill-gas-to-energy projects, mainly due to the current low prices of energy. More than 30 United States projects have ceased operation due to unfavorable economics. However, if oil prices skyrocket as they did during the oil crisis of the late 1970s, the use of landfill gas would truly become economically as well as environmentally beneficial.[41] In the interim, economic benefits from landfill-gas-to-energy projects exist in the form of federal (and in some cases, state) tax incentives and investment credits. However, the tendency of state and local governments to enact legislation and regulatory guidelines of greater stringency than those of federal standards, may stifle the use of options economically available for the recovery and use of landfill gas. Despite all of these obstacles, a landfill gas project must still have the potential to amass significant return on investment from operations, without the benefit of tax credits and other benefits, to attract investment. Most projects, however, are justified financially only through tax credits and favorable purchase price as required by regulation.

Geotechnical Aspects of Landfill Design

Landfill stability is an important aspect of design, particularly in light of the complex, multilayer construction of modern landfills. Failures can occur as slope failures during the construction of a landfill, or slope failures after the landfill has been closed. The critical point of failure is usually the soil/geosynthetic and geosynthetic/geosynthetic interfacial surfaces as well as waste slopes. Landfill failures can have catastrophic results, including damage of leachate- and gas-collection systems, contamination of the surrounding environment, and even—in extreme cases—loss of life. The stability of landfills must be investigated under both static and seismic conditions, as required by RCRA Subtitle D.

Landfill stability is normally analyzed using readily available computer software. This analysis requires knowledge of properties of waste, materials used in the liners and caps (synthetic and natural), and foundation soils. These include such well-known geotechnical properties as unit weight, shear strength, shear moduli, internal friction angle, cohesion, and internal pore pressure, plus geosynthetic material properties such as tensile strength, surface roughness, flexibility, and surface wetness. Waste properties are often difficult to determine because of waste heterogeneity, changes in properties with time, and the difficulty in collecting samples necessary to evaluate these properties.

The operation of the landfill as a bioreactor can create unique geotechnical conditions. Increased moisture content leading to waste saturation can result in positive internal pore pressure and reduced angle of friction (angle at which slope failure occurs at an interface). The impact of waste decomposition on shear strength is also of critical concern. Many studies report finding mud-like conditions at the bottom of wet, deep landfills. Overaggressive leachate recirculation (compounded by the use of impermeable cover soil) has been cited as a contributing factor to the catastrophic slope failure of a landfill in Colombia, South America.

To minimize the probability of landfill failure, it is recommended that side slopes of completed and capped landfills be no greater than 1:3, with 1:4 being preferable. Shallow slopes mean reduced air space for waste disposal; however, the reduced risk of slope failure outweighs economic advantages of increased disposal volume. Use of textured geomembranes can also reduce slippage along interfaces. Finally, proper drainage and gas pressure relief in the cap will reduce pore pressure and reduce the probability of failure. Given the uncertainties of design, significant factors of safety are highly recommended.

Stormwater Management

Many operating and design controls are available to minimize leachate production, including control of the size of the working face, placement of interim cover on the waste, and use of proper stormwater runoff and runon controls. Control of stormwater runon and runoff is also required by Subtitle D of RCRA (Section 258.26). Runon control prevents the introduction of stormwater to the active area of the landfill, thus minimizing the production of leachate, erosion, and contamination of surface water. Reducing runon also limits the production of runoff from the landfill surface.

Runon can be prevented by diverting stormwater from active areas of the landfill. Any facility constructed to control runon must be capable of handling peak volumes generated by a 24-hour, 25-year storm. Typical measures to control runon include contouring the land surrounding the landfill cell or constructing ditches, dikes, or culverts to divert flow.

Runoff that is generated can be collected by swales (see Figure 4-12), ditches, berms, dikes, or culverts that direct contaminated runoff from active areas to storage and treatment facilities, and uncontaminated runoff from closed areas to detention facilities. Runoff from active areas must be collected and at least the volume generated from a 24-hour, 25-year storm must be controlled. Local regulations will dictate the design of uncontaminated stormwater management facilities. For example, in Florida, a detention pond must store the first inch of runoff for 14 days.

Landfill Cap

All Subtitle D landfills have to be capped, regardless of the potential for water intrusion. The purpose is to prevent the production of leachate that can contaminate groundwater. The effect of keeping water out of the landfill is to maintain dry conditions and hinder the process of biodegradation, making most landfills merely storage facilities.

Once the landfill reaches design height, a final cap is placed to minimize infiltration of rainwater, minimize dispersal of wastes, accommodate settling, and facilitate long-term maintenance of the landfill. The cap may consist, from top to bottom, of vegetation and supporting soil, a filler and drainage layer, a hydraulic

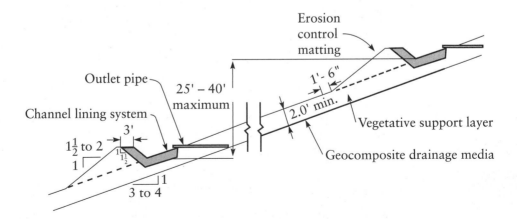

Figure 4-12 Side slope swale in a landfill final cover.

barrier, foundation for the hydraulic barrier, and a gas control layer. Figure 4-13 is a schematic of a recommended top slope cap. EPA regulations require that the top cap be less permeable than the bottom liner.

Slope stability and soil erosion are critical concerns for landfill caps. Typical side slopes are 1:3 to 1:4, and the interface friction between adjacent layers must resist seepage forces and may decrease the contact stresses between layers due to buildup of water and/or gas pressures. Consequently, on slide slopes, composite liner caps (geomembrane placed directly on top of a low permeability soil) are not recommended.

An alternative to the conventional landfill cap design—the *evapotranspiration cover*, also known as the *capillary barrier*—has been introduced. This cover places a thin (6-in.) layer of silt over a 2.5-ft-thick layer of uncompacted soil. The silt layer supports vegetation for transpiration while the soil layer provides moisture storage. Because the uncompacted soil layer remains unsaturated and at lower suction pressure than the saturated silt layer, the flow tends to be driven upward by capillary forces, and percolation into the waste is prevented. Moisture is then removed by transpiration and evaporation. Studies have shown that the evapotranspiration cover is appropriate for most areas of the United States west of the Mississippi River.[44]

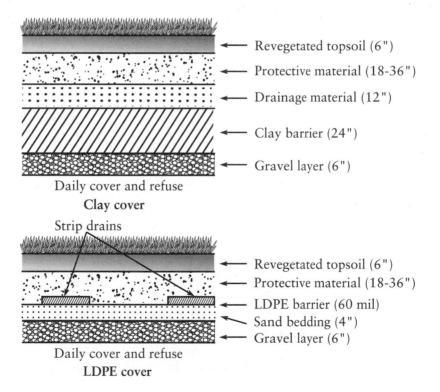

Figure 4-13 *Typical caps used for closing landfills.*

As landfills settle, caps can fail and allow stormwater to penetrate into land-fills. Figure 4-14 shows three modes of cap failure. Although data are scarce, it seems reasonable to assume that the caps are at best short-range covers and that within some years they will become ineffective. Caps are expensive, approaching $200,000 per acre of landfill area, and it is unlikely that they contribute much to the protection of groundwater.

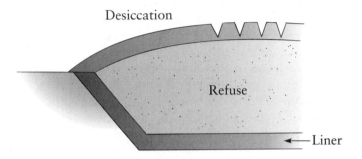

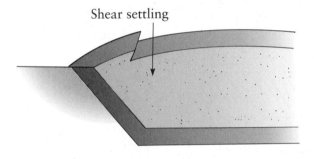

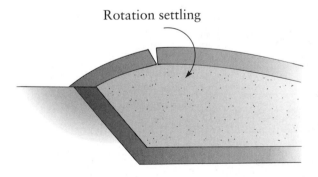

Figure 4-14   Three modes of cap failure.

LANDFILL OPERATIONS

Landfill Equipment

The movement, placement, and compaction of waste and cover in a landfill require a variety of large machines, including tractors, loaders (track and wheel), and compactors. In addition a variety of support equipment is needed, including motor graders, hydraulic excavators, water trucks, and service vehicles.

Originally, track-type bulldozers were used in landfills. Not only did they spread and compact the refuse, but they could be used for placing and spreading cover material as well as preparing the site, building roads, hauling trees, and removing stumps. They typically can achieve waste densities of 800–1000 lb/yd^3 (475–590 kg/m^3).

Specially built landfill compactors (Figure 4-15a) are now almost universally used. They are effective in spreading and compacting large quantities of waste, are quite heavy (exceeding 25 tons), and are equipped with knobbed steel wheels capable of tearing and compacting waste to densities of 1200–1600 lb/yd^3 (710–950 kg/m^3).

Daily dirt cover is excavated and placed on the operating face of a landfill using pans or scrapers (Figure 4-15b). Designed for the sole purpose of scraping up dirt and hauling it to another location, these machines bring the cover dirt to the landfill and then run down the compacted slope, discharging the dirt evenly over the refuse.

The type and quantity of landfill vehicles are determined by the amount and type of waste handled, the amount and type of soil cover, distance of moving waste and cover, weather requirements, compaction requirements, landfill configuration, budget, expected growth, and supplemental tasks anticipated.

Filling Sequences

At an active landfill daily deliveries of waste are placed in *lifts* or layers on top of the liner and leachate-collection system to depths of 20 m (65 ft) or greater. Typical waste placement methods used for landfill construction are shown in Figure 4-16. The area or mound type of landfill construction is commonly used in areas with a high groundwater table or subsurface conditions that prevent excavation. The excavation or trench technique provides waste placement in cells or trenches dug into the subsurface. The removed soil is often used as daily, intermediate, or final cover. Where suitable, waste can also be placed against lined canyon or ravine side slopes. Slope stability and leachate and gas emission control are critical issues for this type of waste placement.

Because it is imperative that the leachate-collection system be protected during landfill operations, waste in the first lift is selected to avoid heavy and sharp objects. This layer is often called the *operational layer*. The waste also must be placed in a manner to keep compactor wheels away from the leachate-collection system. Filling then begins in subsequent lifts, generally in a corner and moving outward to form the next layer. The filling sequence is established at the time of landfill design and

(a)

(b)

Figure 4-15 (a) Landfill compactor used to compact waste placed in a landfill. (b) A pan scraper used to haul and spread daily dirt cover.

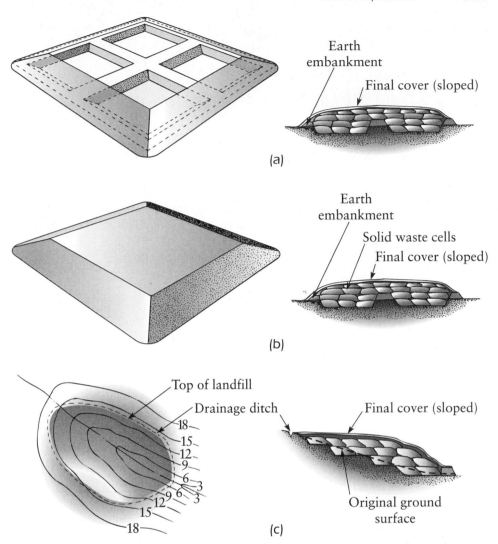

Figure 4-16 *Commonly used landfilling methods: (a) excavated cell/trench, (b) area, and (c) canyon/depression. Source: (49)*

permitting. The working face must be large enough to accommodate several vehicles unloading simultaneously, typically 12 to 20 ft (4–6 m) per vehicle.

As waste is placed in the landfill, heavy equipment is used to compact the waste to maximize use of *airspace*. The degree of compaction expected is a function of several factors, including refuse layer thickness (see Figure 4-17), number of passes made over the waste (see Figure 4-18), slope (flatter slopes compact better by landfill compactors, steeper slopes (maximum 1:3) compact better by track-type tractors), and moisture content (wetter waste compacts more effectively than dry waste).

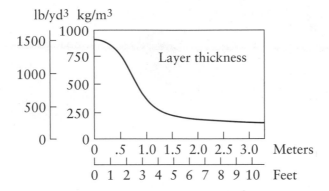

Figure 4-17   Waste density of a landfill as a function of layer thickness.

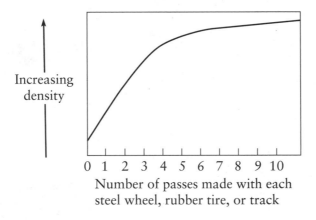

Figure 4-18   Waste density of a landfill as a function of the number of compactor passes.

Daily Cover

Waste is covered at the end of each working day with soil or alternative daily cover such as textiles, geomembrane, carpet, foam, or other proprietary materials. Daily cover is required to control disease vectors and rodents, to minimize odor, litter, and air emissions, to reduce the risk of fire, and to minimize leachate production. The landfill sides are sloped to facilitate maintenance and to increase slope stability; generally, a maximum slope of 1:3 is maintained.

Monitoring

Landfill monitoring is critical to the operation of a landfill. Most commonly, landfill operators monitor the following: leachate head on the liner, leakage through the landfill liner, groundwater quality, ambient air quality (to ensure compliance with the Clean Air Act), gas in the surrounding soil, leachate quality and quantity, landfill-gas quality and quantity, and stability of the final cover.

Control of head on the liner requires the ability to maintain a properly designed leachate-collection system, monitor head on the liner, store or dispose of leachate outside of the landfill, and remove leachate at appropriate rates. Leakage through a single liner is often detected using a lysimeter such as that shown in Figure 4-19. The locations and number of lysimeters are based on the landfill design; however, more than one should be provided to ensure redundancy. The lysimeter should be located below the crest of the landfill liner where the maximum head will be found, the point of maximum leakage potential.

Groundwater monitoring is generally accomplished through the construction and sampling of monitoring wells in the vicinity of the landfill. Optimum design would place a cluster of wells in each location to provide the means to evaluate groundwater quality at multiple depths. These well clusters should be placed up-gradient from the landfill to evaluate background groundwater quality, as well as immediately down-gradient from the landfill to determine the influence of the land-fill on groundwater. Additional monitoring wells are also placed at the property boundary. These wells are sampled quarterly for a variety of organic and inorganic

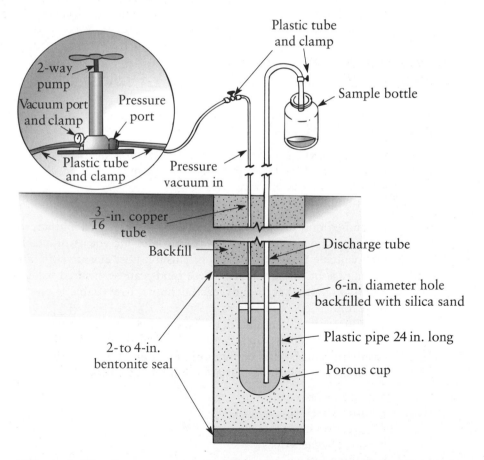

Figure 4-19 Porous cup suction lysimeter for the collection of liquid samples from the landfill. Source: (22)

groundwater constituents as stipulated in RCRA Subtitle D regulations. If down-gradient constituents are found to have a statistically significant increase in concentration as compared with up-gradient concentrations, more complete monitoring must begin. If down-gradient levels continue to exceed drinking water standards, more extensive monitoring and a corrective action plan may be required.

The EPA permits a waiver of groundwater monitoring requirements if it can be demonstrated that there is no potential for the migration of hazardous constituent from the landfill to the uppermost aquifer during the active life of the landfill and the post-closure period. To accomplish this demonstration, computer models are generally used that simulate contaminant movement in the subsurface. The Multimedia Exposure Assessment Model (MULTIMED) was developed by the EPA specifically for this purpose.[45]

Landfill operators must ensure that the concentration of methane gas does not exceed 25% of the lower explosive limit for methane in facilities nor does it exceed the lower explosive limit at the landfill boundary. Gas monitoring probes such as the ones shown in Figures 4-20 and 4-21 are generally placed at the property boundaries and various other locations around the landfill property. The hydrogeologic properties of the site, soil conditions, and the placement of landfill structures determine the locations of the probes.

POST-CLOSURE CARE AND USE OF OLD LANDFILLS

RCRA Subtitle D requires close attention to a landfill for 30 years following closure. The following are required during the 30-year post-closure period:

- maintenance of the integrity and effectiveness of the final cover
- operation of the leachate collection system
- groundwater monitoring
- gas-migration monitoring

A detailed post-closure care report must be prepared and approved prior to waste receipt at the landfill. The report must also include assurance that sufficient funds are available to meet the costs of closure, post-closure care, and corrective action for release of contaminants. These funds can be ensured using a variety of instruments such as surety bonds, cash deposits, irrevocable letters of credit, or private trust funds.

Once a landfill is closed, the land can be reused for many purposes. It is important to consider the final use of a landfill during planning and design phases to ensure that the landfill configuration is compatible with the selected final use. Final use alternatives include, but are not limited to, these:

- golf courses
- natural areas
- recreation parks
- ski/toboggan slopes
- parking lots
- building construction

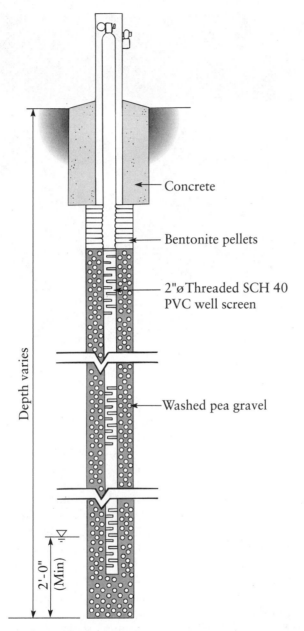

Concrete

Bentonite pellets

2"øThreaded SCH 40
PVC well screen

Washed pea gravel

Depth varies

2'-0"
(Min)

Figure 4-20 *Landfill gas monitoring probe.*

"Mount Trashmore" in the otherwise relatively flat state of Indiana is a popular ski location. The Harborside International Golf Complex near Chicago,
Illinois—formerly the Chicago MSW landfill—hosts two world-class 18-hole golf
courses and a 58-acre practice facility. A 50-acre landfill in urban Cambridge,
Massachusetts, now provides a variety of sports and recreational facilities, including three soccer fields, three softball fields, two play areas, bocce courts, jogging
trails, etc. The project increased the city's open space by 20 percent.

Figure 4-21 Gas extraction well in place.

The reuse of a landfill site is complicated by leachate generation, gas emissions, and the potential for large differential settlement. Differential settlement can be particularly damaging to overlying structures and embankments, as well as any leachate- and gas-control facilities in place. Settlement is the result of added moisture, external loads such as final and daily cover, overlying wastes, and any added structures such as buildings and roads. Settlement typically occurs quite rapidly immediately after closure (first 1 to 12 months) as a result of the rapid dissipation of pore fluid and gas. This period of *primary settlement* is followed by slower settlement over the next 15 to 20 years when *secondary settlement* occurs as a result of waste creep and biodegradation. This long-term settlement may result in compaction of as much as 30% of the waste depth. The rate and magnitude of settlement are difficult to predict due to the heterogeneity of the waste and complex degradation processes. Poor foundations can result in unfortunate situations such as the destruction of the motel shown in Figure 4-22.

Settlement must be considered when designing structures such as roads, parking lots, and structures supported on shallow foundations. Explosive gas can collect between the landfill surface and the supported floor of structures employing deep foundations built on top of closed landfills. To compensate for settlement, permanent structures may need to be equipped with costly foundations and articulated and telescoping pipe connections to maintain proper drainage and integrity of facilities.

Figure 4-22   *An unfortunate motel built on a settling landfill.*

LANDFILL MINING

If significant biodegradation occurs in a landfill, it might be possible to dig up old landfills, separate the nonbiodegradable fraction, and use the dirt and organic soil as a cover material for present landfills. This seems like a reasonable option for communities seeking new landfills, and at one time it appeared that *landfill mining* might be widely practiced. For example, a south Florida landfill mined an old landfill and recovered some metals. However, with the continuing evolution of landfill regulations and control of landfill gas, it has become very difficult to mine landfills. Opening the cap of a landfill allows gas to escape to the atmosphere and permits rainwater to produce contaminated runoff. Finally, mined materials have marginal economic value because they are quite dirty and difficult to clean.

Landfill mining makes economic sense only if it can create new volume for continuing the life of an existing landfill and when this can be done at minimal environmental cost. Only shallow landfills that have few vertical lifts are candidates for landfill mining since they are most likely to have fully biodegraded. Landfill mining operations consist of excavating the buried refuse and screening it to separate out the useful new cover materials. Typically the nonbiodegradable contents have little value and are re-landfilled. Successful operations significantly extend the time to closure for landfills and thus provided an economical alternative to the siting of new landfills.[46]

FINAL THOUGHTS

Landfills are engineering projects that require an unusual mixture of technical skills and public relations acumen, with the latter often outweighing the former. Little is mentioned here relative to the nontechnical problems associated with the planning, design, and operation of landfills, but this is not to imply that these aspects are insignificant.

A popular definition of solid waste is that "it is the stuff everyone wants picked up but no one wants put down." Studies on the psychology and sociology related to the siting of landfills are both fascinating and somewhat frightening. We know entirely too little about public reaction to, and the impact of, such projects, and the design engineer is often placed in a somewhat uncomfortable position of being the Solomon-like judge of public reaction and public good. Nowhere does the human nature of the engineering profession become as important as in the design of the ultimate disposal of a community's residues, and too often the engineer becomes "public enemy number one." Why is this?

Engineering is an old and honored profession. Some of the earliest engineers in the newly founded United States were our best minds and leaders. George Washington taught himself surveying and supervised the construction of roads, canals, and locks. Thomas Jefferson was an inveterate tinkerer, and some of his gadgets are clever even by modern mechanical engineering standards.

Engineers are indeed a special breed. A study performed some years ago evaluated the characteristics of students who entered college intending to study engineering. The study found that those students who continued in the engineering program after two years differed markedly from those who left engineering. Only certain persons choose to become engineers and undertake the rigors of the demanding college curriculum that allows entrance to the field. Engineering students and practicing engineers tend to have a very positive image of themselves and a lively *esprit de corps*. Engineering is *fun*.

Engineers also see themselves as performing a public service. The American Soceity of Civil Engineers motto describing the civil engineering profession as the "people-serving profession" neatly sums up engineers' perception of themselves. Engineers build civilizations. Engineers serve the public's needs.

But while engineers see themselves as successful problem solvers acting in the public interest, the public's perception of engineers is often somewhat different. Well-publicized engineering failures resulting in damage to human and environmental health have encouraged many people to perceive engineers as the creators, not solvers, of problems.

Part of the public attitude toward engineers might be explained by the view that engineers tend to conduct social experiments.[47] Engineers do not have all the answers at hand when a problem arises and often cannot perform full-scale experiments to obtain such answers. For example, when designing a suspension bridge, engineers cannot construct a full-scale model of the bridge to test the design. Instead, they use the best knowledge and designs available and extrapolate, using sound judgment. Sometimes the extrapolation is faulty, and the experiment fails. Engineers learn from such failures, and modify the design process in subsequent projects. What is important is that the experiments occur not in a private laboratory but on the public stage.

Engineering experiments affect humans in many ways. A road, for example, is both a technical and a social undertaking. The road might affect homes and businesses, create traffic noise, or even divert funds from other projects. A new lawn mower is designed and manufactured, and it might cut grass effectively, but at the expense of neighborhood quiet and perhaps even auditory damage to the user. A reservoir is constructed, but at the cost to recreational whitewater rafting, or despoliation of a bottomland virgin ecosystem, or the destruction of an ancestral homestead. A new biological weapon, designed to kill people, is developed with the knowledge that its use would devastate the global ecosystem. How is the engineer to deal with these conflicting public benefits?

The fundamental difference between how engineers see themselves and how the public sees them depends on how well engineers meet public concerns. Sometimes engineers are labeled the "tools of the establishment" or the "despoilers of the environment" or the "diligent destroyers." With the public it's often "us" against "them," and "them" is too often the engineers. Why are the engineers the villains? Why do the engineers so often appear to be the bad guys, or as in the old cowboy movies, the guys wearing the black hats?

The source of the problem is that engineers and the lay public hold different ethical attitudes. Because engineers tend to be utilitarians, they look at the overall and aggregate net benefits, thus diminishing the importance of harm to the individual. Because engineers are positivists, they tend to ignore or dismiss considerations for which reasons of a certain type cannot be given—that is, quantifiable or at least empirical data—thus ignoring intangibles. Engineers value people, of course, but they have a particular view of what is good for people and how this good is to be determined in a given case. Finally, engineers think of themselves as doing applied physical science, not applied social science. The physical science approach to engineering (ignoring the "people-serving profession" motto) allows engineers to think of their work as not being germane to the needs of society. This conflict of ethical outlooks is a root cause of much of the problem with engineers' interaction with the public.

REFERENCES

1. O'Brien, J. 1977. Unpublished data. Durham, N.C.: Duke Environmental Center, Duke University.

2. Hershfield, S., P. A. Vesilind, and E. I. Pas. 1992. "Assessing the True Cost of Landfills." *Waste Management and Research* 10: 471–484.

3. Nelson, A. C., J. Genereux, and M. M. Genereux. 1997. "Price Effects of Landfills on Different House Value Strata." *Journal of Urban Planning and Development*, ASCE, 123, n. 3: 59–67.

4. Vesilind, P. A., and E. I. Pas. 1998. "Discussion of A. C. Nelson, J. Genereux, and M. M. Genereux, A Price Effect of Landfills on Different House Value Strata." *Journal of Urban Planning and Development*, ASCE, 123, n. 3: 59–68.

5. Azar, S. 1998. *The Proposed Eubanks Landfill: The Ramifications of a Broken Promise*. Senior Independent Study. Durham, N.C.: Duke University Department of Civil and Environmental Engineering.

6. Sleats, R., C. Harries, I. Viney, and J. F. Rees. 1989. "Activities and Distribution of Key Microbial Groups in Landfills." In *Sanitary Landfilling: Process, Technology and Environmental Impact*. London: Academic Press.

7. Pohland, F. G., and R. Englebrech. 1976. *Impact of Sanitary Landfills; An Overview of Environmental Factors and Control Alternatives*. N.Y.: Report prepared for the American Paper Institute.

8. Fenn, D. G., K. J. Hanley, and T. V. DeGeare. 1975. *Use of the Water Balance Method for Predicting Leachate Generation from Solid Waste Disposal Sites*. EPA OSWMP, SW-168. Washington, D.C.

9. Remson, I. A., A. Fungaroli, and A. W. Lawrence. 1968. "Water Movement in an Unsaturated Sanitary Landfill." *Journal of the Sanitary Engineering Division*, ASCE, 94, n. SA2.

10. Reindl, J. 1977. "Managing Gas and Leachate Production on Landfills." *Solid Waste Management* (July): 30.

11. Othman, M. A., R. Bonaparte, B. A. Gross, and D. Warren. 1998. *Evaluation of Liquids Management Data for Double-Lined Landfills*. Draft Document Prepared for U.S. Environmental Protection Agency. Cincinnati, Ohio: National Risk Management Laboratory.

12. Thornthwaite, C. W., and J. R. Mather. 1957. *Instructions and Tables for Computing Potential Evapotranspiration and Water Balance*. Publ. Climatology, Drexel Inst. Technol., 10-185.

13. *Water Atlas*. 1998. Water Resources Information Service, USGS. Washington, D.C.

14. McGuinnes, J. L., and E. F. Bordne. 1972. *Comparisons of Lysimeter Derived Potential Evapotranspiration with Computed Values*. Agr. Res. Serv. Tech. Bull. 1452. Washington, D.C.: U.S. Dept. of Agriculture.

15. Schroeder, P. R., C. M. Lloyd, and P. A. Zappi. 1994. *The Hydrologic Evaluation of Landfill Performance (HELP) Model, User's Guide for Version 3*. EPA/600/R-94/168a.

16. Reinhart, D. R., and C. J. Grosh. 1997. "Analysis of Florida MSW Landfill Leachate Quality Data." Report to the Florida Center for Solid and Hazardous Waste Management (March).

17. Reinhart, D. R., and K. Lewis. 1994. "Beneficial Utilization of Landfill Gas." Report to the Florida Center for Solid and Hazardous Waste Management (September).

18. Young, A. 1989. "Mathematical Modeling of Landfill Gas Extraction." *Journal of Environmental Engineering* 115, n 6: 1073.

19. Nyns, E. J. 1992. *Landfill Gas: From Environment to Energy*, First Edition. Luxembourg: Commission of the European Committee.

20. Augenstein, D., and J. Pacey. 1992. *Landfill Gas Energy Utilization: Technology Options and Case Studies*. EPA-600-R-92-116, EPA, Research Triangle Park.

21. Dessanti, D. J., and H. W. Peter. 1984. "Recovery of Methane from Landfill." *Proceedings* Symposium: Energy from Biomass and Wastes VIII: 1053–1090.

22. Tchobanoglous, G., H. Theisen, and S. Vigil. 1993. *Integrated Solid Waste Management: Engineering Principles and Management Issues*, 3rd ed., New York: McGraw-Hill, Inc.

23. Reinhart, D. R., C. D. Cooper, and N. E. Ruiz. 1994. *Estimation of Landfill Gas Emissions at the Orange County Landfill*. Orlando: Civil and Environmental Engineering Department, University of Central Florida.

24. Reinhart, D. R., and T. Townsend. 1998. "Assessment of Leachate Collection System Clogging at Florida Municipal Solid Waste Landfills." Report to the Florida Center for Solid and Hazardous Waste Management (April).

25. *Criteria for Municipal Solid Waste Landfills*. 1988. Washington, D.C.: EPA. July.

26. McBean, E. A., F. G. Pohland, F. A. Rovers, and A. J. Crutcher. 1982. "Leachate Collection Design for Containment Landfills." *Journal of Environmental Engineering Division* ASCE 108, n. EE1: 204–209.

27. McEnroe, B. M. 1993. "Maximum Saturated Depth over Landfill Liner." *Journal of Environmental Engineering* ASCE 119, n. 2: 262–270.

28. Richardson, G., and A. Zhao. 2000. *Design of Lateral Drainage Systems for Landfills*. Baltimore: Tenax Corporation.

29. *Requirements for Hazardous Waste Landfill Design, Construction, and Closure*. 1989. EPA/625/4-89-022. Washington, D.C.: EPA.

30. Lema, J. M., R. Mendez, and R. Blazquez. 1988. "Characteristics of Landfill Leachates and Alternatives for Their Treatment: A Review." *Water, Air, and Soil Pollution* 40: 223–250.

31. Chian, E. S. K., and F. B. DeWalle. 1977. *Evaluation of Leachate Treatment, Vol. 1, Characterization of Leachate*. EPA-600/2-77-186a. Cincinnati: EPA.

32. Cossu, R., R. Stegmann, G. Andreottola, and P. Cannas. 1989. "Biological Treatment." In *Sanitary Landfilling: Process, Technology and Environmental Impact*, ed. by T. H. Christensen, R. Cossu, and R. Stegmann, pp. 251–253.

33. Venkataramani, E. S., R. Ahlert, and P. Corbo. 1974. "Biological Treatment of Landfill Leachates." *CRC Critical Reviews in Environmental Control* 14, n. 4: 333–376. Based on research conducted by W. C. Boyle and R. K. Ham. 1974. "Treatability of Leachate from Sanitary Landfills." *Journal Water Pollution Control Federation* 46, n. 5.

34. King, D., and A. Mureebe. 1992. "Leachate Management Successfully Implemented at Landfill." *Water Environment and Technology* 4, n. 9: 18.

35. Harris, J. M., D. Purchewitz, and D. R. Reinhart. 1999. "Leachate Management for the New Millennium." *Proceedings* 4th annual SWANA Landfill Management Symposium, Denver.

36. Landtec. 1994. "Landfill Gas System Engineering Design: A Practical Approach." Courseware Landfill Gas System Engineering Design Seminar, Orlando.

37. Thorneloe, S. A. 1992. "Landfill Gas Utilization—Options, Benefits, and Barriers." *Proceedings* The Second United States Con-

ference on Municipal Solid Waste Management, Arlington, (June).

38. Wheless, E., S. Thalenberg, and M. Wong. 1993. "Making Landfill Gas into a Clean Vehicle Fuel." *Solid Waste Technologies* 7, n. 6: 32.

39. Sandelli, G. J.. 1992. *Demonstration of Fuel Cells to Recover Energy from Landfill Gas, Phase 1 Final Report: Conceptual Study.* EPA-600-R-92-007, EPA, Research Triangle Park.

40. Pacey, J. G., M. Doorn, and S. A. Thorneloe. 1994. "Landfill Gas Energy Utilization: Technical and Non-technical Considerations." *Proceedings* 17th Annual Landfill Gas Symposium GR-LG 0017, pp. 237–249. Long Beach: Solid Waste Association of North America.

41. Thorneloe, S. A. 1992. "Landfill Gas Recovery/Utilization—Options and Economics." *Proceedings* 16th Annual Conference on Energy From Biomass and Wastes (March). Orlando: Institute of Gas Technology.

42. Stahl, J., E. Wheless, and S. Thalenberg. "Compressed Landfill Gas as a Clean, Alternative Vehicle Fuel." *Proceedings* 16th Annual Landfill Gas Symposium GR-LG 0016, pp. 289–296. Louisville: Solid Waste Association of North America.

43. Waner, J. A., C. A. Phillips, and N. A. Webster. 1994. "Methane Recovery Benefits Study Released." *Water, Environment & Technology* 4, no. 6: 36.

44. Weand, B.L., and V. L. Hauser. 1997. "The Evapotranspiration Cover." *Environmental Protection* 8, n.1: 40–42.

45. Sharp-Hansen, S., C. Travers, P. Hummel, and T. Allison. 1990. *A Subtitle D Landfill Application Manual for the Multimedia Exposure Assessment Model (MULTIMED).* Athens, Ga.: EPA.

46. Reinhart, D.R., and T. Townsend. 1997. *Landfill Bioreactor Design and Operation.* Boca Raton: Lewis Publishers.

47. Martin, M. W., and R. Schinzinger. 1989. *Ethics in Engineering.* New York: McGraw-Hill.

48. Franklin Associates. 1999. *Characterization of Municipal Solid Waste in the United States: 1998 Update.* EPA 530, Washington, D.C.

49. Brunner, D. R., and D. J. Kelly. 1972. *Sanitary Landfill Design and Operation.* Washington, D.C.: EPA Office of Solid Waste Management.

ABBREVIATIONS USED IN THIS CHAPTER

BDL = below detection limit
BOD = biochemical oxygen demand
COD = chemical oxygen demand
CWA = Clean Water Act
CO = carbon monoxide
CO_2 = carbon dioxide
DOC = Direct Osmotic Concentration
EU = European Union
EPA = Environmental Protection Agency
GCL = geosynthetic clay liner
GPAD = gallons per acre per day
HDPE = high-density polyethylene
HELP = Hydrologic Evaluation of Landfill Performance (a computer model)

ID = inside diameter
LCS = leachate collection system
LDS = leachate detection system
LULU = locally undesirable land use
MSW = municipal solid waste
MULTIMED = Multimedia Exposure Assessment Model
MVR = Mechanical Vapor Recompression
MW = megawatt
NIMTO = Not In My Term of Office
NPDES = National Pollution Discharge Elimination System
NOPE = Not On Planet Earth
NO_x = nitrogen oxides

OA = osmotic agent
POTW = Publicly Owned Treatment Works
PVC = Polyvinyl Chloride
RCRA = Resources Conservation and Recovery
    Act

RO = Reverse Osmosis
TDS = Total Dissolved Solids
VCD = Vapor Compression Distillation
VOA = volatile organic acid
VOC = volatile organic compound

PROBLEMS

4-1. For your own community (assume appropriate quantities of refuse, densities, etc.) calculate how many years it would take to fill a football field 20 ft deep. Include the necessary earth cover.

4-2. Estimate the life of a landfill for a user population of 10,000. The available area for the area-type landfill is 10 acres. The water table is estimated at 20 ft below the ground surface. The pit must have side slope of 1:3, and the final surface must have a slope of 1:4. Assume that the soil occupies 20% of the compacted volume.

4-3. Assume the following fraction, by weight, of material in a refuse:

Component	Percent, by weight
Food waste	10
Yard waste	20
Mixed paper	40
Corrugated cardboard	10
Glass	10
Ferrous (cans)	10

a. Estimate the in-place density of this refuse, using the bulk density values listed in Table 4-1.

b. If the following materials are recovered in a materials recovery operation, estimate the bulk density of the remaining refuse going to the landfill:

Component	Percent recovered
Garbage	0
Yard waste	0
Mixed paper	50
Glass	60
Ferrous (cans)	95

c. If, for this community, the life of the landfill was originally calculated as 10 years, what is the expected life if the materials listed above are recovered?

4-4. You are the engineer hired by a community trying to site a modern (Subtitle D) landfill. At the public hearing one of the neighbors of the proposed landfill asks you if her well water might become contaminated by the landfill leachate.

a. How would you answer her? Take your time and give a full answer.

b. The town hired you to site the landfill. Is the town your client, or do you have higher professional responsibility?

4-5. A landfill initially receives 1000 tons/day of waste. This receipt rate increases 10% per year. The landfill is closed at the end of the fourth year. Using the EPA exponential model, calculate the yearly gas production for 20 years after opening. What is the peak gas production in m^3/yr?

4-6. You are a municipal engineer who has just received a request from the city director of planning to comment on a proposed large multistory shopping center. You note that it is to be built on a site that you vaguely remember as having been a landfill many years ago. Nobody in the office remembers for sure when this was and what it was used for, when it closed, or how long it had been used. You do know that it has been closed for over 25 years. Write a letter to the planning director expressing your concerns and recommendations.

4-7. How does a modern (Subtitle D) landfill differ from the older existing landfills? List and explain all items that need to be considered in Subtitle D landfills that were not important in older landfills before the passage of RCRA legislation.

4-8. A community of 100,000 people decides to construct a landfill that is to have a volume of 2,000,000 cubic yards. How long can this landfill be expected to last? Use approximate numbers.

4-9. Determine the required spacing between pipes in a leachate-collection system using a geonet and the following properties. Assume the most conservative design and that all stormwater from a 25-yr, 24-hr storm enters the leachate-collection system.

     Geonet transmissivity, θ = 10 cm²/sec

     Design storm (25-year, 24-hour) = 8.2 in.

     Drainage slope = 2%

     Maximum design head on liner = 6 in.

4-10. Repeat Example 4-6 assuming that the highest flow rate in each collection leg is 750 cubic feet per minute at 115°F.

4-11. A landfill is proposed for the area shown on the map in Figure 4-24. Why might this not be a good location? Where might be a better place? What are the advantages and disadvantages of the new location? Some relevant facts: Groundwater at the proposed landfill site is 5 ft below surface; the soil is sand from −30 feet to +20 feet and sandy loam above 20 feet.

4-12. A town of 100,000 people wants to construct a landfill on the site shown in Figure 4-25. Design such a landfill. Estimate the life of the landfill, and specify the equipment to be used, the method of operation, and other considerations necessary for acceptable operation. Draw in roads and any permanent installations on the map.

4-13. Describe, from a microbiological perspective, what goes on inside a sanitary landfill.

4-14. A town has hired a consulting firm to find a new landfill site. After they find the ideal site, Megafill Inc., located 500 miles from the town, offers to transport and dispose of the waste for only $20 per ton more than the expected tipping fee at the new landfill.

   a. What are some of the arguments in favor of, and against, the town signing up with Megafill?

   b. To better understand the site, Megafill offers to fly you, the city's public works director, and the city council to the site in their corporate jet. Do you accept the offer? Why or why not?

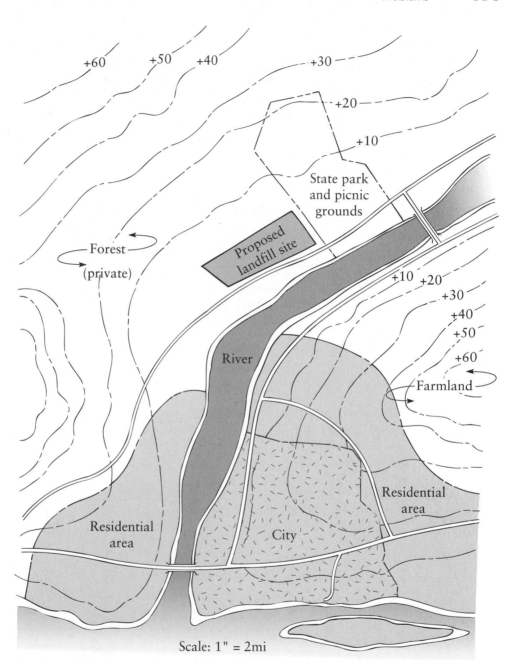

Figure 4-24 See Problem 4-11.

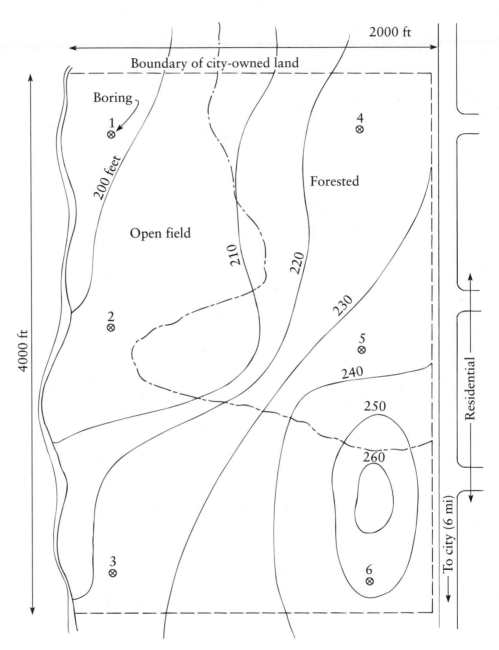

Figure 4-25 See Problem 4-12.

Processing of Municipal Solid Waste

In almost all engineering disciplines the designs of the processes used to produce a desired end product are based on the reasonable assumption that the nature of the raw material is known and can be defined accurately and precisely. This condition lends a feeling of confidence to the analysis of unit operations and breeds sophistication in the design of the process. Unfortunately, the MSW processing profession is not so blessed. Solid waste processing and materials recovery facilities must be able to accept almost all manner of solid waste. Some types of solid waste are easy to process, but occasionally materials that are difficult and/or dangerous to handle are also found in refuse. As a result, solid waste processing operations have large factors of safety, and must be designed for extraordinary contingencies. This requirement often results in overdesign and underutilization in order to be able to process all of the feed material.

In this chapter solid waste handling prior to further materials separation or processing is discussed. None of the unit operations discussed here actually accomplishes materials separation, but rather prepares the refuse for the separation operations that follow.

REFUSE PHYSICAL CHARACTERISTICS

Municipal refuse, either in its original or shredded state, has some material properties that make its processing and conveying hazardous and difficult. Although the art and science of solids conveying and storage are well studied in the mining and chemical engineering fields, the application of most of that knowledge is inappropriate for a heterogeneous, unpredictable, and time-variable material such as refuse. Hickman[1] lists the materials characteristics that must be considered in the design of any storage, conveying, or processing equipment. Among these are the following:

Particle Size

Because of the nature of MSW, particle size is difficult to define. Sieves define size by only two dimensions; thus a piece of wire could pass a sieve and still prove troublesome in conveying. The problems of defining particle diameter is discussed in Chapter 2, and the process of size separation by sieving is discussed further in Chapter 6.

Bulk Density

Shredded MSW, when stored in a storage pit, can achieve densities as high as 250 lb/ft³ (400 kg/m³). The variation in density has been found to be significant, as shown in Figure 5-1. This variation in density is an important factor in designing storage facilities.

Angle of Repose

The angle of repose is the angle, to the horizontal, that the material will stack without sliding. Sand, for example, has an angle of repose of about 35°, depending on the moisture content. Because of variable density, moisture, and particle size, the angle of repose of shredded refuse can vary from 45° to greater than 90°.

Material Abrasiveness

Refuse consists of many types of abrasive particles, including sand, glass, metals, and rocks. Removal of this abrasive material is often necessary before some operations, such as pneumatic conveying, can become practical.

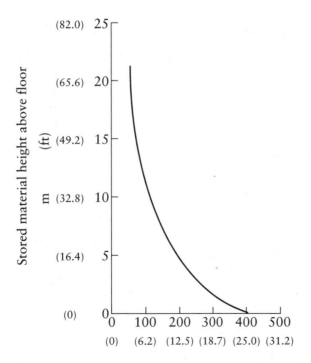

Figure 5-1   Variation of bulk density in MSW. Source: (1)

Moisture Content

All of the foregoing properties are influenced by moisture content. The extent of this effect depends on the material. When the moisture level exceeds 50%, the high organic fraction can undergo spontaneous combustion if the material is allowed to stand undisturbed.

STORING MSW

The storage of MSW has long been a serious problem, especially at large combustor facilities. Most waste-to-energy combustors must be continuously fired and require sufficient storage for at least two days to allow for the unavailability of refuse over weekends. Similarly, if materials recovery systems operate through the weekend, they must store material in sufficient quantities to even out the fluctuations in supply.

Two major considerations in the design of MSW storage facilities are public health and fire. Rats and other rodents can inhabit storage areas unless special precautions are taken. The odor of slowly decomposing garbage can be overwhelming and can cause public relations problems downwind.

Spontaneous combustion is possible with the storage of MSW. The rule of thumb is that two days of storage is the safe maximum, with a week being dangerous. A fire in a storage pit is not only difficult to extinguish, but the resulting wet refuse after the fire is extinguished presents new disposal problems.

All storage facilities should be constructed as *first-in/first-out* systems. Unfortunately, this is not a simple task, and many of the existing storage systems tend to result in long-term storage of some fraction of the refuse.

A common storage system consists of a pit with an overhead bridge crane. Garbage and transfer trucks back up to the pit and discharge solid waste directly into it. An overhead crane, with an operator either directly on the crane or centered over the pit, is used to both spread the load in the pit and to retrieve the solid waste. The crane can drop the solid waste into a feed chute, onto a conveyor belt, or directly into a transfer vehicle.

Another common storage system relies on a large tipping floor. Solid waste is deposited onto the floor and is then stacked as high as 20 ft (7 m) by a front-end loader. In some facilities a concrete or steel push wall is incorporated into the design. The front-end loader can also meter the solid waste onto a metal pan conveyor or directly into a transfer vehicle.

The design of better storage facilities requires not only a knowledge of the theory of materials flow, but also a means of experimentally evaluating the flow rate of solid material in a storage chamber. The use of velocity probes, especially for solid waste, is clearly unacceptable. A number of potentially effective techniques are stereophotogrammetry, radio pills (transmitters that move with the solids in the bin), radiological tagging (e.g., with cesium 137), and X-ray methods.[2] With non-homogeneous materials such as refuse, the radio pill or photogrammetry seem to be most applicable.

CONVEYING

Six basic types of conveyors are used for refuse: (1) rubber-belted conveyors, (2) live bottom feeders, (3) pneumatic conveyors, (4) vibratory feeders, (5) screw feeders, and (6) drag chains. The first three types are used primarily to move refuse; the last three are used to feed or meter refuse to a load sensitive device such as a combustor.

Rubber-belted conveyors have been used to move unshredded raw material, and are especially acceptable for less abrasive and less rugged loads such as source-separated recyclables. Rigid metal interlocking belts (commonly referred to as *metal pan conveyors*) have been successfully applied to raw refuse conveying. Skirts (sides) are sometimes used for bulky material. A typical conveyor with skirts is shown in Figure 5-2. A trough-type rubber belted conveyor without skirts is shown in Figure 5-3.

A conveyor with inclines greater than 20° will usually experience *tumble back* by some material. This can be considered advantageous if the conveyor is feeding a shredder, since the movement of the refuse can even out the feed to the shredder. One shredder manufacturer recommends angles from 38° to 40° to achieve a more even feed rate to the shredder.[3] In other applications a very steep incline is required, and rubber belt conveyors with flights can be used.

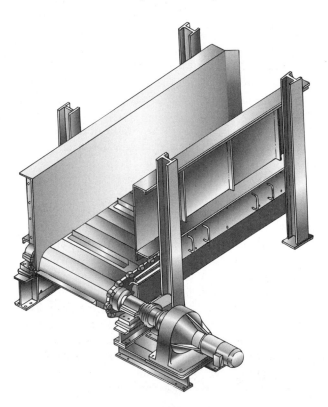

Figure 5-2 Typical feed conveyor. (Courtesy Allis-Chalmers)

The power requirements of belt conveyors can be estimated by a number of empirical equations, such as the following:[4]

$$\text{horsepower} = \frac{LSF}{1000} + \frac{LTC}{990} + \frac{TH}{990} + P$$

where L = length of conveyor belt, ft
S = speed of belt, ft/min
F = speed factor, dimensionless
T = capacity, tons/h
C = idle resistance factor, dimensionless
H = lift, ft
P = pulley friction, horsepower

The first two terms on the right-hand side of the equation represent the power necessary to move the load horizontally, while the third term represents vertical movement. The last term represents power loss due to friction. Although the units in this equation are obviously spurious, the constants in the denominators are used to convert everything to power.

For specific materials, empirical evidence suggests some maximum loadings on conveyor belts without skirts as shown in Table 5-1. These capacities are for a belt with a 20° trough, or angle of the edge of the belt with the horizontal, as shown in Figure 5-3, and a belt speed of 100 ft/min (30.5 m/min).

Table 5-1   Rubber Conveyor Belt Capacities for Selected Materials, at a Belt Speed of 100 Feet/Minute

| Material | Belt width (inches) | | |
| | 18 | 36 | 60 |
	Capacity, tons/hour		
Glass bottles	6.0	28.3	83.4
Plastic bottles	0.8	3.7	10.8
Aluminum cans	0.8	3.7	10.8
Newsprint	3.9	18.1	53.1
MSW from truck	8.7	40.0	117

Source: (5)

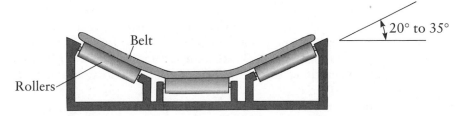

Figure 5-3 Conveyor commonly used for MSW. Capacities for this conveyor are shown in Table 5-1.

Live bottom hoppers are related to vibrating feeders and are used to move MSW out of holding bins or transfer trailers. As the name implies, the bottom of the hopper has sliding interlocking beams that move at set speeds. By moving them slowly forward and then rapidly backward, the motion slowly moves the burden forward. Live bottom feeders have been installed in facilities where refuse has to be moved a short distance such as in transfer stations. Figure 5-4 is an illustration of a typical live bottom feeder.

Pneumatic conveyors have been used mainly for collecting raw bagged MSW in hospitals and other large buildings and in feeding shredded organic fractions to boilers as supplemental fuel. The required air velocities in pneumatic tubes can be estimated as

$$v_m = v_a - v_f$$

where v_m = material velocity, ft/min
v_a = velocity of the air stream, ft/min
v_f = the *floating velocity*, or terminal velocity when falling in still air, ft/min.

The floating velocity can be calculated by an empirical equation such as

$$v_f = 3250\sqrt{(\text{SG})d}$$

where v_f is the floating velocity in ft/min, d is the aerodynamic diameter of a representative particle in inches, and SG the specific gravity of the material (relative to water).

The material velocity must be sufficient to be able to dislodge stuck particles and to even out the flow, and can be estimated as

$$v_m = 585\sqrt{W}$$

where W is the bulk density of the material in lb/ft^3. Because of the problems of measuring specific gravities in heterogeneous mixtures, the specific gravity can be estimated as

$$\text{SG} = 0.1(W)^{2/3}$$

The total recommended air velocity is thus

$$v_a = 1030\sqrt[3]{W}\sqrt{d} + 585\sqrt{W}$$

where v_a is the recommended air velocity in ft/min. The problem of accurately estimating the representative diameter of a particle makes the practical use of this equation difficult. Experience has shown, however, that maintaining air velocities of about 4500 ft/min (1400 m/min) is sufficient for maintaining the materials flow in vertical tubes. Table 5-2 lists some recommended air velocities for common materials. Operational experience has shown that a materials-to-air concentration of 0.1 (e.g., 0.1 kg of paper/1.0 kg of air) is reasonable.[6] The friction loss within a duct due to materials flow is less than 10%, well within a factor of safety commonly used in fan design.

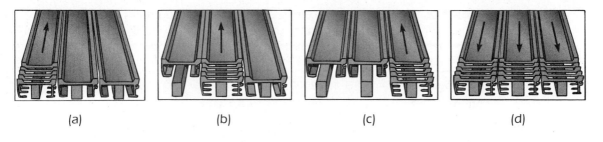

(a)   (b)   (c)   (d)

(e)

Figure 5-4 Typical live bottom (walking floor). (a) The first slat group slips under the load to the rear of the load. (b) The second slat group slips under the load. (c) The third slat group slips under the load. (d) All slat groups and the load advance forward together. (e) A truck equipped with a walking floor has just emptied its contents. (Courtesy Keith Walking Floor)

Table 5-2 Recommended Air Velocities for the Pneumatic Conveying of Some Representative Materials

Material	Minimum air velocity (ft/min)[a]
Coal, powdered	4000
Cotton	4500
Iron oxide	6500
Shavings	3500
Vegetable pulp	4500
Paper	5000
Rags	4500

[a] To obtain m/sec, multiply ft/min by 0.00508.
Source: (7)

Pneumatic conveyors suffer from wear problems, especially if glass is being conveyed. Elbows in pneumatic lines can be expected to abrade quickly, and sacrificial pieces must be used and frequently replaced.

Vibrating feeders are advantageous because they also even out materials flow. These devices are used to move small quantities of rigid material. For example, a vibrating conveyor can be used to feed glass to a hand-sorting table. A 2-cm stroke at a frequency of 900 strokes/min is common.

Screw conveyors are used to meter shredded refuse into a furnace because the screw serves as an air lock and the feed rate of fuel can easily be adjusted by changing the rotational speed of the screw. The volume of material moved by screw conveyors can be estimated by recognizing that the capacity of the conveyor in the *flooded* condition (i.e., all the space between the blades is full, as might occur when a screw conveyor is used in the bottom of a hopper) is

$$Q = CNRV$$

where Q = delivery of refuse, m³/min
$\quad C$ = the efficiency factor
$\quad N$ = number of conveyor leads
$\quad R$ = rotational speed of screw, rpm
$\quad V$ = volume of refuse between each pitch, m³

The number of leads means the number of blades that are wrapped around the conveyor hub. A common wood screw, for example, has one lead. Theoretically, if the distance between adjacent conveyor blades is constant, the forward motion of the material being conveyed is directly proportional to the number of leads. The units for the above equation can be any volume and time, such as cubic meter and seconds. These terms are defined further in Figure 5-5.

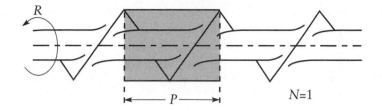

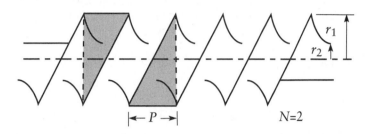

Figure 5-5   Screw conveyor (Note: Top screw has N=1, and bottom screw has N=2).

The volume within each pitch can be calculated approximately as

$$V = P\pi(r_1^2 - r_2^2)$$

where P = pitch (distance between adjacent conveyors' blades if the number of leads, $N = 1$), m

r_1 = radius to the conveyor tip, m

r_2 = radius to conveyor hub, m

The dimensionless efficiency factor, C, is a function of the amount of slippage that occurs. Ideally, a screw conveyor operates by allowing the material to slide freely on the blade and to thus prevent radial rotation. Any rotation by the material (sticking to the blade) lowers efficiency. In live bottom bins, where the conveyor screws are not in individual troughs, considerable slippage is expected.

If the screw conveyor is not flooded, its capacity cannot be determined theoretically because the rate is influenced by a large number of variables.[8]

EXAMPLE
5-1

A live bottom bin has eight screw conveyors, each with $r_1 = 15$ cm, $r_2 = 6$ cm, $N = 1$, $P = 50$ cm, and $R = 10$ rpm. Assuming that $C = 0.5$, calculate the total material flow.

$$V = P\pi(r_1^2 - r_2^2) = 50(3.14)(225 - 36) = 29{,}673 \text{ cm}^3$$

For each conveyor,

$$Q = CNRV = (0.5)(10)(29{,}673)(10^{-6}) = 0.15 \text{ m}^3/\text{min}$$

or total flow is $8 \times 0.15 = 1.2$ m^3/min.

Drag chain conveyors are used to move solid waste in applications such as waste-to-energy plants. A drag chain conveyor consists of an open- or closed-top metal rectangular pan. A chain runs the length of the pan along each side. Across the chain at 5- to 10-foot intervals are metal or wood flights. The chain drags the flights, which move the refuse. On the bottom of the metal pan can be slide door openings to chutes. Opening the door allows the refuse to fall into the chutes. Refuse that does not fall into the chutes can be returned to the beginning of the drag conveyor.

As a general rule, refuse should be conveyed and transferred as little as possible. It is thus good engineering design to eliminate, or at least minimize, points of transfer and conveying within a facility.

COMPACTING

One problem in the disposal of MSW is the low density of the material, which requires large volumes for its collection, handling, and final disposal. Compacting MSW can thus lead to significant cost savings. The structure of refuse can be pictured as an assemblage of particles interspaced with open air spaces called *voids*. Because these voids are large, and since many of the particles are absorbent, any moisture is absorbed in the material and is not in the voids. The total volume of material is made up of the solids plus the voids, or

$$V_m = V_s + V_v$$

where V_m = volume of material
V_s = volume of solids (including the moisture)
V_v = volume of voids

The *void ratio* is defined as

$$e = \frac{V_v}{V_s}$$

and the *porosity* is

$$n = \frac{V_v}{V_m}$$

By weight the total material is made up of the solids plus moisture, or

$$W_m = W_s + W_w$$

where W_m = weight of material, including moisture
W_s = weight of solids
W_w = weight of moisture

The *bulk density* is defined as

$$\rho_b = \frac{W_m}{V_m}$$

Note that this is on a *wet basis*. The entire sample is weighed *as is* and its volume calculated. Most densities in compaction literature are expressed in terms of bulk density, mainly because it is easy to measure and can be readily used in comparative studies. Bulk densities can also be expressed on a *dry basis* if the sample is then dried and the weight of the moisture subtracted. The compaction of refuse in this text and in most literature is expressed as the increase in bulk densities (wet basis).

When MSW is compacted, the density is increased as a result of the crushing, deforming, and relocating of individual items in the refuse. Hollow containers, such as bottles and cans, begin to collapse at different pressures, depending on their orientation and strength. For example, cans collapse at pressures of 10 to 30 psi (0.1 to 0.3 N/m^2) and glass bottles at 5 to 35 psi (0.05 to 0.35 N/m^2).[9]

The compaction of some materials is irreversible, in that when the pressure is released, the material does not spring back to its original volume. MSW, however, contains many items that contribute to reversible compaction. At normal compaction pressures, 20% expansion can occur within a few seconds after the release of pressure, and this expansion can be as much as 50% after a few minutes.[10] The greater the pressure, the greater will be the bale integrity (its resistance to falling apart). A typical compression curve, using a small sample of refuse in a laboratory press, is shown in Figure 5-6.

SHREDDING

Strictly speaking, *shredding* is one form of *size reduction*, others being such processes as *cutting, shearing, grinding, crushing,* and other imaginative terms, many of which originated in mining engineering. In solid waste work, however, shredding is the generic term for size reduction even though, strictly speaking, shredding is only one method of size reduction. In this text, as in solid waste literature, shredding encompasses all the processes used for making little particles out of big particles.

Many types of shredders are presently on the market, and almost all of them were developed originally for an application and feed material other than refuse. Most of our present refuse shredding technology comes from the mining industry, which has for many years used shredders for ore processing. The application of this technology to refuse, however, is not an easy matter, because these devices were developed for homogeneous feeds having well-established breakage characteristics.

Use of Shredders in Solid Waste Processing

The first applications of shredders to MSW were to facilitate disposal, with little consideration for materials recovery. The pioneering work on shredding for disposal was done by Robert Ham and his colleagues at the University of Wisconsin. They found that shredded MSW had a more uniform particle size, was fairly homogeneous, and compacted more readily than unshredded waste, mainly because the

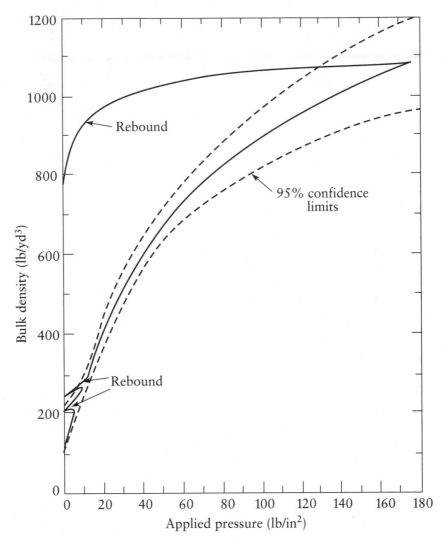

Figure 5-6 Compression curve for a sample of MSW in a laboratory. The rebound curves occur once the compressive pressure is released. Source: (9)

larger voids had been eliminated.[11] After shredding, MSW looks not unlike confetti, and has a light, bulky nature. In fact, the overall density of the material is decreased by over 50%, from 350–400 lb/yd³ (200–240 kg/m³) to 125–150 lb/yd³ (75–90 kg/m³). Shredding reduces required landfill volume since shredded refuse compacts better within the landfill. In addition, the landfill will have more uniform settlement and this helps maintain the integrity of the top cap.

Experience with landfilling of shredded MSW indicates that shredded refuse does not require an earth cover. Extensive testing has shown that the conditions that make an earth cover necessary in a conventional landfill no longer exist with shredded refuse. Earth cover in a shredded refuse landfill is considered unnecessary because of the following reasons:[12]

- Odor—The shredded refuse is so well mixed, and retains its aerobic character when spread in reasonably thin layers, that odor is not a problem.
- Rats—There are no large food particles in shredded refuse that could support a rat population.
- Insects—The drier refuse, regularly covered with new layers, suppresses insect breeding, and all of the maggots are killed during shredding.
- Blowing paper—Small pieces are not caught by the wind and do not blow away, while large pieces of paper and plastics found in a conventional landfill would be readily transported by wind.

Because of the lack of cover, leachate from shredded refuse is produced sooner, and this leachate is at a higher pollutional concentration than leachate from normal landfills. But because the shredded MSW is allowed to be open to the air, significant drying takes place, and the total leachate production is minor. Because of the superior compaction characteristics of shredded MSW, shredders have also been used before high-compression baling.[13]

Although the advantages of shredding prior to landfilling seem impressive, most regulatory agencies insist that daily cover still be applied even if the refuse is shredded. Such a requirement effectively eliminates the cost advantage of shredding refuse prior to landfilling.

A second use of shredding is in the production of *refuse-derived fuel*, or *RDF*. The breaking apart of the various constituents within MSW results in a shredded waste that has a more uniform heating value and requires less excess air, thus saving on air pollution control equipment and costs.

A third use of shredders is in the processing of yard waste as well as demolition debris, branches, and other organic material to produce a mulch that can then be composted or used as is as a ground cover. Shredders used for this purpose are, strictly speaking, *grinders*, consisting of a tub in which grinding gears rub the feed particles against the inside wall and break them up into smaller pieces. A typical grinder is shown in Figure 5-7.

Perhaps the most important use of shredders is in materials recovery. Mixed municipal solid waste is composed mostly of materials that are physically attached, and it is not easy or simple to separate them. Consider as an example a typical kitchen appliance. The electric motor has perhaps 10 or more materials, both plastics and metals, and the housing is of perhaps two different types of plastic or metal. Disassembling this device to recover the various materials would be prohibitively expensive. The alternative is to shred it up into small pieces, and at least some of these small pieces would then be of a single material, which could be separated (the topic of the next chapter).

Types of Shredders Used for Solid Waste Processing

The list of size-reduction equipment used in both the mining and chemical industries is surprisingly long. One well-known mining and ore dressing handbook, for

Figure 5-7   Tub grinder for yard waste.

example, lists over 50 different devices that could be applied to solid waste shredding.[4] A review of size-reduction equipment widely used in chemical engineering applications lists 21 different devices.[14]

The application of a specific device for MSW has not been a simple matter, however, since the material is significantly different from ore and other homogeneous feeds encountered in these industries. For example, coal can be counted on to shatter upon impact, and thus a shredder that would process coal would also shatter glass bottles rather well. On the other hand, the same shredder probably would not shred metal cans, which must be cut or torn apart within a shredder.

One of the first size-reduction devices used for solid waste processing was the *hammermill*, consisting of a central rotor on which are pinned radial hammers that are free to swing on the pins. The rotor is enclosed in a heavy-duty housing, an integral part of the shredding operation. In the *horizontal hammermill* the rotor is supported by bearings on either end and the feed is by gravity (free drop) or conveyor (force fed) (Figure 5-8). A discharge grate placed below the rotor determines the size of the product, since a particle cannot pass through this grate until it is smaller than the grate opening in two dimensions. Some hammermills are symmetrical so that the direction of the rotor can be changed to alternate wear surfaces without necessitating hammer maintenance after each run.

The *vertical hammermill*, as the name implies, has a vertical shaft, and the material moves by gravity down the sides of the housing (Figure 5-9). These mills usually have a larger clearance between the housing at the top of the mill, and progressively smaller clearances toward the bottom, thus reducing the size of the material in several steps as it moves through the machine. Since there is no discharge grate, the particle size of the product must be controlled by establishing a proper clearance between the lower hammers and the housing (Figure 5-10).

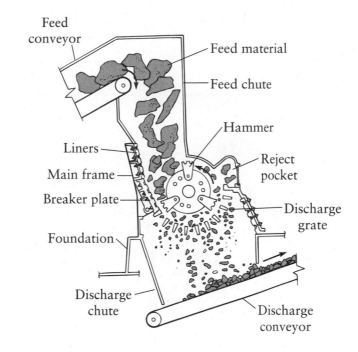

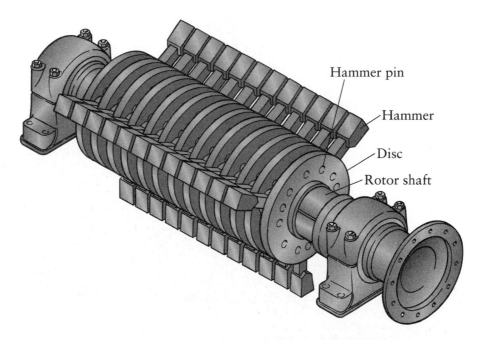

Figure 5-8 *Horizontal hammermill shredder. Source: (15)*

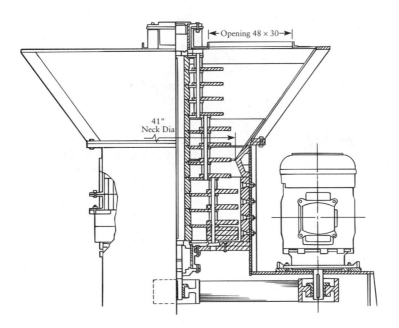

Figure 5-9   Vertical hammermill shredder. Source: (15)

Figure 5-10 *Inside a vertical hammermill shredder.*

Another type of shredder used for MSW is known as a *hog*, originally used in pulp and paper manufacturing for wood chops. The hog is used to shred green waste. Care must be taken to ensure that no abrasive items such as glass or metals are involved.

Slow-speed *shear shredders* were originally used to slice whole tires prior to disposal. If a whole tire is placed in a landfill, it can create problems by eventually floating to the surface due to its low density (if the space within the tire is still filled with air). Some states, such as California and Florida, require tires to be cut into pieces prior to landfilling. Shear shredders have been increasingly used for processing of mixed solid waste as well because of their lower energy requirements, lower rates of wear, and most importantly, the reduced chance of explosions. A typical shear shredder is shown in Figure 5-11.

Finally, a *flail mill* consists of arms with elbows that beat at the plastic bags to open them so that the contents can be processed further. Often glass bottles are broken in flails, but other than that, minimal size reduction occurs.

(a)

(b)

Figure 5-11   Shear shredder. (a) Close-up of the shredder blades. (b) In operation shredding tires. (Courtesy Shred-Tech)

Describing Shredder Performance by Changes in Particle-size Distribution

One of the most important design and operational parameters to be considered in size reduction is the change in particle-size distribution of the feed and the final product. The effect of shredding can differ for various material components in solid waste. Figure 5-12 is a graphical description of 13 different categories showing the size distribution after shredding in a hammermill.[9] The wide variation in size is obvious, and is one of the primary attributes of shredding, which allows for subsequent separation of the various material components. The discussion below is focused on describing the composite curve, which is not a picture of a homogeneous material but is made up of various components within the different particle-size categories.

Laboratory measurement of particle size is in itself difficult, since the material is in odd shapes. A piece of wire, for example, presents a difficult problem in classification because it is clearly quite small in two dimensions (and thus can escape further reductions in size in the shredding operation), but the effect of its length on subsequent separation operations—such as air classification—can be troublesome.

The method of measuring particle size can also influence the results of any given study. The common procedure for measuring particle-size distribution is by sieving, yet the shape of both the particles and the sieve openings can affect the number of particles that can pass through an opening.[9] In addition to shape factors, problems with providing an adequate duration of sieving, wear and tear on the sieve and the material, variations in the sieve apertures, and errors in observation and sampling all suggest that there may be problems involved in comparing size distribution data obtained at various laboratories.[16] Particle-size measurement is discussed more fully in Chapter 6.

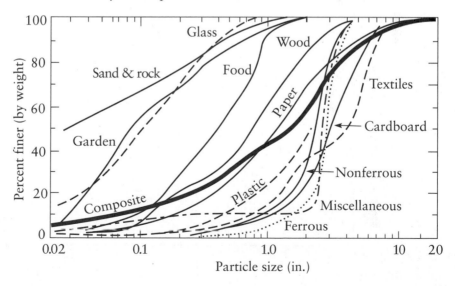

Figure 5-12 *Size reduction of various MSW components after shredding.*
Source: (10)

The size distribution of particles generally cannot be expressed by any single-valued function and is instead expressed by an equation describing the distribution of various size fractions. The general nomenclature[17] used for these equations is defined in Figure 5-13, a plot of the cumulative weight Y less than size x plotted versus particle size x. The particle sizes are broken into an arbitrary number of intervals (usually according to sieve sizes) with the top size (or grade) called number 1, on down to the nth grade. The ith interval has within it W, weight fraction of the material.

In the breakage process, as the curve is shifted to the left (greater number of smaller particles), some portion of the materials in the ith fraction remain (are not broken), and some portion originates from some larger sizes, or some general interval j. Y_j is the cumulative fraction by weight less than size j.

A number of equations have been proposed for describing the particle-distribution curve. Gaudin,[14] for example, suggested the following equation to describe the particle-size distribution for brittle materials:

$$Y = \left(\frac{x}{q}\right)^p$$

where Y is the cumulative fraction of material by weight less than size x, and q and p are constants specific to the material processed and the conditions under which the breakage occurs. In this case q is the theoretical maximum size and p defines the slope of the line on log–log coordinates.

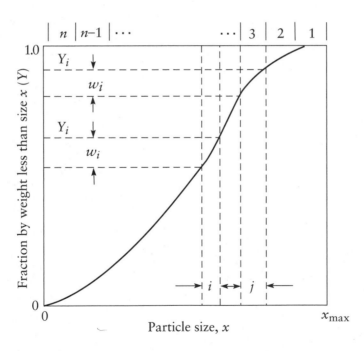

Figure 5-13   Cumulative particle-size distribution curve.

The most widely used particle-size descriptor is the Rosin-Rammler model,[18] first proposed in 1933, and stated as:

$$Y = 1 - \exp\left(\frac{-x}{x_0}\right)^n$$

where n is a constant and x_0 is the *characteristic particle size*, or just *characteristic size*, defined as the size at which 63.2% ($1 - 1/e = 0.632$) of the particles (by weight) are smaller. The Rosin-Rammler equation is a generalized expression for sigmoidal curves, such as those in Figure 5-14. Note that the constant n is the slope of the line $\ln(1/(1 - Y))$ versus x on log–log coordinates, since the linear form can be derived as:[38]

$$Y = 1 - \exp\left(\frac{-x}{x_0}\right)^n$$

$$\ln\left(\frac{1}{1-Y}\right) = \left(\frac{x}{x_0}\right)^n$$

$$\log\left[\ln\left(\frac{1}{1-Y}\right)\right] = n\log\left(\frac{x}{x_0}\right)$$

$$\log\left[\ln\left(\frac{1}{1-Y}\right)\right] = n\left[\log x - \log x_0\right]$$

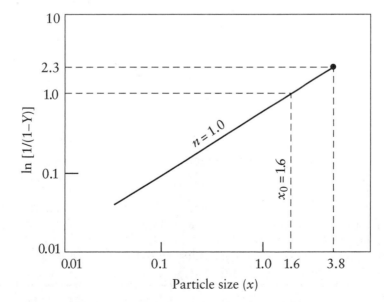

Figure 5-14 Rosin-Rammler particle-size distribution curve.

The value of x_0 is also defined as that size wherein $(1/(1 - Y)) = 1.0$, or equivalently where $1/e$ of the particles are larger than x_0. The equation suggests that for a specific value of x_0, as the constant n increases due to changes in machine or feed characteristics, the value Y decreases, meaning that a coarser product is obtained. Conversely, for a given n, a larger x_0 also defines a coarser particle size of the product.

The Rosin-Rammler equation is plotted on log–log coordinates in Figure 5-14 to illustrate the definition of both n and x_0. Table 5-3 presents a compilation of some Rosin-Rammler coefficients for various refuse shredders.

Table 5-3   Rosin-Rammler Exponents for Shredded Refuse

	n	x_0
Washington, DC	0.089	2.77
Wilmington, DE	0.629	4.56
Charleston, SC	0.823	4.03
San Antonio, TX	0.768	1.04
St. Louis, MO	0.995	1.61
Houston, TX	0.639	2.48
Vancouver, BC	0.881	2.20
Pompano Beach, FL	0.587	0.67
Milford, CT	0.923	1.88
St. Louis, MO	0.939	3.81

Source: (19)

The characteristic size can be calculated from a specification such as 90% passing a given size, common in ore comminution practice. Example 5-2, after Trezek and Savage,[20] illustrates the procedure.

EXAMPLE 5-2

Suppose that a sample of refuse must be shredded so as to produce a product with 90% passing 3.8 cm. Assume that $n = 1$. Calculate the characteristic size.

Since $Y = 0.90$, $\ln(1/(1 - Y)) = \ln(1/(1 - 0.90)) = \ln 10 = 2.3$. Plot $x = 3.8$ cm versus $\ln(1/(1 - Y)) = 2.3$ on log–log paper, as shown in Figure 5-14. For $n = 1$, the slope of the line is 45°, which can be drawn. Lines for any other slope (n) can similarly be constructed by measuring the slope with a ruler. The characteristic size x_0 is then found at $\ln(1/(1 - Y)) = 1.0$, as $x_0 = 1.6$ cm.

The conversion from x_0 to 90% passing is possible by recognizing that

$$x_0 = \frac{x}{\left[\ln\left(\frac{1}{1-Y}\right)\right]^{1/n}}$$

If $Y = 90\%$, this expression reduces to

$$x_0 = \frac{x_{90}}{2.3^{1/n}}$$

where x_{90} = screen size where 90% of the particles pass. If the value of n is 1.0,

$$x_{90} = 2.3\, x_0$$

These expressions are convenient for design purposes, as illustrated later in this chapter.

The characteristic size and the slope n can also be useful in measuring the effectiveness of a shredder in achieving breakage.[21] By sampling and sieving, a particle-size distribution of the shredder output can be obtained and the characteristic size and slope n calculated. Such calculation is facilitated by the use of *Rosin-Rammler Paper*, first developed by the Bureau of Mines in 1946. Figure 5-15 shows a sample of such paper, and its use is illustrated by Example 5-3.

EXAMPLE
5-3

A product from a MSW shredder was sieved with the following results. Calculate the Rosin-Rammler characteristic size and the slope n.

Sieve size x(in.)	Percent retained on sieve	Percent finer than Y
6	11	89
3	9	80
1	17	63
0.5	21	42
0.1	22	20
fines	20	—

Note that if 11% of the material is retained on the first sieve of 6 inches, 89% of the material has to be finer (smaller) than this size. If 9% is retained on the second sieve, $11 + 9 = 20\%$ of the material has been retained and 80% has to be finer than 3 in.

These values of x, particle size, and Y, percent finer than (or percent passing), are plotted as shown in Figure 5-15. The characteristic size x_0 is read off the graph at $Y = 63.21$ passing as 1.6 in. The slope n is calculated by measuring with a ruler as $3.0/5.5 = 0.55$.

The use of such calculations are shown in Figure 5-16, which represents actual data from a shredder operation. The concern that prompted the study was hammer wear and whether or not the operation of the shredder deteriorated over time. As can be seen from the figure, the characteristic size and the slope n did not change much over the nearly 15,000 tons of refuse processed.[20]

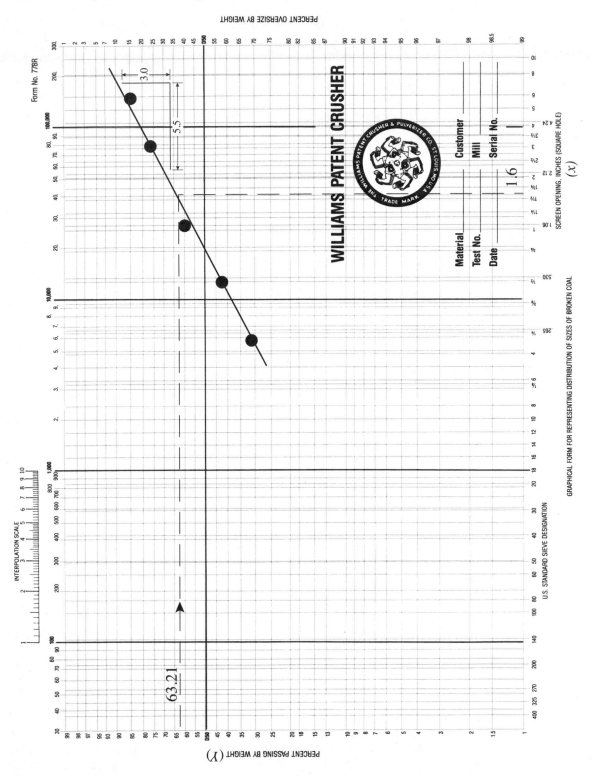

Figure 5-15 Rosin-Rammler paper for plotting particle-size distribution.

Another breakage model that assumes only a single fracture of materials is the Gaudin-Meloy model,[21] written as

$$Y = 1 - \left(1 - \frac{x}{x'_0}\right)^r$$

where x'_0 is again called the *characteristic size* of the feed material, but note the distinction between x'_0 and the Rosin-Rammler x_0. The size ratio r is defined as the ratio of the size of the broken piece to the size of the original piece.

A number of other particle-size distribution functions have been suggested, [22, 23, 24] but none of these appears to improve the required precision over the models discussed above. They merely add complexity to a problem where the lack of precision in measurement of particle size makes further sophistication questionable.[25] The π breakage theory is an exception since it provides a complete description of the product from a shredding operation. The explanation of this theorem is complex, and is therefore included as an appendix to this chapter.

Power Requirements of Shredders

Only limited information is available about the power requirements of shredders. Various estimates indicate that the hammermill is grossly inefficient, with only about 0.1% to 2.0% of the energy supplied to the machine appearing as increased surface energy of the product solids.[26] Part of the explanation for such low efficiencies lies in the plastic deformation and viscoelastic flow that accompanies shredding; these deformation processes require many times more energy than the creation of new surface area.

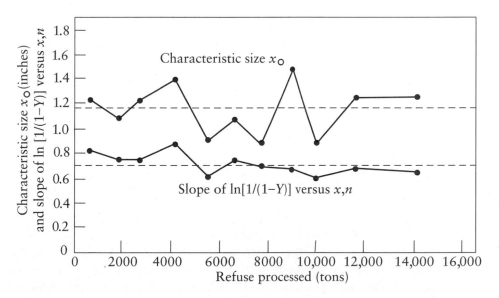

Figure 5-16 *Data from an evaluation study of a vertical hammermill shredder. Source: (34)*

The efficiency of a shredding operation depends on how the energy is applied and on how the material reacts to it. For example, a roll mill, which crushes materials but does very little shearing, would waste energy trying to shred a newspaper. Also, since fracture in brittle materials occurs progressively, with flaws building up within the particle until the particle breaks up, the rate of application of the force is important. Because there is a time lag between the application of the force and eventual fracture, a machine that applies load rapidly is inherently inefficient, since more energy is required than if the force could be applied more slowly. It follows that for some high-speed machines, a slower speed could result in less energy use, at least to the point at which the rotor inertia is too low and the energy requirement once again increases with decreasing speed. Higher speeds, nevertheless, produce the finest product particle size, but this requires more energy.

Energy requirements also increase with feed rate. Plotting the feed rate versus energy use in kilowatts results in straight lines for both secondary and primary shredding.[27] For low feed rates there is simply not enough material going through, and the machine is loafing. As the feed rate increases, the efficiency increases (lower specific energy use), and eventually the machine becomes overloaded. The amount of moisture in the feed refuse also influences energy use, and the minimum energy requirement appears to be in the 35 to 40% moisture range.[20] Moisture also affects particle size, with drier feed resulting in smaller particles in the product.

Accurate estimates of the energy requirements for size reduction have not been possible, even for homogeneous materials. Several empirical relationships have been suggested. Those that consider only the fracture process are all based on the assumption that

$$\frac{dE}{dL} = -CL^{-n}$$

which states that the energy dE required to achieve a small size change dL in a unit mass of material is inversely proportional to the size of the article L. The symbols n and C are constants. If $n = 1$, for example, the equation can be integrated to yield

$$E = C \log\left(\frac{L_1}{L_2}\right)$$

where L_1/L_2 is the size-reduction ratio and E is the work done to reduce the particles from size L_1 to size L_2. C, as before, is a constant. Physically, this expression states that the work done is proportional to the number of new smaller particles created from the larger ones. This is known widely as *Kick's law*.[26]

If $n = 2$, the integrated form of the equation yields

$$E = C\left(\frac{1}{L_2} - \frac{1}{L_1}\right)$$

which is known as *Rittinger's law* and physically represents the assumption that the energy required is proportional to the amount of new surface formed by the size-reduction process.[26] Rittinger's law seems to agree better with the results of rough grinding operations, whereas Kick's law is a better approximation of fine grinding.

If $n > 1$ but is otherwise undefined, the integrated form of the equation is

$$E = \frac{C}{n-1}\left(\frac{1}{L_2{}^{n-1}} - \frac{1}{L_1{}^{n-1}} \right)$$

The *Bond work index*[28] is based on the assumption that the work done in crushing and grinding is directly proportional to the total length of new cracks formed in the material being reduced in size.[29] In this case it is suggested that $n = 1.5$, and the general equation becomes the *Bond law*:

$$E = E_i \frac{\sqrt{L_F} - \sqrt{L_P}}{\sqrt{L_F}} \sqrt{\frac{100}{L_P}}$$

where E = specific work (kWh/ton) required to reduce a unit weight of material with 80% finer than some diameter L_F in micrometers, to a product with 80% finer than some diameter L_P, in micrometers. In the above equation, E_i = work index, a factor that is a function of the material processed. This value is also the theoretical work required to reduce a unit weight from infinite size to 80% finer than 100 μm; units are (kWh/ton).

The expression above is more conveniently expressed as

$$E = 10\, E_i \left(\frac{1}{\sqrt{L_P}} - \frac{1}{\sqrt{L_F}} \right)$$

The work indices for some common industrial materials are tabulated in Table 5-4. The 10 in the equation is the square root of 100 μm, hence the units of E_i are kWh/ton.

Table 5-4   Work Indices for Common Industrial Materials

Material	Work index (kWh/ton)
Coal	11.4
Glass	3.1
Granite	14.4
Slag	15.7

Source: (29)

The Bond work index can also be estimated by the dimensionally incorrect empirical relationship[29] as

$$E_i = 2.59\, \frac{C_S}{SG}$$

where E_i = work index, (kWh/ton)
C_s = impact crushing resistance, ft-lb/in. of thickness required to break
SG = specific gravity

Health and Safety

One need only reflect on the types of materials people thoughtlessly throw away into refuse cans to realize the serious health and safety aspects of shredding MSW. Although the health and safety aspects of size reduction deserve considerable study, only a brief summary of the problem is presented here. Specifically, shredders can be hazardous because of noise, dust, and explosions.

The noise levels around a 3-ton/h hammermill range from 95 to 100 dBA with much of the noise produced being a low-frequency rumble. (dBA is a standard method of noise measurement, and stands for decibels on the A scale of the sound-level meter. This scale is an attempt to duplicate the hearing efficiency of the human ear.)[30] In addition to a high constant noise level, materials recovery facilities processing MSW produce considerable impact noise, which is difficult to measure, and the effect of which on human beings is poorly understood. The existing federal Occupational Safety and Health Act (OSHA) standard limits noise to 90 dBA over an 8-h working day. The corresponding limit set by the EPA is 85 dBA. It seems likely that shredder operators will need to wear ear protection, and noise reduction should be considered in the design of future resource recovery facilities.

Dust can cause several problems: It can be a vector for the transmission of pathogenic microorganisms, it can itself have a detrimental effect on health by affecting the respiratory system, and it can explode. This last problem is discussed in the next section.

OSHA standards presently limit dust inhalation to 15 mg/m^3 of total dust over an 8-h day. Limited studies of dust production in resource recovery facilities have shown that dust levels are from 7 to 13 times higher than the OSHA standard. This finding, which would not surprise shredder operators, dictates the use of face masks while working.[31]

Plate counts at shredding operations have indicated that the total bacterial counts during the shredder operation are as much as 20 times greater than the ambient, which contains about 880 organisms/m^3 of air. Studies have shown that in resource recovery facilities where shredders are used, coliform counts can jump from 0 to 69 per cubic foot, and fecal streptococci from 0 to over 500 per cubic foot.[32] These indications clearly show the potential danger of disease transmission by the air route during shredding of MSW.

The high temperature and metal-to-metal contact in shredders has caused numerous explosions at existing shredder installations. Actually, small explosions such as those due to the breakage of aerosol cans occur regularly, and shredders are designed to accept these without suffering damage. Larger explosions, however, can damage the shredder and have resulted in fatalities.

Two types of explosions are generally recognized—dust explosions and those caused by explosive materials such as gunpowder and partially filled gasoline cans. Some operators argue that the dust will never become explosive by itself and that all explosions are caused by combustible materials. This contention remains to be investigated.

Dust can cause explosions when the concentration of a combustible dust is sufficiently high and there is adequate oxygen and a spark. Below a certain dust concentration, the heat of combustion is not sufficient to propagate combustion, and an explosion cannot occur. If the concentration of oxygen is reduced to below about 10%, explosions should not occur. Other than continuously flooding the chamber with an inert gas, however, it is difficult to keep oxygen out of the shredder. The only realistic means of preventing explosions is to maintain a high level of surveillance on what is being fed to the shredder. Accordingly, almost all shredder installations have people scanning the feed conveyor for such potentially dangerous items as cans of paint thinner, gasoline tanks from cars, lawn mower engines, and so on.

Even keen surveillance, however, is not foolproof. In one instance a gas tank was removed from the conveyor belt, but some of the gasoline spilled on the refuse. This occurred just at the change of shifts and the new crew was not informed of the spilled gas. When the shredder was again fed, the gasoline vaporized and ignited from space heaters above the workers, sending a fireball back into the shredder.

Two methods presently used to reduce the damage when explosions occur in refuse shredders are venting and flame suppression. All shredders are constructed with blowout doors, so that the pressure within the shredder housing can escape. Some are equipped with flame suppressors, which are designed to be released as soon as the pressure builds up to a critical level. This system will work for most types of explosions, but it is not fast enough for dynamite, gunpowder, and other explosives.

Hammer Wear and Maintenance

Because of the relatively unsophisticated and brute-force nature of the shredding process, the wear and tear on shredders can be substantial. One of the major maintenance headaches (and expenses) at shredding facilities is the wear of the hammers. The pattern of wear on hammers is illustrated in Figure 5-17. The wear is almost exclusively on the bottom edge of the hammer's crushing face, since this is the area of impact as the material is crushed against the grate. The wear seems to be due mostly to abrasion, although severe impact with very hard objects can also contribute to wear. Hammer wear can be reduced by both hard-facing the hammers with abrasion-resistant alloys and by slowing the speed of shredding.

As the hammers wear down, the shredding performance decays. The percent of materials passing the sieve (Y) can be related to the tons of refuse processed (T) by the equation

$$Y = b_0 + b_1 \exp(-b_2 T)$$

where b_0, b_1, and b_2 are all constants.[33] For example, for 0.185-in. particle size, the relationship is

$$Y = 8 + 35e^{-0.0081T}$$

Hammer wear for horizontal hammer mills can be expected to be in the range 0.05 to 0.10 lb/ton when shredding MSW. Experience has shown, however, that much higher rates of hammer wear are also possible.

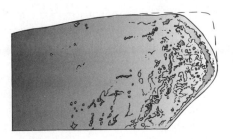

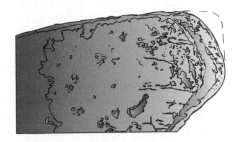

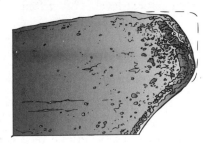

Figure 5-17 Hammer wear in hammermill shredders. (Courtesy of G. J. Trezek)

Figure 5-18 shows hammer wear for a vertical shredder used for raw refuse.[34] The hammers close to the top of the shredder (Figure 5-10) are used for material breakage and do not suffer much wear, while the hammers close to the bottom show much higher rates of wear.[34]

Shredder Design

Shredders are not, in the strict sense, designed by a consulting engineer or the purchaser. They are actually selected much as water pumps are selected for a specific application. Such specifications as the speed, motor horsepower requirements, and the rotor inertia must all be specified.

The rotor inertia is usually expressed as WR^2, where W is the mass of the rotor assembly and R the radius to the hammermill tips. This is not an accurate measure of rotor inertia, but is simply a convenient parameter for comparative purposes. A wide range of WR^2 is offered by manufacturers, from about 50,000 to 150,000 lb-ft^2 (2000 to 6000 kg-m^2).[15]

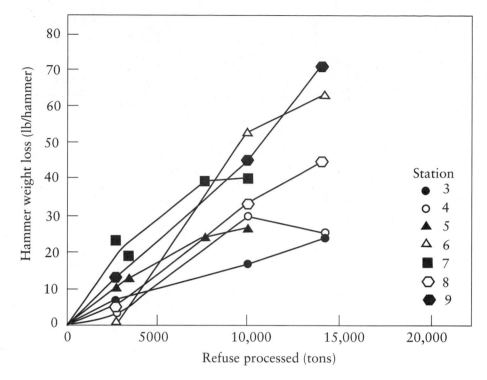

Figure 5-18 *Hammer wear distribution depends on location in a vertical shredder.* Source: (34)

The motor horsepower is designed on the basis of starting horsepower. If the motor inertia WR^2 is expressed as lb-ft^2, the rotor speed N in rpm, and starting time t in seconds, then the torque T is

$$T = \frac{WR^2 N}{9.6\, gt}$$

where g is the gravitational constant, 32.2 ft/sec^2. The horsepower is then calculated as

$$\text{horsepower} = \frac{2\pi TN}{33,000}$$

where T is torque, in ft-lb.

Typically, a shredder used for MSW has a width/diameter ratio greater than 1.0, a hammer weight of 150 lb (70 kg), a rotor inertia of 35,000 lb-ft^2 (1500 kg-m^2), a hammer tip speed of 14,000 ft/min (4260 m/min), four rows of hammers, and a starting time of 30 s. As a rule of thumb, a MSW shredder must be designed for at least 15 kWh/ton. For horizontal hammermills the grate openings can be changed to achieve different size distribution of the product. Similarly, final clearance in vertical mills can be changed to produce various-sized products. The characteristic size x_0 (see page 193) and the specific energy can be related to the grate spacing or exit clearance as shown in Figure 5-19.

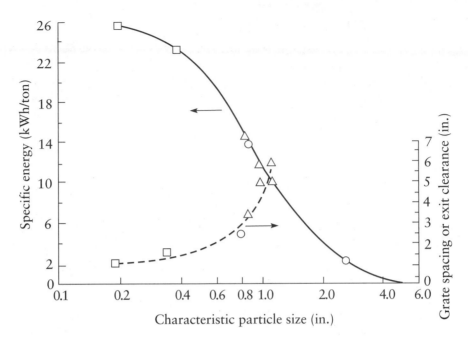

Figure 5-19 Relationship between grate spacing, energy use, and product size. Source: (20)

EXAMPLE
5-4

For $x_0 = 0.64$ in. (1.62 cm) and $n = 1$ find the grate spacing or clearance and the motor power requirements for a feed of 5 tons/h.

Using Figure 5-19, enter at $x_0 = 0.64$ and find the grate opening as 2.54 in. and specific energy requirement as 18 kWh/ton, which translates to 90 kW (120 hp).

The Bond work index may be used for shredder design by noting first that the Rosin-Rammler equation can be written as

$$Z = \exp\left(\frac{-x}{x_0}\right)^n$$

where Z = cumulative fraction *greater than* some stated size x
$\quad x_0$ = characteristic size
$\quad n$ = constant

That is, $Z = 1 - Y$, where Y is previously defined as the cumulative fraction finer than some size x.

At $Z = 0.2$ (meaning that 20% of the feed is larger), then $x = L_P$, the screen size through which 80% of the product passes. Solving for L_P,

$$0.2 = \exp\left(\frac{-L_P}{x_0}\right)^n$$

$$L_P = x_0 \, (1.61)^{1/n}$$

Substituting into the Bond work index equation,

$$E = \frac{10 \, E_i}{\left[x_0(1.61)^{1/n}\right]^{1/2}} - \frac{10 \, E_i}{L_F^{1/2}}$$

This equation now allows for an estimation of shredder power requirements, given a definition of the required product, and an estimation of the work index E_i.

Although no data on refuse shredder power consumption versus particle size are yet available, it is possible to back-calculate for the Bond work index. Table 5-5 is such a tabulation, and shows that on average, the E_i for refuse is about 430 (kWh/ton).

Table 5-5   Bond Work Index for Shredding of Refuse and Refuse Components

Shredder location	Material shredded	Bond work index, E_i (kWh/ton)
Washington, DC	Refuse	463
Wilmington, DE	Refuse	451
Charleston, SC	Refuse	400
San Antonio, TX	Refuse	431
St. Louis, MO	Refuse	434
Houston, TX	Refuse	481
Vancouver, BC	Refuse	427
Pompano Beach, FL	Refuse	405
Milford, CT	Refuse	448
St. Louis, MO	Refuse	387
Washington, DC	Glass	8
Washington, DC	Paper	194
Washington, DC	Steel cans	262
Washington, DC	Aluminum cans	654

Source: (19)

EXAMPLE
5 - 5

Assuming that E_i = 430 kWh/ton, x_0 = 1.62 cm, n = 1.0, and L_F = 25 cm (about 10 in., a realistic estimate as shown in Figure 2-6), find the power requirement for a shredder processing 5 tons/h.

$$E = \frac{10(430)}{\left[16,200(1.61)^{1/1}\right]^{1/2}} - \frac{10(430)}{(250,000)^{1/2}} = 18\,\text{kWh/ton}$$

or

$$18\,\frac{\text{kWh}}{\text{ton}}\left(5\,\frac{\text{tons}}{\text{h}}\right) = 90\,\text{kW}$$

Note that this is the same value as found in Example 5-4.

PULPING

Wet pulping, although a well-developed process in the pulp and paper industry, has been applied to solid waste processing by only one manufacturer. Their unit, pictured in Figure 5-20, is a tub 12 ft (3.6 m) in diameter with a high-speed cutting

Figure 5-20   Pulper used for processing MSW.

blade on the bottom driven by a 300-hp motor. The raw refuse is pulped and all pulpable and friable materials are reduced in size so as to fit through the holes immediately below the cutting blade. The resulting slurry has a solids content of about 4%. Pieces of metal and other nonbreakable materials are ejected from the pulper through an opening on the side of the tub, washed, and put through a ferrous recovery system. The slurry can be centrifuged to remove the organics. No such facilities are operating in the United States, the last one at Dade County, Florida, being closed in the late 1980s.

ROLL CRUSHING

Roll crushers are used in resource recovery operations for the purpose of crushing brittle materials such as glass while merely flattening ductile materials such as metal cans—hence allowing for subsequent separation by screening. Roll crushers were first employed in materials recovery facilities for the reclamation of metals from incinerator residue, and have found use in processing partially source-separated refuse composed of glass containers and aluminum and steel cans.

A variation of the roll crusher is the roll crusher/perforator. Curbside recycling programs typically collect PETE (soda bottles) and HDPE (milk containers) from residents. Even though instructed not to, some residents screw the lids back onto the empty containers. When these containers are baled, they then do not compress and can cause problems with the finished bales. To solve this problem, some processors have placed roll crushers/perforators in front of the baler. Puncturing the container prior to baling eliminates the problem.

Roll crushers work by capturing and forcing the feed through two rollers operating in opposite directions, exactly like the wringers on older washing machines. The first objective of roll crushing is to capture the pieces that are to be crushed. This capture depends on the size and characteristics of the particles, and the size, gap, and characteristics of the rollers. To illustrate the importance of this capture, imagine attempting to force a basketball through a clothes wringer. A small rubber ball, on the other hand, could be captured readily and flattened.

The variables involved in the analysis of roll crushing are shown in Figure 5-21. The diameters of the two rolls are D while the diameter of the particle to be crushed is d. The normal force between the particle and the rollers is N, and the tangential force is T. If the resultant force R is pointed downward, the particle will be captured and crushed. If it points upward, the particle will ride on the rollers.

The vertical component of N is

$$N_v = N \sin \frac{n}{2}$$

where n is the angle between the two tangential forces and $n/2$ is the angle between the horizontal and the line connecting the centers of the feed particle and the roller. Similarly, the vertical component of the tangential force is

$$T_v = T \cos \frac{n}{2}$$

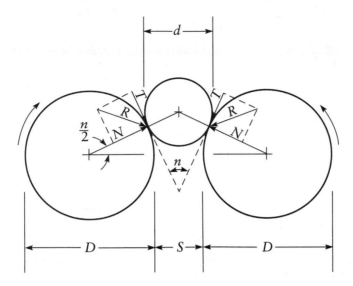

Figure 5-21   Roll crusher.

At the point where crushing is possible, $N_v = T_v$, or

$$N \sin \frac{n}{2} = T \cos \frac{n}{2}$$

or

$$\tan \frac{n}{2} = \frac{T}{N}$$

In this instance the angle n becomes known as the *angle of nip*. Since T/N = the coefficient of friction (φ), the necessary condition for crushing to take place is

$$\tan \frac{n}{2} \leq \varphi$$

From Figure 5-21,

$$\frac{S}{2} + \frac{D}{2} = \frac{D}{2} + \frac{d}{2} \cos \frac{n}{2}$$

where S is the separation between the rollers. It follows that

$$\cos \frac{n}{2} = \frac{D+S}{D+d}$$

EXAMPLE
5-6

It is intended to crush pieces of glass of nominal diameter 5 cm (feed) to maximum diameter 0.5 cm (product). The coefficient of friction between steel and glass is 0.4. Find the diameter of the rollers that will just capture and crush this material.

$$\tan \frac{n}{2} = 0.4$$

$$\frac{n}{2} = 21.8$$

$$\cos \frac{n}{2} = 0.928 = \frac{D+S}{D+d} = \frac{D+0.5}{D+2}$$

and $D = 18.8$ cm.

The capacity of a roll crusher can be estimated as the maximum volume squeezed through the rollers (Figure 5-22):

$$C = kMDWS\rho$$

where C = capacity, tonnes/h
k = dimensional constant = 60 if all dimensional units are as below
M = speed of rollers, rpm
D = diameter of rollers, m
W = length of rollers, m
S = separation, m
ρ = density of material, g/cm^3

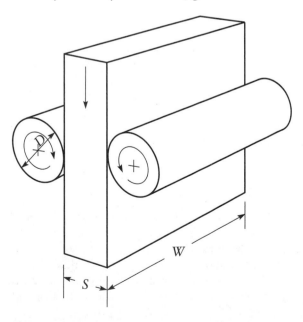

Figure 5-22 Capacity of a roll crusher.

E X A M P L E
5 - 7

For rollers in Example 5-6, find the capacity if $M = 60$ rpm, $W = 0.5$ m, and $\rho = 2.5$ g/cm^3.

$$C = kMDWS\rho = (60)(60)(0.189)(0.5)(0.005)(2.5) = 4.2 \text{ tonnes/h}$$

A related use of roll crushers is as *can flatteners*. Run at slow speed, can flatteners are placed in processing facilities after magnetic separation of aluminum from steel cans, and are economical if the cans are to be shipped long distances.

GRANULATING

For some materials, such as plastic bottles, the high energy and cost of hammermills are not warranted. Size reduction can be achieved far better using *granulators*, which are slow-speed shears that cut instead of shatter. Granulators can be economically effective if the plastic is to be shipped long distances since the granulated plastic has a far higher density than compacted bottles.

FINAL THOUGHTS

People who have had extensive experience in the storage and handling of MSW are confirmed cynics. A new piece of equipment, no matter how highly touted by the manufacturer, is received with a wry, knowing smile. The question always is: "Yes, but will it work with MSW?" Too many times they have seen the unsuccessful transfer of technology and hardware from another application. Always the problem is the same—solid waste is a heterogeneous and unpredictable material, and equipment designed for a simpler feed cannot handle MSW. With an increasing need for MSW processing, we will see more and more equipment specifically designed for refuse and proven in the field using the real stuff. Only then will these cynics be convinced.

APPENDIX
THE PI BREAKAGE THEOREM

Breakage of particles can be described by the breakage function, $B(x,y)$, which is defined as the fraction by weight of products that have a size less than x when particles of original size y are broken once. For example, $B(20,500) = 0.4$ means that 40% by weight of an original particle of size 500 falls below size 20 after a single breakage. The breakage function can also be normalized, so that $B(x/y)$ describes the particle-size distribution regardless of the size of the original particle. For example, $B(0.04) = 0.26$ means that 26% by weight of the product falls in sizes below 0.04 of the original size.

Much of the original development in breakage theory is credited to Epstein,[35] with the matrix application discussed below by Broadbent and Callcott.[36] Trezek at the University of California at Berkeley was the first to apply this technique to MSW,[30] but the development shown here is originally developed by Vesilind, Pas, and Simpson.[37]

The basic assumption in this development is that during the breakage process, a fraction of the particles entering as feed are not broken, and exist as part of the product. For any given particle size, therefore, the product consists of some particles that were originally that size and were not broken, and some that are the result of larger pieces breaking into smaller ones, the latter being referred to as complement.

A second assumption is that the size distribution of the product due to the breakage of any single particle can be described by a continuous function. In other words, a large particle will break into many smaller pieces, which—when analyzed by sieving—will yield a smooth particle-size distribution. In the matrix analysis technique developed by Broadbent and Callcott the distribution of particles following such breakage is described by the function

$$B(x,y) = \frac{1 - \exp(-x/y)}{1 - \exp(-1)}$$

where y is the particle size being broken and $B(x,y)$ is the cumulative fraction of the product equal to or smaller than x. Although this equation was chosen in the Broadbent-Callcott analysis, other equations may describe the particle-size distribution as well or better, and may be found to be superior when this analysis is applied to MSW.

Consider now a series of sieves, where the sieve sizes are related by a constant factor a so that the range of sizes passing the largest sieve is 1 to a, the next sieve is a to a^2, the next sieve a^2 to a^3, and so on to a^{n-1} to a^n. (Such geometric gradation is not necessary for this analysis, but it happens to be convenient for illustrative purposes.) The feed particle-size distribution can then be described by the fractions f falling into these size ranges or grades, so that fraction f_1 of the feed is comprised of particles between size 1 and size a, f_2 is between a and a^2, and so on. Obviously,

$$\sum_{i=1}^{n} f_i = 1.0$$

Further, assume that the particle-size distribution within each grade can be described by the geometric mean, or

$$a^{1/2}, a^{3/2}, a^{5/2}, \ldots, a^{(2n-1)/2}$$

Assuming now, as discussed above, that the fraction of the feed within each grade that actually breaks is π so that the amount of the various fractions that are broken are

$$\pi f_1, \pi f_2, \pi f_3, \ldots, \pi f_n$$

The fraction π of the topmost grade (f_1) breaks into smaller pieces according to

$$B(x,y) = \frac{1 - \exp(x / y)}{1 - \exp(-1)}$$

so that of the π fraction breaking, the cumulative fraction of those particles equal to or finer than those in the top grade after breakage is

$$B_{11} = \frac{\left[1 - \exp\left(\dfrac{-a^{1/2}}{a^{1/2}}\right)\right]}{\left[1 - \exp(-1)\right]}$$

Note that $B_{11} = 1.0$, since $y = a^{1/2}$, the particle size being broken and $x = a^{1/2}$, the size of the product.

Some fraction of the particles in the top grade break into a smaller size $a^{3/2}$. The cumulative fraction of particles equal to or smaller than this size originating in the top grade of particle size $a^{1/2}$ is

$$B_{21} = \frac{\left[1 - \exp\left(\dfrac{-a^{3/2}}{a^{1/2}}\right)\right]}{\left[1 - \exp(-1)\right]}$$

and so on. The breakage of the top grade is summarized as follows:

Grade	Cumulative fraction of original (πf_1) particles equal to or finer than the grade after breakage
1 to a	$B_{11} = \dfrac{1 - \exp\left(-a^{-1/2}\big/a^{1/2}\right)}{1 - \exp(-1)}$
a to a^2	$B_{21} = \dfrac{1 - \exp\left(-a^{-3/2}\big/a^{1/2}\right)}{1 - \exp(-1)}$
a^2 to a^3	$B_{31} = \dfrac{1 - \exp\left(-a^{-5/2}\big/a^{1/2}\right)}{1 - \exp(-1)}$
.	.
.	.
.	.
a^{n-1} to a^n	$B_{n1} = \dfrac{1 - \exp\left(-a^{-(2n-1)/2}\big/a^{1/2}\right)}{1 - \exp(-1)}$

Of the topmost feed grade, πf_1 was designated for breakage ($(1 - \pi f_1)$ did not enter the breakage process). The cumulative mass of the original feed that is equal in size to, or smaller than, $a^{1/2}$ after breakage is $B_{1,1}(\pi f_1)$. If $b_{11} = 1.0$, as we have assumed, all the particles of the product are equal to or finer than $a^{1/2}$.

Similarly, the products of the breakage of the topmost grade are now distributed throughout the smaller grades such that $B_{2,1}$ represents the cumulative fraction of particles equal to or finer than $a^{3/2}$, $B_{3,1}$ is the cumulative fraction equal to or finer than $a^{5/2}$, and so on. The fraction of the product particles in the top grade after breakage is thus $b_{11} = B_{1,1} - B_{2,1}$. This can also be interpreted as the fraction of particles, which, although broken, remained in the top grade. Similarly, the fraction of particles in the next smallest grade is $b_{21} = B_{2,1} - B_{3,1}$, and so on.

The example given above is for the breakage of particles that were originally in the topmost grade only, with a geometric size $a^{1/2}$. But breakage occurs within all grades, and this relationship must apply equally well for any other grade of the feed. Summing up the particles in the product that originated from breakage of larger particles, we get the following equations:

Grade	Product resulting from breakage
1 to a	$P_1 = \pi b_{11} f_2$
a to a^2	$P_2 = \pi b_{21} f_1 + \pi b_{22} f_2$
a^2 to a^3	$P_3 = \pi b_{31} f_1 + \pi b_{32} f_2 + \pi b_{33} f_3$
.	.
.	.
.	.
a^{n-1} to a^n	$P_n = b_{n1} f_1 + b_{n2} f_2 + \ldots + b_{nn} f_n$

where the first b subscript refers to grade of product and the second subscript defines the origin of that fraction. For example, the product in the second grade (a to a^2) is now made up of feed particles that originated as the top grade (1 to a) and were broken (the quantity $\pi b_{21} f_1$) and the particles within the second grade that, although broken, remained there as $\pi b_{22} f_2$. The product of the third grade is composed of particles that originated on the first grade ($\pi b_{31} f_1$), the second grade ($\pi b_{32} f_2$), and the third grade ($\pi b_{33} f_3$). To state it another way, the parameter b_{ij} is defined as the fraction of material in size interval j that falls into the size interval i after breakage. Thus the products of size interval 1 are distributed so that b_{21} falls into size interval 2, b_{31} falls into interval 3, and so on. The sum of the values of b_y over all the values of i is 1.

These equations can be summarized in a single equation using matrix notation. If $\mathbf{P}'$ is the product vector, $\mathbf{f}$ is the feed vector, and $\mathbf{B}$ is defined as the breakage matrix, derived from Equation 5-1, then

$$\mathbf{P}' = \mathbf{B}(\pi)\mathbf{f}$$

Recall that not all of the particles in any one size fraction were broken in the process. Hence the product contains a fraction that originally existed in the feed and were not broken, $(1 - \pi) f_n$. The final product within each grade consists of these original particles, plus those that resulted from the breakage process. Hence

Grade	Total product
1 to a	$P_1 = \pi b_{11} f_1 + (1 - \pi) f_1$
a to a^2	$P_2 = \pi (b_{21} f_1 + b_{22} f_2 + (1 - \pi) f_2$
a^2 to a^3	$P_3 = \pi (b_{31} f_1 + b_{32} f_2 + b_{33} f_3 + (1 - \pi) f_3$
.	.
.	.
.	.
a^{n-1} to a^n	$P_n = \pi (b_{n1} f_1 + b_{n2} f_2 + \dots + b_{nn} f_n) + (1 - \pi) f_n$

These equations can similarly be expressed in matrix form as

$$\mathbf{P} = \pi \mathbf{B} \mathbf{f} + (1 + \pi) \mathbf{f}$$

In Equation 5-3 the feed vector $\mathbf{f}$ and the product vector $\mathbf{P}$ can be obtained from sieve analysis, and the breakage matrix $\mathbf{B}$ is defined by the quantities $b_1, b_2, b_3, \dots$ and can be calculated. The only unknown quantity is thus the scalar value π, which could well be a valuable index for describing the shredding operation. The validity of Equation 5-3 depends on the validity of the original assumptions. If π is indeed constant for a given shredding operation, any one size particle is as likely to be broken as any other. This is not altogether true, since particles tend to become more resistant to fracture as they become smaller.

The question of the breakage function is still open to argument. Trezek et al. investigated the applicability of several models and decided that either the Gaudin-Meloy function or a modified Broadbent-Callcott function gave excellent results.[20] The latter was written as

$$B_{(x)} = \frac{1 - \exp\left[-\left(x/x_0\right)^n\right]}{1 - \exp(-1)}$$

where n is a positive index requiring back-calculations and varying from 0.845 to unity. The term x_0 is the *characteristic size* as previously defined. Figures 5-23 and 5-24 show the results of some experiments using these two functions. The first figure is an estimate of primary shredding, and the second shows the results from secondary shredding. In both cases either breakage function seems to predict the particle-size distribution adequately.

Another refinement to the model is to eliminate the original assumption that the fraction of any particle size that enters the breaking process is constant for all sizes—in other words, that π is constant. To do this, we can argue that π can be replaced by a matrix $\mathbf{S}$ such that it characterizes the size-reduction process, and is a diagonal matrix with entries $S_1, S_2, S_3, \dots S_n$ along the diagonal and zeros elsewhere. The final product equation can be expressed as

$$\mathbf{P} = \mathbf{B} \mathbf{S} \mathbf{f} + (\mathbf{I} - \mathbf{S}) \mathbf{f}$$

where $\mathbf{I}$ is a unit matrix.

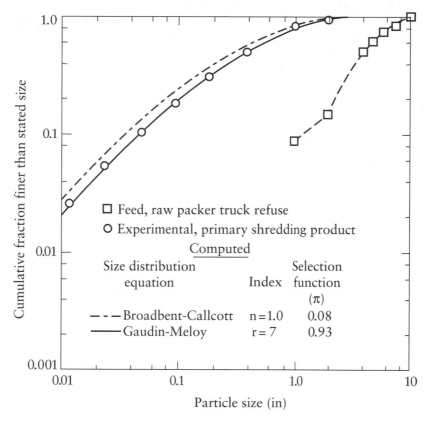

Figure 5-23 *Experimental and calculated particle-size distribution for primary shredding. Source: (20)*

A radioactive tracer has been used to measure the breakage process and to evaluate some of the assumptions made in the foregoing analysis.[39] In this study the breakage function $B(x, y)$ did not vary with the time of grinding, given a certain feed size. It was also found reasonable to normalize $B(x,y)$ to $B(x/y)$. Finally, they found by back-calculating the values for S and $B(x/y)$ that they were within reasonable agreement with experimental data.

EXAMPLE
5-8

For the following feed, calculate the fraction of particles in each size range if it is assumed that $\pi = 0.5$ and the breakage function is $B(x/y) = (1 - \exp(-x/y))/(1 - \exp(-1))$.

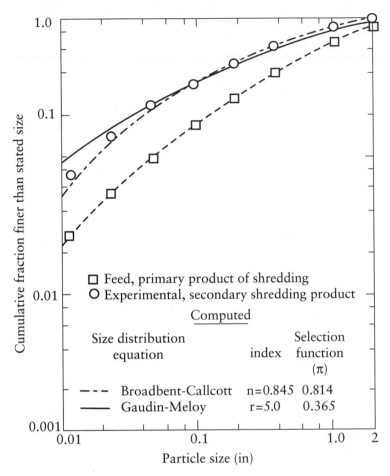

Figure 5-24 Experimental and calculated particle-size distribution for secondary shredding. Source: (20)

Average size in sieve (mm)	Fraction of feed
100	0.80
75	0.20
50	0
25	0

The basic equation is

$$P = \pi Bf + (1 - \pi)f$$

where P is the product vector, B is the breakage matrix, and f is the feed vector. Written in longhand:

$$
\begin{bmatrix} P_1 \\ P_2 \\ P_3 \\ P_4 \end{bmatrix} = (\pi) \begin{bmatrix} b_{11} & 0 & 0 & 0 \\ b_{21} & b_{22} & 0 & 0 \\ b_{31} & b_{32} & b_{33} & 0 \\ b_{41} & b_{42} & b_{43} & b_{44} \end{bmatrix} \begin{bmatrix} f_1 \\ f_2 \\ f_3 \\ f_4 \end{bmatrix} + (1 - \pi) \begin{bmatrix} f_1 \\ f_2 \\ f_3 \\ f_4 \end{bmatrix}
$$

The breakage matrix can be calculated from the breakage function:

$$
B(x,y) = \frac{1 - \exp(-x/y)}{1 - \exp(-1)}
$$

$$
B_{100,100} = \frac{1 - \exp(-100/100)}{1 - \exp(-1)} = 1.0
$$

$$
B_{75,100} = \frac{1 - \exp(-75/100)}{1 - \exp(-1)} = 0.83
$$

This means that of the particles in size interval 1 (100 mm) that enter the breakage process, 83% break down to size 75 mm or smaller. Hence the faction of the particles that remain in the size interval 1 (even after breakages) is b_{11} = 0.17. Similarly,

$$
B_{50,100} = \frac{1 - \exp(-50/100)}{1 - \exp(-1)} = 0.62
$$

or 62% of the material that broke becomes 50 mm or smaller. Hence 0.83 − 0.62 = 0.21, or 21% of the original size material is now in size interval 2, or b_{21} = 0.21. Similarly, $B_{25,100}$ = 0.35 and b_{31} = 0.62 − 0.35 = 0.27, and the last size interval must have a contribution of 0.35. Note that these fractions (0.17, 0.21, 0.27, and 0.35) sum to 1.0.

The material that was at 75 mm also breaks, so that

$$
B(50,75) = \frac{1 - \exp(-50/75)}{1 - \exp(-1)} = 0.77
$$

or b_{22} = 0.23. Also,

$$
B(25,75) = \frac{1 - \exp(-25/75)}{1 - \exp(-1)} = 0.45
$$

and b_{32} = 0.77 − 0.45 = 0.32, and b_{43} = 0.45. There was no feed material at the 50- and 25-mm sizes, and therefore b_{33} = b_{43} = b_{44} = 0. In summary,

$$
\begin{bmatrix} P_1 \\ P_2 \\ P_3 \\ P_4 \end{bmatrix} = (\pi) \begin{bmatrix} 0.17 & 0 & 0 & 0 \\ 0.21 & 0.23 & 0 & 0 \\ 0.27 & 0.32 & 0 & 0 \\ 0.35 & 0.45 & 0 & 0 \end{bmatrix} \begin{bmatrix} 0.8 \\ 0.2 \\ 0 \\ 0 \end{bmatrix} + (1 - \pi) \begin{bmatrix} 0.8 \\ 0.2 \\ 0 \\ 0 \end{bmatrix}
$$

The solution to the simultaneous equations is

$P_1 = 0.47$
$P_2 = 0.20$
$P_3 = 0.15$
$P_4 = 0.18$

Note that the sum of the product fractions equals 1.0, as it should.

REFERENCES

1. Hickman, W. B. 1976. "Storage and Retrieval of Prepared Refuse." *Proceedings* ASME Natl. Waste Process Conf., Boston.

2. Resnick, W. 1976. "Flow Visualization Inside Storage Equipment," *Proceedings* International Conference on Bulk Solids—Storage, Handling and Flow. London: Powder Advisory Centre.

3. *Planning and Specifying a Refuse Shredding System.* Appleton, Wis.: Allis-Chalmers.

4. Taggart, A. F. 1945. *Handbook of Mineral Dressing.* New York: John Wiley & Sons, Inc.

5. Mitchell, J. R. 1971. "Designing for Batch and Continuous Weighers." *Chemical Engineering* (Feb. 28): 177.

6. Tanzer, E. K. 1968. "Pneumatic Conveying for Incineration of Paper Trim." *Proceedings* ASME National Incineration Conference, New York.

7. Madison, R. D. (Ed.). *Fan Engineering.* Buffalo, N.Y.: Buffalo Forge Company.

8. Bates, L. 1976. "Performance Features of Helical Screw Equipment." *Proceedings* International Conference on Bulk Solids—Storage, Handling, and Flow. London: Powder Advisory Centre.

9. Ruf, J. A. 1974. "Particle Size Spectrum and Compressibility of Raw and Shredded Municipal Solid Waste." Ph.D. thesis, University of Florida.

10. American Public Works Association. 1972. *High Pressure Compaction and Baling of Solid Waste.* EPA-OSWMP SW-32d. Washington, D.C.

11. Ham, R. K. 1975. "The Role of Shredded Refuse in Landfilling." *Waste Age* 6, n. 12: 22.

12. Reinhard, J. J., and R. K. Ham. 1974. *Solid Waste Milling and Disposal on Land Without Cover.* EPA (NTIS PB-234 930 and PB-234 931). Washington, D.C.

13. *Baling Solid Waste to Conserve Sanitary Landfill Space.* 1974. EPA-OSWMP. Washington, D.C.

14. Gaudin, A. M. 1926. "An Investigation of Crushing Phenomena." *Transactions* AIME 73: 253.

15. Franconeri, P. 1976. "Selection Factors in Evaluating Large Solid Waste Shredders." *Proceedings* ASME National Waste Processing Conference. Boston.

16. Trezek, G. J. 1974. *Significance of Size Reduction in Solid Waste Management.* EPA-600/2-77-131, Cincinnati.

17. Austin, L. G. 1971–72. "Introduction to the Mathematical Description of Grinding as a Rate Process." *Power Technology* 5, n. 1.

18. Rosin, P., and E. Rammler. 1933. "Laws Covering the Fineness of Powdered Coal." *Journal,* Institute of Fuel 7: 29–36.

19. Stratton, F. E., and H. Alter. 1978. "Application of Bond Theory to Solid Waste Shredding." *Journal of the Environmental Engineering Division*, ASCE, 104, n. EE1.

20. Trezek, G. I., and G. Savage. 1975. "Report on a Comprehensive Refuse Comminution Study." *Waste Age* (July): 49–55.

21. Gaudin, A. M., and T. P. Meloy. 1962. "Model and a Comminution Distribution Equation for Single Fracture." *Transactions*, AIME, 223: 40–43.

22. Evans, I., and C. D. Pomeroy. 1966. *The Strength, Fracture, and Workability of Coal.* Elmsford, N.Y.: Pergamon Press, Inc.

23. Harris, C. C. 1969. "The Application of Size Distribution Equations to Multi-event Process." *Transactions*, AIME, 244: 187–190.

24. Bennet, J. G. 1936. "Broken Coal." *Journal*, Institute of Fuel 10, n. 22.

25. Gawalpanchi, R. R., P. M. Berthouex, and R. K. Ham. 1973. "Particle Size Distribution of Milled Refuse." *Waste Age*, pp. 34–45.

26. Coulson, J. M., and J. F. Richardson. 1955. *Chemical Engineering*. Elmsford, N.Y.: Pergamon Press, Inc.

27. Diaz, L. F. 1975. "Three Key Factors in Refuse Size Reduction." *Resource Recovery and Conservation* 1: 111–113.

28. Bond, F. C. 1952. "The Third Theory of Comminution." *Transactions*, AIME, 193: 484.

29. Perry, J. H. (Ed.). 1963. *Chemical Engineering Handbook*. New York: McGraw-Hill Book Company.

30. Trezek, G. J., D. M. Obeng, and G. Savage. 1972–73. *Size Reduction in Solid Waste Processing*. EPA, Grant No. R801218, Second Year Progress Report.

31. Drobny, N. L., H. E. Hull, and R. F. Testin. 1971. *Recovery and Utilization of Municipal Solid Waste*. EPA-OSWMP SW-Ioc. Washington, D.C.

32. Diaz, L. F., et al. 1976. "Health Considerations Associated with Resource Recovery." *Compost Science* (Summer): 18–24.

33. *Final Report on a Demonstration Project at Madison, Wisconsin, 1966–1972*. 1973. EPA-OSWMP. Washington, D.C.

34. Vesilind, P. A., A. E. Rimer, and W. A. Worrell. 1980. "Performance Characteristics of a Vertical Hammermill Shredder." *Proceedings* 1980 National Waste Processing Conference. Washington, D.C. ASME.

35. Epstein, B. 1948. "Logarithmic-Normal Distribution in Breakage of Solids." *Industrial and Engineering Chemistry* 40: 2289.

36. Broadbent, S. R., and T. C. Calcott. 1956, 1957. "Coal Breakage Processes." *Journal*, Institute of Fuel 29: 524 and 30: 13.

37. Vesilind, P. A., E. I. Pas, and B. Simpson. 1986. "Evaluation of the Pi Breakage Theory for Refuse Components." *Journal of the Environmental Engineering Division*, ASCE 112, n. 6.

38. Vesilind, P. A. 1980. "The Rosin-Rammler Particle Size Distribution." *Resource Recovery and Conservation* 5: 275–278.

39. Gardner, R. P., and L. G. Austin. 1962. "The Use of Radioactive Tracer Technique and a Computer in the Study of the Batch Grinding of Coal." *J. Inst. Fuel* 35: 174.

ABBREVIATIONS USED IN THIS CHAPTER

dBA = decibels on the A scale of the sound level meter
EPA = Environmental Protection Agency
HDPE = high-density polyethylene
MSW = municipal solid waste

PETE = polyethylene terephthalate
RDF = refuse-derived fuel
OSHA = Occupational Safety and Health Act
SG = specific gravity

PROBLEMS

5-1. A pneumatic conveyor is to move wood chips with a maximum diameter of 1 in. Estimate the air velocity required.

5-2. Why are *first in/first out* storage methods necessary in the processing of MSW?

5-3. Two shredders are used to process nearly identical municipal solid waste. The products are sieved, and the data are presented as Rosin-Rammler plots. From these plots it is concluded that the characteristic size (x_0) and n value are as follows:

Shredder	Characteristic size (in.)	n value
"Trash Mauler"	0.34	1.0
"Gobbler"	0.46	1.2

a. Which shredder did a better job of reducing the size of the solid waste? How do you know?

b. Which shredder yielded a more uniform product? How do you know?

c. Using a Rosin-Rammler plot, show the curves for each shredder. Assume a linear curve on the Rosin-Rammler plot.

5-4. Calculate the shredder energy requirements for shredding the feed to the product as shown in Figure 5-25. Use the Bond work index method. Make any assumptions required.

5-5. Shredded solid waste was run through a set of sieves with the following results:

Sieve size (inches)	Fraction of feed retained on sieve
10	0.05
8	0.20
4	0.32
2	0.18
1	0.20

What is the Rosin-Rammler characteristic size (x_0) of this product? Use a graph to obtain your answer.

5-6. Aluminum beer cans are to be processed in a roller mill to a maximum thickness of 0.5 cm. Calculate the size of rollers required.

5-7. Estimate the theoretical maximum capacity of a single lead screw conveyor. The dimensions are measured as pitch = 20 in., radius to conveyor tip = 18 in., and radius to conveyor hub = 6 in. The speed is measured as 30 rpm. How much will the final answer be sensitive to a 10% error in measurement for each of the variables?

5-8. A product from a shredder follows the Rosin-Rammler particle size distribution, with $n = 1.0$, and a characteristic size (x_0) of 2 cm. What is the 90% passing size?

5-9. Show, using a mathematical derivation, that the characteristic size (x_0) in the Rosin-Rammler equation is defined as the point where 63% is finer than that particle size.

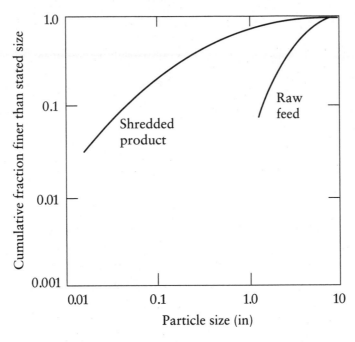

Figure 5-25   See Problem 5-4.

5-10. A shredded refuse is classified by particle size as follows:

Sieve size (cm)	Percent by weight passing
10	85
5	65
2	48
1	40

What is the characteristic size (x_0) as defined by a Rosin-Rammler plot?

5-11. A shredder has a feed and product as shown:

	Percent by weight finer than	
Size (in.)	Feed	Product
4	80	95
2	15	65
1	5	25
0.5	0	10

a. What is the characteristic size (x_0) of the feed? the product?
b. Do both distributions fit the Rosin-Rammler particle-size distribution function?
c. Draw and label typical product particle-size distribution curves if

i. The feed becomes wetter (higher moisture content)

ii. The shredder is run at a higher speed.

d. What is the effective power requirement (kWh/ton) if the Bond work index is 400?

5-12. You are reviewing an engineering report recommending the construction of a materials recovery facility. A shredder for processing raw MSW (right off the truck) is recommended as a part of this facility. You turn to the cost estimate and discover that the only two costs associated with the shredder are annualized capital cost and the power cost. What other costs would you recommend be included in this calculation?

5-13. A horizontal hammermill shredder is to process 8 tons/hour of MSW to a characteristic size (x_0) of 0.7 inch. What grate spacing and horsepower would be required to accomplish this?

5-14. A horizontal hammermill shredder is to process 8 tons/hour of MSW to a characteristic size (x_0) of 0.7 inch. Estimate the required horsepower using the Bond work index method. Assume that the feed characteristic size is 4 inches, and the particle-size distribution follows the Rosin-Rammler model, with $n = 1$. (*Note*: You will need log–log graph paper).

5-15. The city of Durham, North Carolina, has a population of 100,000, and the daily per capita production of MSW is 4.0 pounds. A horizontal hammermill shredder is to be installed to process the raw waste to a product that is 90% finer than 6 cm. The shredder is to operate 8 hours per day. Assume the n in the Rosin-Rammler equation is 1.0, the Bond work index is 400 kWh/ton, and the raw refuse is 80% passing 20 cm. Estimate the size of the electric motor needed to power this shredder.

5-16. A shredder is to reduce the particle size of 80% passing 10 cm to 80% passing 2 cm, at a solids flow rate of 10 tonnes (10,000 kg) per hour. Size the motor for this shredder.

5-17. A hammermill shredder is to process 8 tons/hour of mixed municipal solid waste to a characteristic size of 0.7 inch. What grate spacing and horsepower would be required to accomplish this? Use the Bond work index method and assume that the feed characteristic size is 4 inches and that the feed particle-size distribution follows the Rosin-Rammler model, with $n = 1.0$.

5-18. Describe the approaches used to control the destructive force of explosions in hammermills.

5-19. A horizontal shredder processing municipal solid waste was found to produce shredded product as shown below:

Particle size (in.)	Percent finer than (%)
10	95
5	85
2	50
1	34
0.5	20

What is the characteristic size for shredded product?

5-20. What is the *Bond Law* used for? Why is it useful

Materials Separation

This chapter is devoted to various means of separating selected components from mixed municipal refuse and/or previously separated recyclables. Materials recovery facilities (MRFs, pronounced "murphs")* that process mixed waste are called *dirty MRFs*, while those that process partially separated material (the recyclables) are called *clean MRFs*.

All the separation devices included in this chapter are based on a principle of *coding* and *switching*. Some property of the material is used as a recognition code, such as magnetic/nonmagnetic or large/small, and switches—such as magnets or screens—are then used to achieve separation. All materials separation devices (including human beings) operate on the same principle: First, there must be a recognizable code to differentiate the materials in question, and then this code must be used in a switching device that physically separates the materials.

GENERAL EXPRESSIONS FOR MATERIALS SEPARATION

In separating various pure materials from a mixture, the separation can be either *binary* (two output streams) or *polynary* (more than two output streams). For example, a magnet capturing ferrous material is a binary device, whereas a screen with a series of different sized holes, producing several products, is a polynary separation device.

Binary Separators

A schematic of a binary separator is shown in Figure 6-1. The input stream is composed of a mixture of x and y, and these are to be separated. The mass per time (e.g., tons/hour) of x and y fed to the separator is x_0 and y_0, respectively. The mass per time of x and y exiting in the first output stream is x_1 and y_1, and the second output stream is x_2 and y_2.

Assume that the device was intended to separate the x into the first output stream and y into the second. If the separator is totally effective, then all of x goes to the first output, and all of y to the second. In practice this is seldom achievable, and the first stream is contaminated with some y, and the second with some x. The

*and small MRFs are then SMRFs?

Input   Output

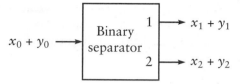

Figure 6-1 *Binary separator.*

effectiveness of the separation can then be expressed in terms of recovery. The recovery of component x in the first output stream is R_{x_1}, defined as

$$R_{x_1} = \left(\frac{x_1}{x_0}\right) 100$$

where recovery is expressed as a percentage, and x_1 and x_0 are in terms of mass/time. Similarly, the recovery of y in the second output stream is expressed as

$$R_{y_2} = \left(\frac{y_2}{y_0}\right) 100$$

Since the mass balance holds:

$$x_0 = x_1 + x_2$$

then

$$R_{x_1} = \left(\frac{x_0 - x_2}{x_1 + x_2}\right) 100$$

The effectiveness of a separator cannot be judged only on the basis of recovery. Consider for a moment what would happen if the binary separator were run so as to achieve $x_2 = y_2 = 0$. In other words, all of the feed is exited as output number one. In that case the recovery of x is 100%, but the device is not performing its desired function, since no separation occurs. A second operational parameter is therefore required, and this is usually an expression of purity, stated as

$$P_{x_1} = \left(\frac{x_1}{x_1 + y_1}\right) 100$$

where P_{x_1} is the purity of the first output stream in terms of x, expressed as percent. Similarly, the purity of the second output stream in terms of y is

$$P_{y_2} = \left(\frac{y_2}{x_2 + y_2}\right) 100$$

Usually, both purity and recovery are needed for a complete and accurate description of binary separation performance.

At times the input and output streams are more conveniently expressed in terms of concentrations instead of mass/time. The equations for calculating the effectiveness of the separation can in this case be shown to be the following:

$$R_{x_1} = \frac{[x_1]([x_0] - [x_2])100}{[x_0]([x_1]-[x_2])}$$

where x_1 is the concentration of x, in output stream 1, x_0 is the concentration of x in the input stream, and x_2 is the concentration of x in output stream 2, with all concentrations expressed as percentages. A similar expression can be written for component y (with 1 and 2 subscripts reversed, of course).

The purity of x would be

$$P_{x_1} = \frac{[x_1]\rho_x}{[x_1]\rho_x +[y_1]\rho_y} 100$$

where ρ_x and ρ_y are the densities of x and y, respectively.

Often a binary separator is designed to extract one type of material from a waste stream. For example, a magnet draws off ferrous materials as the desired output. Such an output is often called the *product* or *extract*, and the second output is the *reject*. Literally, the magnet extracts the ferrous materials and rejects the rest. In subsequent discussions of materials separation in this text, one output is referred to as the extract and the other as the reject.

Polynary Separators

Two types of polynary systems are possible, as shown in Figure 6-2. In the first case x_0 and y_0 are the two components in the feed, and the separator has more than two

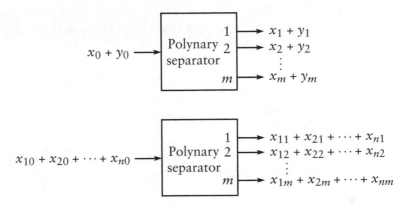

Figure 6-2 Polynary separators.

output streams, with x and y appearing in all of them, but in different amounts. In such a system the recovery of x in the first output stream is

$$R_{x_1} = \left(\frac{x_1}{x_0}\right) 100$$

as before, where x is in mass/time units. Similarly, the purity of x in the first output stream is

$$P_{x_1} = \left(\frac{x_1}{x_1 + y_1}\right) 100$$

The recovery of x in the mth output stream is

$$R_{x_m} = \left(\frac{x_m}{x_0}\right) 100$$

The second type of polynary separator is the most general case where the feed contains n components $(x_{10}, x_{20}, x_{30} \ldots x_{n0})$ and these are to be separated into m outputs. The notation is shown in Figure 6-2. x_{11} is the x_1 that ended up in the first output, x_{21} is the x_2 that ended up in the first output, and so on. The recovery of x in the first output is thus

$$R_{x_{11}} = \left(\frac{x_{11}}{x_{10}}\right) 100$$

where x_{11} is the x_1 that ended up in the first output and x_{10} is the x_1 in the feed. The purity of this stream in terms of x_1 is

$$P_{x_1} = \left(\frac{x_{11}}{x_{11} + x_{21} + \ldots + x_{n1}}\right) 100$$

with all x terms having units of mass/time.

Effectiveness of Separation

Because of the inconvenience of using two measures of separation effectiveness (recovery and purity) to define the operation of a materials separator, a single-value parameter would be useful. One such parameter sometimes used in literature, but not recommended, is overall recovery, defined as

$$OR_{x,y} = \left(\frac{x_1 + y_1}{x_0 + y_0}\right) 100$$

This parameter is useful only for process design, such as sizing conveyor belts. Since this term is not a measure of separation effectiveness (i.e., 100% overall recovery can be achieved by simply bypassing or turning off the separator), it should not be used in describing the operation of materials separation.

Rietema[1] reviewed these efforts and suggested a measure of separation effectiveness. Stated for a binary separation with input of x_0 and y_0, Rietema defined effectiveness as

$$E_{x,y} = 100 \left| \frac{x_1}{x_0} - \frac{y_1}{y_0} \right| = 100 \left| \frac{x_2}{x_0} - \frac{y_2}{y_0} \right|$$

Another means of obtaining a single value of binary separator performance, developed by Worrell[2] and modified by Stessel,[3] is to multiply the fraction of x in the first output stream by the fraction of y in the second output stream and take the square root of the product, or

$$E_{x,y} = \left(\frac{x_1}{x_0} \frac{y_2}{y_0} \right)^{1/2} 100$$

Both Rietema's and the Worrell-Stessel definitions of effectiveness are rational in that if perfect separation occurs (all of x_0 goes to the first output, so that $x_1 = x_0$, and $y_2 = y_0$), the effectiveness is 100%. That is, the device is 100% effective in performing its intended operation. Likewise, if no separation occurs ($x_0 = x_1$ and $y_0 = y_1$), then both measures of effectiveness are zero.

EXAMPLE
6 - 1

A binary separator has a feed rate of 1 tonne/h. It is operated so that during any 1 hour, 600 kg reports as output 1 and 400 kg as output 2. Of the 600 kg the x constituent is 550 kg, while 70 kg of x ends up in output 2. Calculate the recoveries and the effectiveness of the separation using the methods discussed above.

The recovery of x in the first output is

$$R_{x_1} = \left(\frac{x_1}{x_0} \right) 100 = \frac{(550)100}{550 + 70} = 88\%$$

The purity of this output stream is

$$P_{x_1} = \left(\frac{x_1}{x_1 + y_1} \right) 100 = \frac{(550)100}{600} = 92\%$$

Using Rietema's definition of effectiveness,

$$E_{x,y} = 100 \left| \frac{x_1}{x_0} - \frac{y_1}{y_0} \right| = 100 \left| \frac{500}{620} - \frac{50}{380} \right| = 67\%$$

and according to the Worrell-Stessel effectiveness equation,

$$E_{x,y} = 100 \sqrt{\frac{550}{620} \frac{330}{380}} = 88\%$$

PICKING (HAND SORTING)

The most primitive method for the separation of materials from waste, and historically the first, is *hand sorting* or *picking*. Ever since civilization began, scavengers have been an integral part of society. Selectively accepting other people's waste, collecting and processing it, and selling it at a profit is a time-honored profession and, in recent times, quite a profitable one. The first hand-sorting facility in the United States was built by Colonel Waring for New York City in 1898.[4] The refuse from 116,000 people was sorted, and over 2 1/2 years of operation about 37% of the refuse was recovered, a major part of which was rags. The recovered material yielded an income of about $1 per ton. The income from this and other plants was not sufficient to maintain them, however, and the job of scavengering was given back to private entrepreneurs, who paid the city about $1 per ton for the privilege.

In modern society scavenging at landfills and trash cans is discouraged because of health considerations and the potential for accidents. The trade lives, however, in the person of the *pickers* at materials recovery facilities.

Pickers (or more properly *hand sorters*) have two major functions. First, they recover any items of value that need not be processed. Commonly, corrugated cardboard, bundles of newspaper, and large pieces of metal (reinforcing bars, etc.) are recovered by the pickers. This is known as *positive sorting*. Their second function is to remove all those items that could cause damage to the rest of the processing system, such as explosives, as discussed in the previous chapter. This type of sorting is called *negative sorting*.

Most of the time the functions of hand pickers (salvage and protection) are combined. For example, one large processing facility recovered a piece of titanium 60 cm in diameter and 10 cm thick off the conveyor belt leading to the shredder. Not only is this a valuable piece of metal, but it would have completely destroyed the shredder if it had been allowed to go in.

The *coding* and *switching* functions in hand sorting are simple to define. The material is recognized visually (coding) by such properties as color, reflectivity, and opacity; verified by sensing its density; and removed (separated) by hand picking. Hand sorting is usually done on the conveyor belt after the bags have been mechanically opened in a trommel or a bag-opening flail mill. At a clean MRF the material may arrive in the loose form or in paper bags, and no opening is needed.

At some facilities no such preprocessing is used, and the sorting operation is hence highly inefficient. Typically, the conveyor belt is loaded, and the material is leveled out by a skimmer. The pickers stand on either side of the conveyor belt and remove the selected materials. Experience has shown that pickers can salvage up to about 1000 lb/h/person. However, the quantity sorted is highly dependent on the density of the material. For example, the picker removing cardboard removes far more material by weight than the picker removing film plastic.

A picking belt should be no more than 24 in. (60 cm) wide for one-sided picking, or 36 to 48 in. (90 to 120 cm) wide for pickers on both sides, and should not move faster than 30 to 40 ft/min (about 9 m/min), depending on the number of pickers.[5] If at all possible, the picking operation should be done in daylight.

Artificial light, especially fluorescent bulbs, give off a narrow band of light, and this makes identification (coding) of the various components difficult. Large skylights should be installed if outside operation is impractical.[6] Picking material from MSW is a dirty and dangerous profession, not recommended for the squeamish. A typical picking belt is pictured in Figure 6-3.

SCREENS

Screening is a process of separation by size. A series of uniform-sized apertures allows smaller particles to pass while rejecting the larger fraction. A particle can pass a screen if it is smaller than the opening in at least two dimensions. Material that passes through the holes is called the *extract*, and material that does not pass through the holes is called *reject*.

Screening in material recovery operations has been used commonly toward the end of a series of unit operations and is intended primarily for glass removal, since glass would by then have been crushed to fine particles. Screens have also been used for reclaiming a high-organic (garbage) fraction from shredded waste[7] and as rough sorters at the beginning of materials recovery facilities. The breakage and removal of much of the glass in primary screening has been found to be highly beneficial in reducing wear on downstream shredders.

Figure 6-3   Pickers in a clean MRF.

Screens, like other separation devices, cannot be expected to attain 100% recovery. In other words, some undersized material (smaller than the screen apertures) will report as reject and not be removed as extract. (In screening, remember that the *reject* does not pass through the holes, while the *extract* does pass through the holes.) The equation expressing the recovery of undersize materials from a screen is based on the recovery equation for a binary material separation operation,

$$R_{x_1} = \left(\frac{x_1}{x_0}\right) 100$$

where R_{x_1} = screen recovery, %

$\quad\quad x_1$ = amount of material recovery as extract, that is, the material falling through the holes, mass/time

$\quad\quad x_0$ = amount of undersized material that *could have* fallen through the holes, mass/time

All of the oversize material entering as feed report as reject, not falling through the apertures, which is only logical, since oversize is defined by screening. However, much of the material in refuse is flexible, and a large plastic bag may fall through a hole size of 4 inches. By our definition, the plastic bag is then part of the extract, or the undersize fraction. In binary separation, if we define the extract (exit stream 1 in Figure 6-1) as x_1, then $y_1 = 0$, and the above definition of recovery is adequate to describe screen operation. In other words, the purity of the extract is always 100%. Theoretically, it is possible to achieve very high recoveries with screening, but only at the cost of limiting the throughput. To maintain adequate throughput, most screens are operated between 85% and 95% recovery.[8]

A misleading and yet often used expression of screen effectiveness is

$$R_{x_1} = \frac{x_1}{x_0 + y_0}$$

Recovery expressed in this way is more correctly termed a *split*, or that fraction of the feed that exited from output stream number 1. This expression is not a fair indication of how well the screen performed in separating the small particles that it was theoretically capable of separating.

Trommel Screens

By far the most popular screen for processing municipal refuse is the revolving screen or the trommel, which is an inclined cylinder, mounted on rollers, with holes in the side, as shown in Figure 6-4. The drums roll at slow speeds, 10 to 15 rpm, thus using very little power. The main advantage of the trommel screen is its resistance to clogging. Some of the material within the screen might tend to hang on, but will eventually drop off. Trommel screens can also be equipped with spikes to break open plastic bags. This has been used in mixed waste material recovery facilities at the beginning of the process line.

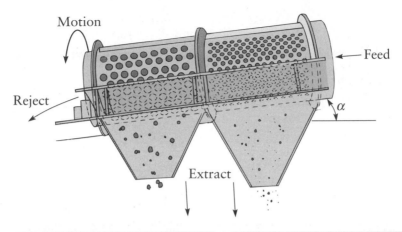

Figure 6-4 Trommel screen.

The trommel screen works by allowing the refuse in the screen to tumble around until the smaller pieces find themselves next to the apertures and fall through. The tumbling motion may be of two kinds, as shown in Figure 6-5:

1. *Cascading*: The charge is lifted up by the circular motion of the screen and then tumbles down on top of the layer heading upward.
2. *Cataracting*: The speed of the screen is sufficiently great to actually fling the material into the air, where it will drop along a parabolic trajectory back to the bottom of the screen.

Cataracting produces the greatest turbulence, and the trommel should achieve the greatest efficiency. As the drum speed is increased further, a third type of motion is

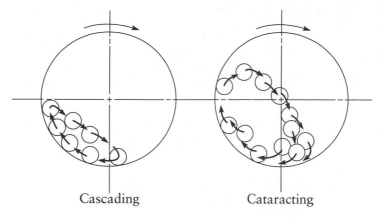

Cascading   Cataracting

Figure 6-5 Two types of particle paths in a trommel screen.

eventually attained—*centrifuging*. In this case the material adheres to the drum and never drops off, resulting in low recovery.

With reference to Figure 6-6, consider a particle p in contact with the inside of the screen. The centrifugal force acting to press it against the inside wall is c and the w_1, a component of the gravitational force w, acts to pull it away. The angle between the vertical and the line Op is α_1, and $w_1 = w \cos \alpha_1$. If $c > w_1$, the particle remains in contact with the screen. However, if $w_1 > c$, the particle will fall off. If c remains greater than w_1 as α_1 decreases to zero (particle at the top), the particle never does drop off but remains on the wall through the rotation. At the point of separations $c = w \cos \alpha$, where α is the angle at which separation occurs.

The centrifugal force is

$$c = \frac{w}{g}\left(r\omega^2\right)$$

where ω = rotational velocity, rad/sec
      r = radius, cm
      g = acceleration due to gravity, cm/sec^2

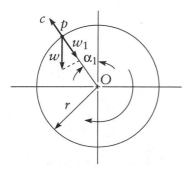

Figure 6-6 Definition of terms for trommel screen analysis.

Combining the two equations,

$$w \cos \alpha = \frac{w}{g}(r\omega^2)$$

$$\cos \alpha = \frac{r\omega^2}{g}$$

and since $\omega = 2\pi n$, where n is the speed of the screen in revolutions per second,

$$\cos \alpha = \frac{4\pi^2 r n^2}{g}$$

From this relationship it is clear that the angle α at which the particle leaves the wall of the screen and begins its free flight varies with both r and n. The critical point is at $\alpha = 0$, or $\cos \alpha = 1$, so that the *critical speed* is

$$n_c = \sqrt{\frac{g}{4\pi^2 r}}$$

If $w_1 > c$ at α, the particle will lose contact with the wall and begin its flight in a parabolic path until it once again hits the screen (Figure 6-7). The equation of the parabola at the origin p_1, is

$$y = x \tan \alpha - \frac{gx^2}{2 V_1^2 \cos^2 \alpha}$$

where V_1 = initial velocity of the particle p_1 as it leaves the wall
      x, y = coordinates

The equation of the circular path of the screen is

$$x^2 + y^2 = (2r \sin \alpha)\, x + (2r \cos \alpha)\, y = 0$$

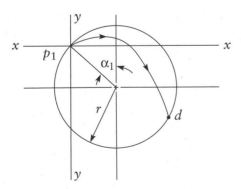

Figure 6-7   Flight path of a particle leaving the inside wall of a trommel screen.

The simultaneous solution of these equations gives the coordinates of the point d, where these two curves intersect. This point is at the coordinates

$$x = 4r \sin \alpha \cos^2 \alpha$$

$$y = -4r \sin^2 \alpha \cos \alpha$$

The development above is for a single particle within a screen. This is, of course, an unrealistic assumption, and the motion of particles has to be analyzed taking into account the action of other particles.

The critical speed and the fraction of the screen occupied by the refuse are related as shown in Figure 6-8. The plot is in terms of bulk volume, defined as

$$F = \frac{S}{V}$$

where F = bulk volume fraction
S = volume occupied by the solids and the air spaces between the solids
V = total volume inside the trommel screen

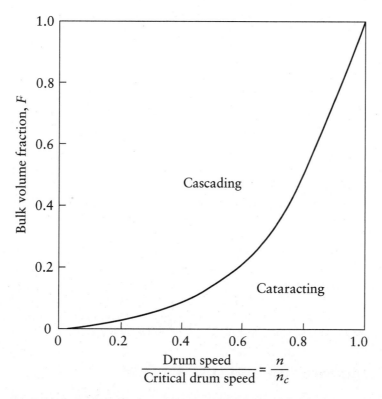

Figure 6-8 Effect of volume occupied by solids on trommel flow characteristics. Source: (9)

If the screen is totally full ($F = 1.0$), only cascading is possible for speeds less than critical (there is no space for the particles to fall back through the air). At lower volume fractions, cataracting becomes possible at speeds lower than critical. At the limit, with only one particle in the screen, cataracting occurs at even low speeds since there is no particle-particle interference.

EXAMPLE
6-2

Assume a 2.7-m-diameter trommel. Calculate the critical speed.

$$n_c = \sqrt{\frac{g}{4\pi^2 r}} = \left[\frac{980}{4(3.14)^2(270/2)}\right]^{1/2} = 0.43 \text{ rotation/sec}$$

or 26 rpm.

The recovery obviously varies with the speed. A 9-ft-diameter (2.7-m) trommel screen operates at its highest effectiveness at about 45% of the critical speed. The rule of thumb is that recovery is greatest at a speed where the load rides one-third of the distance to the top of the screen.

Another operating variable, not considered above, is the slope of the drum. Within limits, as the slope is increased, the solids retention time is decreased, and the percent of product recovered is decreased because the particles have less chance of finding a hole through which to drop. Where trommel screens have different-sized holes along the drum, the slope affects the amount of material within each size range captured. Figure 6-9 shows some laboratory results with shredded MSW. The trommel for these tests had a section of 1-in. holes.[10] The recovery was measured as the percent of particles less than 1 in. captured. These data for underloaded conditions show that recovery drops off rapidly with angle of incline. A small slope, however, would have a low throughput, and a proper balance between recovery and throughput must be achieved.

For unshredded MSW optimum trommel performance can be obtained if the solids retention time is between 30 seconds and 1 minute,[11] and the material makes 5 to 6 revolutions within the drum.[12] A pilot plant obtained 95 to 100% recovery of 3-in. shredded refuse at a rate of 2 tons/h. At 2.5 tons/h, the recovery dropped to 91%. The 4-ft-diameter, 6-ft-long trommel rotated at 18 rpm.

Figure 6-10 shows how the power requirements for a trommel screen vary with loading. At very low load the extra power demand is zero, and increases as the trommel lifts more and more of the charge. When the trommel fills up and no separation occurs, the extra power demand is again zero.

Reciprocating and Disc Screens

The second type of screen used is a disc screen, which consists of rotating discs that move solid waste across the screen. These screens are very rugged and can process large quantities of solid waste (Figure 6-11).

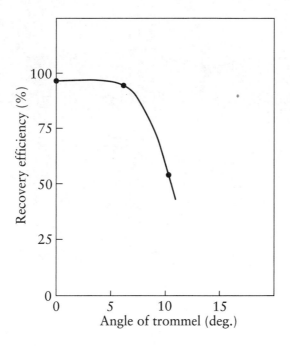

Figure 6-9 Trommel screening of MSW. Angle of the trommel affects efficiency of solids recovery. Source: (10)

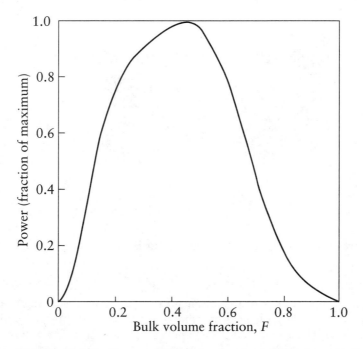

Figure 6-10 Power requirements of a trommel screen as a function of load. Source: (9)

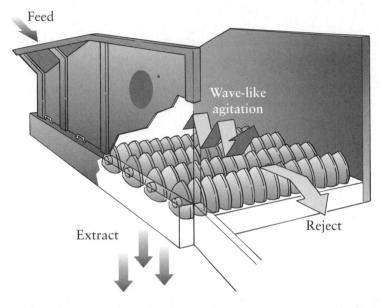

Figure 6-11 *Disc screen. (Courtesy of Bulk Handling Systems Inc.)*

A third type of screen is an inclined or horizontal shaking screen. This screen, however, is readily plugged by rags, paper, and other objects, and is limited in its application to cleaner feeds. One application of such a screen has been the removal of the small pieces of glass so as to produce uniformly sized pieces that might be color sorted. Figure 6-12 is an estimate of the capacity of a vibrating screen when separating glass from shredded refuse.

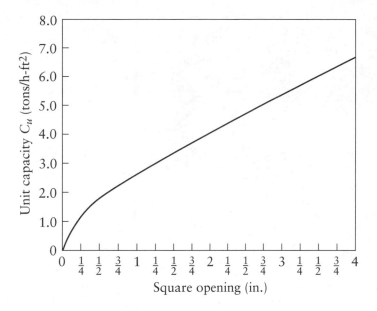

Figure 6-12 *Unit capacity of a vibrating screen in separating glass from MSW. Source: (8)*

Disc and reciprocating screens in materials recovery facilities can be used for removing glass after crushing because the glass would be in small pieces, while the size (in two dimensions) of other materials—such as cans, plastics, and paper—would not be reduced. Screens have also been used for reducing the amount of inert (noncombustible) material in facilities that produce refuse-derived fuel (RDF) since again the shredded paper would be much too big to fit through the holes.

FLOAT/SINK SEPARATORS

The *code* for all float/sink separators is the settling (or rising) velocity of a solid particle within a fluid. One of the earliest float/sink separators was the process of winnowing in which the farmer threw the mixture of grain seed and chaff into the wind, and the wind was able to carry off the chaff without suspending the grain (Figure 6-13). More accurately, the settling velocity of the grain was much higher than the settling velocity of the chaff, and the grain thus fell quickly to the earth.

In solid waste separation a number of operations use the float/sink principle, including air classification (not unlike winnowing), heavy-media separation, jigging, flotation, and others. In all cases the theory of operation is the same.

Figure 6-13 Winnowing, from an old wood cut. Source: (12)

Theory of Operation

The motion of a solid particle suspended in a fluid, such as an air or water, is governed by three forces:

F_E = some external force such as gravity or centrifugal force
F_B = buoyant force
F_D = drag force

The motion of the particle is described by Newton's Law, acceleration = force/mass.

$$\frac{dv}{dt} = \left(F_E - F_D\right)\frac{1}{\rho_s V}$$

where V = particle volume, m^3
ρ_s = particle density, kg/m^3
dv/dt = particle acceleration, m/sec^2

A particle falling in a fluid under gravity has two phases in its motion: acceleration and terminal velocity. The latter is attained when the three forces—F_E, F_D, and F_B—are balanced and the acceleration $dv/dt = 0$. Thus during a steady fall (terminal velocity),

$$F_B = F_D + F_E$$

These three forces can be expressed as

$$F_E = \rho_s V a$$
$$F_B = \rho V a$$
$$F_D = \frac{C_D v^2 \rho A}{2}$$

where ρ_s = density of the solid particle, kg/m^3
$\quad\quad V$ = volume of the particle, m^3
$\quad\quad a$ = acceleration due to some external force, m/sec^2
$\quad\quad \rho$ = density of the fluid, kg/m^3
$\quad\quad C_D$ = drag coefficient
$\quad\quad v$ = differential velocity between the particle and the fluid, m/sec
$\quad\quad A$ = projected area of the particle, m^2

Assuming for the sake of convenience that the particle is a perfect sphere,

$$A = \frac{\pi d^2}{4} \text{ and } V = \frac{\pi d^3}{6}$$

where d is the particle diameter. Further assuming that the external force causing the acceleration is gravitational, $a = g$, the equation reduces to

$$v = \left[\frac{4(\rho_s - \rho)gd}{3C_D\rho} \right]^{1/2}$$

which is the well-known *Newton's law*. Assuming laminar flow conditions where the drag coefficient, $C_D = 24/N_R$, where N_R = the Reynolds number,

$$N_R = \frac{v\rho d}{\mu}$$

where v = velocity, m/sec
$\quad\quad \rho$ = density, kg/m^3
$\quad\quad d$ = diameter, m
$\quad\quad \mu$ = viscosity, kg/sec – m

Substituting yields the familiar *Stokes law*:

$$v = \frac{d^2 g(\rho_s - \rho)}{18\mu}$$

Although this expression is not applicable for air classifiers—because the Reynolds number in air classification is about 10,000, placing the flow well into the turbulent flow regime—it yields some clues as to efficiency of float/sink classification. The objective is to have as large a difference in the settling velocities (v) as possible, and it is clear that the diameter (d) plays an important role since v is a function of d^2. Often, with such uncontrolled, irregular, and unpredictable material as shredded refuse, the diameter is difficult to define. One solution to this problem is to define an *aerodynamic diameter* (or *hydrodynamic diameter* if the fluid is water), which can be back-calculated using known velocities. Using this technique, the aerodynamic diameters for shredded MSW light fraction have been found to be about 40% of actual diameters as defined by screening.[13]

Alternatively, a modified drag coefficient might be used. Although C_D is approximately 1.0 for disk-shaped particles under ideal conditions, experimental evidence suggests that $C_D \approx 2.5$ for refuse particles falling in air. Example 6-3 illustrates this concept.

EXAMPLE
6-3

Assuming that the drag coefficient is 2.5, calculate the air velocity necessary to suspend 2-cm (screened) particles of shredded aluminum. Note that $\rho_s = 2.70$ g/cm^3 and $\rho = 0.0012$ g/cm^3.

$$v = \left[\frac{4(2)(1.2 - 0.0012)980}{3(2.5)(0.0012)}\right]^{1/2} = 1004 \text{ cm/sec}$$

In some cases instead of adjusting the drag coefficient or back-calculating an aerodynamic diameter, it might be more convenient to define an *effective diameter*, which might be the average dimension of the particle as it is presented to the fluid stream. Since

$$A = \frac{\pi d^2}{4}$$

the effective diameter can be defined as

$$d_a = \left(\frac{4A}{\pi}\right)^{1/2}$$

This may be reasonable where flat objects, such as pieces of paper, are suspended. In other cases the effective diameter could equally well be defined by the volume as

$$d_v = \left(\frac{6V}{\pi}\right)^{1/3}$$

A further discussion of particle-size analysis is included in the appendix to Chapter 2.

Another problem with applying Newton's law to float/sink classification is that the analysis assumes a single particle settling in an infinite fluid (no boundary conditions). This is obviously not practically possible, and some accommodation must be made for the problems of interparticle actions and the effect of the walls.

In the turbulent regime the effect of the wall can be accounted for by a correction factor[16] given as

$$m = 1 - \left(\frac{r}{R}\right)^{3/2}$$

where r = radius of the sphere
R = radius of the tube
m = correction factor

This correction factor can be used to adjust the terminal velocity to take into account the effect of walls, so that

$$v = m\left[\frac{4d(\rho_s - \rho)g}{3C_D\rho}\right]^{1/2}$$

When the particles are sufficiently concentrated, they act as a body with little interparticle movement. This can occur in a float/sink separator when a large slug of feed enters the throat section and is carried upward in mass with the fluid stream. The velocity of the suspension, v_c, can be expressed relative to the velocity of a single particle, v, as defined by the Newton equation.[14] Under laminar flow conditions (Reynolds number less than about 500):

$$\frac{v_c}{v} = f_1\left(\epsilon, \frac{d}{D}\right)$$

where D is the diameter of settling or flotation column and ϵ = void fraction. For turbulent conditions,

$$\frac{v_c}{v} = f_2\left(\frac{vd\rho}{\mu}, \epsilon, \frac{d}{D}\right)$$

Experimental evidence[14] suggests that for laminar flow conditions,

$$v_c = v\epsilon^{[4.65 + 19.5\, d/D]}$$

For turbulent conditions d/D is not significant,[50] and

$$v_c = v\epsilon^{4.65}$$

Finally, the discussion above assumes that the particles are all made of rigid materials and that the fluid stream does not change their shape. Obviously, many of the materials in shredded refuse, such as paper and plastics, are flexible and porous, and the shape they present to the fluid in the separation is unpredictable and dynamic.

To summarize, the code in float/sink separators is the aerodynamic (or hydrodynamic) velocity of the particle in the fluid. But this is not the material property that should be used as the code for separation. The particles should be separated by their *density* only, and not on the basis of diameter, shape, rigidity, and surface roughness. What is needed is a method of using the float/sink process such that the only basis for separation is density and not the irrelevant material characteristics.

Such an objective can be achieved by first considering the settling column shown in Figure 6-14. In such a simple classifier, if Stokes's law holds, the terminal settling velocity of particles with densities greater than the fluid would be (based on the Stokes law):

$$v \propto (\rho_s - \rho)d$$

where v = terminal settling velocity
    d = particle diameter
    ρ_s = particle density
    ρ = fluid density

With reference to Figure 6-14, because the terminal velocity is governed by both size and density differences, the separation achieved is into three phases: the small lights, the large heavies, and a mixture of large lights and small heavies. Starting from rest, these particles accelerate until they attain terminal velocity. This velocity is reached by having the drag force increase from zero (at rest) until it becomes

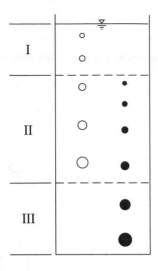

Figure 6-14 Vertical stratification of settling particles, based on difference in settling velocities.

equal to the net gravitational force. At any instant during the acceleration phase the motion of the particle can be described as

$$F_E - F_B - F_D = \frac{V\rho_s}{g}\frac{dv}{dt}$$

where F_E = force due to gravity
$\quad F_B$ = buoyant force
$\quad F_D$ = drag force
$\quad V$ = volume of particle
$\quad \rho_s$ = density of particle
$\quad g$ = acceleration due to gravity
$\quad v$ = particle velocity
$\quad t$ = time

At time zero, with $F_D = 0$, and recalling that $F_E = V\rho_s$ and $F_B = V\rho$, where ρ = fluid density, the initial acceleration is

$$a_0 = \frac{dv}{dt} = \left(1 - \frac{\rho}{\rho_s}\right)g$$

Thus particles of the same density (or material), regardless of their size, have the same initial acceleration. Of course, the moment resistance enters the picture, this no longer holds, because particle size is a factor in drag forces. For any two particles of different density, the denser particle has a greater initial acceleration than does the less dense particle. Although this difference is small, and short-lived, it is the basic principle on which the separation on the basis of density is achieved.

Consider now two particles of the same density but of different size: A = large and B = small. In a still fluid, with both particles starting from rest, the large particle will accelerate faster, but since its terminal velocity is greater, it will still be accelerating when the small particle has attained its terminal velocity. This is shown schematically in Figure 6-15. The distance traveled by either of the particles in any given time t is the area under the curve, or

$$\int_0^t v\,dt$$

A third particle, C, which is of a greater density but smaller size, has a terminal velocity equal to the large particle A. It would be difficult to separate these two particles in a device where both density and size are important. It is possible, however, to take advantage of the different accelerations and achieve separation by imposing a series of short accelerations starting from rest. The distance traveled during any one short spurt is small, but the difference of travel, when summed over many accelerations, is sufficient to have the more dense particle A travel further than the less dense (but larger) particle B, thus achieving stratification and eventual separation.

Figure 6-15 illustrates that the denser but smaller particle B has—until time t_1—a higher settling acceleration than does the larger but lighter particle C. After time t the lighter particle has a higher acceleration, and catches up to the smaller particle at time t_2. (The difference between the areas under the curves to time t_1 equals the area from t_1 to t_2.) If the heavy but small particle B is to be separated from the lighter but larger particle C, the acceleration time has to be well below t_2. This time is a design variable that determines the size of particle to be separated as part of the more dense fraction.

For any float/sink process, if the separation is to be on the basis of density only and not on the basis of such other particle characteristics as size, the objective is to *never allow the particles to attain their terminal velocity*. The particles should be

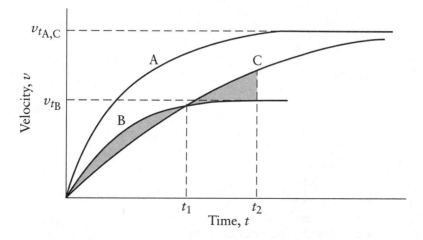

Figure 6-15 *Acceleration of particles with different densities and sizes. Particle A is high density and small, while C is low density and large, and they have equal settling velocities. Particle B is also high density, but smaller than A.*

continually accelerated and decelerated, never allowing them to reach terminal velocity. Some of the types of float/sink separators discussed below take full advantage of this opportunity.

Jigs

A *jig* is a device that achieves the separation of less dense from more dense particles by using the differences in their abilities to penetrate a shaken bed. One very simple type of jig is the miner's pan used in gold mining days. This pan is filled with dirt and gravel from a stream bottom and shaken until the grains of gold penetrate the pan content and lodge on the bottom, appearing as the remainder of the pan contents is poured off. The separation of the sand and pebbles from the gold grains and nuggets is not by size but by density. The modern jig works in essentially the same way.

Figure 6-16 shows a diagram of jig. The feed may be thought of as comprising a less dense fraction, a more dense fraction, and a middling fraction (an intermediate mixture, which contains contaminants as well as some light and heavy fraction, depending on the efficiency of the operation). The mixture in the jig is subjected to pulsating forces produced by a plunger (or diaphragm, air, or other mechanism) in the water medium so that the entire bed is lifted up and settles back. The particles as they are settling never have a chance to attain their terminal settling velocities, and the separation in a jig is on the basis of density only.

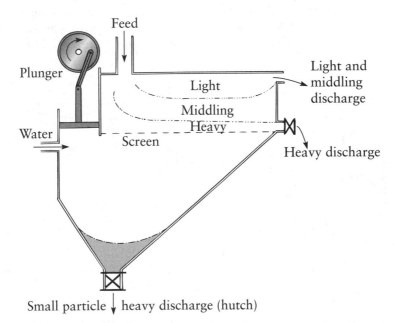

Figure 6-16   A common plunger jig.

As the bed expands, the heavy particles of sufficiently large size and proper shape crash through the bed since the bed is in the *quick* condition and offers little resistance to such settlement. As the pulsations continue, the larger, more dense pieces end up at the bottom, and the less dense particles move to the top of the bed. A screen allows the dense particles to fall to the bottom hopper while the less dense particles (with or without the middling) are drawn off the top.

In addition to the common plunger jig shown in Figure 6-16, which is also called the *Harz jig*, many other variations on the same theme are available commercially. Most of these vary the motion of the plunger so that the pressures imparted on the bed have different frequency diagrams, thus achieving specific bed movement.

Air Classifiers

The objective of air classification is to separate the less dense, mostly organic materials from the more dense, mostly inorganic fraction, using air as the fluid. The basic premise is that the less dense materials will be caught in an upward current of air and carried with the air, while the more dense fraction will drop down, unable to be supported by the air currents. The light fraction entrapped in the air stream must be separated from the air. Commonly, this is done with a *cyclone*, but it can be accomplished equally well with a large box or bag into which the particles drop while the exit air is filtered and escapes. The air can be either pushed or pulled, and the fan can be placed either before or after the cyclone. Except for smaller installations, placement of the fan so as to suck the material through the blades is not recommended, because of the wear and tear suffered by the blades. The various arrangements for air classifiers are shown in Figure 6-17.

Air classification is more successful if the particles in the air classifier do not adhere to each other. One means of breaking up the clumps of material is to use vibration. The *vibrating air classifier* is shown in Figure 6-18. Vibrating air classifiers combine the separation achieved due to the vibration with air entrainment. The feed vibrates along a sloping surface, with the light material shaken to the top, where the air stream carries it around a U-shaped curve.

Similar in function, but slightly different in construction, is a series of classifiers more properly labeled *air knives*, where the air is blown horizontally through a vertically dropping feed. The aerodynamically light particles will be carried with the air stream while the heavy ones will have sufficient inertia to resist a change in direction and drop through the air stream. This technique also allows for separation into more than two categories as shown in Figure 6-19. One other use of an air knife has been to help keep light contamination from carrying over during magnetic separation. Air is blown opposite the direction of travel of the metal under a magnet. This helps separate the lights from the metals and keeps the lights from being carried over onto the metals conveyor.

Considerable confusion exists in the terminology for the product (output) streams of air classifiers. Some authors call the material rising with the air stream

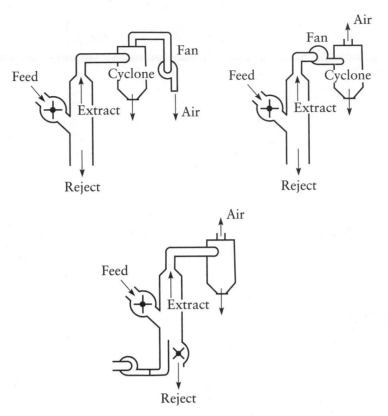

Figure 6-17   Three basic arrangements for air classification.

the *light fraction* and the material falling the *heavy fraction*. These terms are misleading, since they imply ideal separation. Better terms for air classifiers are *overflow* and *underflow*, but these imply vertical geometry. In keeping with the terminology introduced previously, in this text the material suspended by and removed by the air stream is called the *extract*, and the material not so removed is the *reject*.

Theoretical Best Performance of Air Classifiers

Because the *code* for air classifiers is aerodynamic velocity, it should be possible to estimate the effectiveness of air classification by measuring the aerodynamic velocities of the particles to be separated. One such method is the *drop test*, in which representative particles to be separated are dropped from a reasonable height in the absence of air turbulence, and the time needed to reach the ground is measured. Figure 6-20 shows the results of one such test for four different components: paper, plastic, aluminum, and steel.

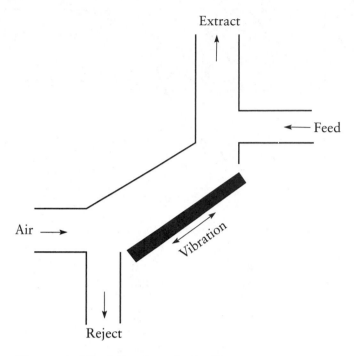

Figure 6-18 *Vibrating air classifier.*

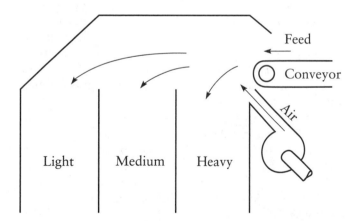

Figure 6-19 *Horizontal air knife.*

The theoretical effectiveness of air classification (that is, if all the particles exited the air classifier just as they were expected to) can be estimated from such a test. For example, using the results shown in Figure 6-20, as the air velocity of a classifier is increased from zero, the first material to start to report to the extract (float up) is paper. At an air velocity of 1000 ft/min, all of the paper reports to the

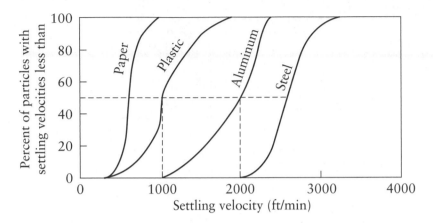

Figure 6-20   Typical results from a drop test.

extract. But at this velocity, about 50% of the plastic also reports to the extract. If the air velocity is increased beyond 1000 ft/min, some of the aluminum would start to rise with the extract. At an air velocity of 2000 ft/min, all of the paper and plastic would be captured as the extract, but also 50% of the aluminum. The steel fraction would not report to the extract until the air velocity exceeded 2000 ft/min. At 3000 ft/min, almost all of the feed would report to the extract.

If the objective in the air classification of this mixture is to produce a refuse-derived fuel (RDF) consisting of only paper and plastic, the air velocity cannot be allowed to exceed 1000 ft/min, or the RDF will contain some aluminum. If the recovery of the paper and plastic is important and some contamination is allowed, then running the air classifier at an air speed of 2000 ft/min would result in 100% recovery of the paper and plastic fraction. If the objective is to produce a pure metal (aluminum and steel) fraction as the reject (drop to the bottom), running the classifier at 2000 ft/min would result in a pure aluminum/steel mixture, but fully 50% of the aluminum would be lost to the extract.

This test gives the theoretical settling velocities of the particles if the drop distance is great relative to the distance needed to accelerate to terminal settling velocity. For such materials as paper and sheet plastic, this is not a problem. With materials that have high velocities, however, the acceleration distance might be significant and a simple drop test might give misleading results. Hasselriis has shown that it is possible to calculate the settling velocity from drop test data using the following relationship.[15]

$$y = \frac{v_s^2}{g} \ln(\frac{1}{2})[e^{v_s^{gt}} - e^{v_s^{gt}}]$$

where y = distance traveled, ft
v_s = terminal settling velocity, ft/sec
g = acceleration due to gravity, 32.2 ft/sec^2
t = time to travel distance y, s

This equation requires an iterative solution since it cannot be readily solved for the terminal settling velocity.

Figure 6-21 is a plot showing the theoretical performance of an air classifier as estimated from a drop test, and the actual performance. In general, the actual performance should be worse than the theoretical best, which should be viewed as an ideal goal.

In the absence of a drop test the terminal velocities for various materials can be calculated using an empirical equation:[16]

$$v_s = 1.9 + 0.092\,\rho_s + 5.8\,A$$

where v_s = terminal (falling) velocity, ft/sec
  ρ_s = particle density, lb/ft^3
  A = particle area (e.g., for a plate, A = length $\times$ width)

The size limits to this equation are between 0.0625 and 1.00 in.^{2}, which is generally considerably smaller than shredded MSW.

If the area function in the preceding equation is eliminated, little loss in accuracy results.[17, 32] In other words, the terminal velocity seems to be most significantly affected by density and only slightly by area and other variables (within the limits used in the experiments). The model, using only density as the independent variable, would then be

$$v_s = 1.91\rho_s^{1/2}$$

This model shows reasonable agreement with steel, aluminum, paper, and balsa wood as the materials.

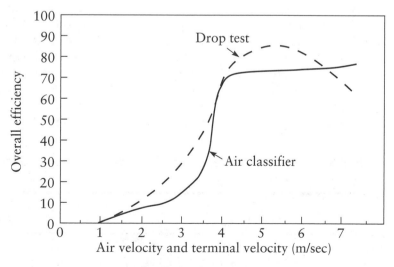

Figure 6-21 Theoretical air classification based on a drop test versus actual air-classifier performance.

Air Classifier Performance

Table 6-1 shows some typical results of air classifier performance with shredded refuse as feed.

Table 6-1 Typical Air-Classification Results for Shredded MSW

| | Percent by weight | |
	Shredded refuse	Extract from air classifier
Noncombustible		
Rocks and dirt	0.3	0
Ferrous metal	7.8	0.08
Nonferrous metal	1.0	0.05
Glass and other	7.8	1.82
Total	16.9	1.95
Combustible		
Paper	52.2	78.8
Wet garbage	11.8	0.1
Yard and garden	6.7	8.6
Other	12.2	10.6
Total	82.9	98.1

Source: (18)

The fuel characteristics of the extract in Table 6-1 are shown in Table 6-2. Also tabulated are the fuel properties of two other fuels prepared from MSW by air classification. The first is fuel composed of cubettes that were prepared from air-classified MSW, and the second is a proprietary product known as Eco-Fuel, which is air classifier extract chemically treated to make it biologically stable. Additional information on the properties of refuse-derived fuel is found in Chapter 7.

Paper and cardboard seem to be insensitive to particle size, whereas the heavy fraction is somewhat more sensitive, as shown in Table 6-3. Moisture content does not seem to influence greatly the recovery of lights, although a drop of perhaps 5% in recovery is expected when the moisture content doubles.[18]

Table 6-2 Fuel Characteristics of Air-Classification Shredded and Processed Refuse

	I Air-classified MSW	II Cubettes	III Eco-Fuel
Moisture (%)	15	15	10
Ash (%)	8.3	6	11
HHV, Btu/lb	6930	6800	6900

Sources: I: (16), II: (17), III: (19)

Table 6-3   Effect of Particle Size on Recovery in an Air Classifier

| | Nominal particle size, in. (90% passing) | | |
	1.9	3.1	24.3
Moisture	27.0	32.1	24.3
Paper and cardboard, percent recovered	92.3	90.4	92.8
Grit, percent recovered	74.8	87.7	86.2
Plastic, percent recovered	73.2	70.6	92.7
Percent heavies in extract	1.5	1.7	3.5

Source: (18)

Air-classifier performance can also be expressed in terms of the air:solids ratio, as shown in Figure 6-22. As the air flow is increased for a given solids feed rate, higher recoveries are experienced. Conversely, lower feed rates for a given air velocity enhance recovery.[20] The capacity of a classifier should be sufficient to withstand sudden surges.[18]

Although the effect of feed rate on air-classifier performance has not been fully studied, it seems reasonable that performance would deteriorate with increased feed rate. If a light particle is fed into the throat section, it must accelerate to attain the speed of the air stream. If heavier particles are present, these interfere with this acceleration and hence increase further the congestion within the throat. It is reasonable that the feed rate and average residence time of particles within the throat section are related as

$$t = \left(\frac{V}{Q}\right)^n$$

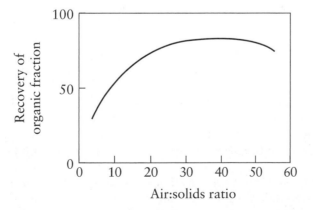

Figure 6-22 Air-classifier performance as described by the air:solids ratio.

where t = average residence time of particles
$\qquad V$ = volume within throat section
$\qquad Q$ = feed flow rate
$\qquad n$ = factor that accounts for the congestion within the throat; $n > 1.0$

It is possible to achieve performances that are better than theoretical by taking advantage of the same principles as are used in the operation of the jig. The air classifier is a float/sink separator, functioning because some particles have a lower aerodynamic velocity and others a higher aerodynamic velocity. To achieve separation on the basis of density only, and not on the aerodynamic velocity, air classifiers have been modified to force the particles to continually accelerate and decelerate. Four designs have been tested to take advantage of this possibility. These are shown in Figure 6-23 as the (a) zig-zag classifier, (b) baffled classifier, (c) constricted air classifier, and (d) pulsed flow classifier. All four of these modifications use the principle of never allowing particles to attain their terminal velocities, and enhancing separation on the basis of material density. In addition all four create high turbulence within the air classifier throat section, and this promotes the separation of particles that might be stuck together.

The *zig-zag air classifier* consists of a column with a series of 90° or 60° turns. Laboratory work with smoke tracers has demonstrated that at nominal air velocities (flow/cross-sectional area of tube) of 2.5 m/sec (500 ft/min) a central air core is formed, with turbulent vortices at the corners that touch the central core.[21] As the air velocity is increased to 3.5 m/sec (750 ft/min), the corner vortices disappear altogether, indicating fully turbulent conditions. The fraction of material reporting as heavies and lights can obviously be altered by changing the flow characteristics in the classifiers.

Within the turbulent vortices, the clumps are broken up, and the light particles are transferred to the upward air stream. The heavy particles drop from vortex to vortex until they exit at the bottom. The dropping action also helps break apart any agglomerated particles. The heavy particles tend to slide down the lower sides of the throat until they are hit by the upward rush of air at the corner (Figure 6-24). If the downward velocity of the particle is sufficiently great relative to the air-stream velocity, the particle takes path C, as shown on the figure. This allows it to continue its downward movement on the wall of the next segment. If, however, the air-stream velocity is sufficiently great, the particle may be caught within the air stream and experience trajectory A. If the particle moves along trajectory B, it has a 50% chance of going up or down.[22]

A creative alternative is to make the air classifier behave like a jig by pulsing the fluid and causing the particles to constantly be accelerating and decelerating. The *pulsed flow air classifier* has a straight throat, but the flow of air to the classifier is pulsed so that the particles never are allowed to attain their terminal settling velocities.[3]

All four types of modified classifiers—zig-zag, baffled, constricted, and pulsed flow—have the added advantage that the turbulence caused by the air flow helps to break up the clumps of refuse, stuck together because of mechanical accidents (e.g., pieces of metal stuck into paper) or adhesive forces (e.g., grease and oil acting as glue).

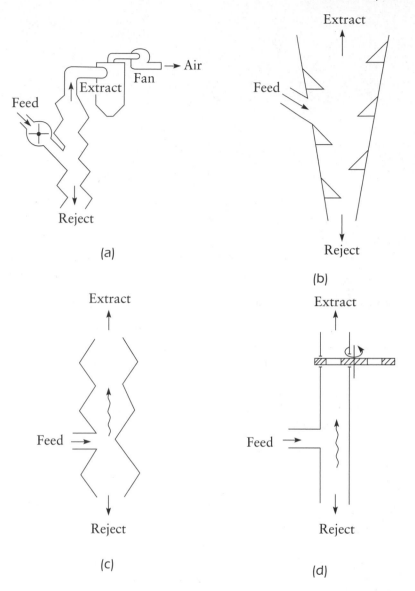

Figure 6-23 Various air classifiers: (a) zig-zag, (b) baffled, (c) constricted, and (d) pulsed flow.

Removing the Separated Particles from the Air Stream

Air classifiers can operate successfully only if the extracted material is removed from the air stream once they have been separated. This operation can be accomplished by using a large chamber, but such settling chambers are inadequate both from the standpoint of efficiency and because of the large space requirements. A more efficient means of removing the suspended particles from the air stream is to use a *cyclone*.

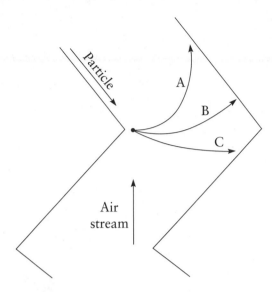

Figure 6-24   Behavior of a particle in a zig-zag air classifier.

The operation of the cyclone is illustrated in Figure 6-25. The air and solid particles enter the cyclone chamber at a tangent, setting up a high-velocity rotational air movement within the chamber. The solid particles, having greater mass, move outward toward the inside wall, are slowed down on contacts and eventually drop to the bottom of the chamber under the influence of gravity. The air exits through the central tube, free of solids.

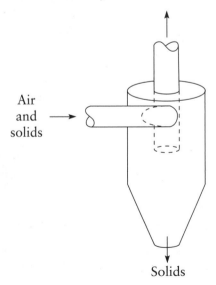

Figure 6-25   Cyclone used for air cleaning.

The objective of a cyclone is to move particles to the outside by centrifugal action. The radial velocity of a particle, v_R, is important, as illustrated in Figure 6-26. Assuming laminar conditions (a bad assumption), the particle terminal radial velocity is governed by Stokes's law, except that the gravitational acceleration must be replaced by centrifugal acceleration, defined as

$$a = r\omega^2$$

where a = centrifugal acceleration, m/sec^2
  r = radius, m
  ω = rotational velocity, rad/sec

so that

$$v_R = \frac{d^2(\rho_s - \rho)\omega^2 r}{18\mu}$$

where d = particle diameter, m
  ρ = density of air, kg/m^3
  ρ_s = density of particle, kg/m^3
  μ = air viscosity, N-s/m^2

Recognizing that $r\omega^2 = v_{tan}^2/r$, where v_{tan} is the tangential velocity, the radial velocity can be expressed as

$$v_R = v_s \frac{v_{tan}^2}{gr}$$

which is the basic design equation for cyclones. The objective is, for any feed into the cyclone, to have the highest possible v_R, and this can be achieved by having a high tangential velocity (v_{tan}), a high particle settling velocity (v_s) and a small r. The latter observation leads to using banks of small-diameter cyclones instead of a large

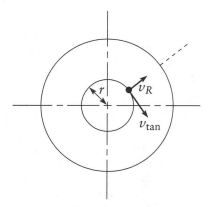

Figure 6-26 Radial movement of a particle within a cyclone.

single unit. For a given volume the bank of small-diameter cyclones is far more efficient in removing particles from the air stream. Unfortunately, with refuse-derived fuel, a small diameter will result in clogging, and large diameters are absolutely essential.

This analysis may be somewhat simplified by assuming that $\rho_s \gg \rho$. Thus

$$v_R \approx \frac{d^2 \rho_s \omega^2 r}{18\mu}$$

At any radius r the tangential velocity is

$$v_{tan} = r\omega$$

and

$$\omega = \frac{v_{tan}}{r}$$

where ω is the rotational velocity in rad/sec. The time t needed for one rotation is thus

$$t = \frac{r}{2\pi v_{tan}}$$

The distance traveled by a particle during one rotation is

$$S \approx v_R t \approx \frac{d^2 \rho_s \omega^2 r}{18\mu} \left(\frac{r}{2\pi v_{tan}} \right)$$

where S is the radial distance traveled by a particle during one rotation. (This is necessarily approximate, since the radius and hence the v_{tan} changes as the particle moves.)

Since $v_{tan} = \omega r$,

$$S \approx \frac{d^2 \rho_s v_{tan}}{36\mu\pi}$$

The objective is to increase S, the radial distance traveled by the particles during a single rotation. The faster the particle collides with the wall, the faster it is removed from the air stream. From the above equation S can be increased by increasing the particle size d, increasing its density ρ_s, increasing the tangential velocity v_{tan}, and decreasing the fluid viscosity μ. Experience with cyclones for the removal of shredded light fraction has shown that large factors of safety are needed.

It is useful to review the above analysis of separation in a cyclone. It starts with the assumption that Stokes's law holds, which is patently false in the highly turbid environment of a cyclone. When this analysis is used in air pollution control, such as the removal of fine particulates (about 5 μm in diameter) from combustion, the particles are so small that they are not influenced by the turbulence and tend to obey Stokes's law even in the presence of macroturbulence. In theory the boundary layer around the particles is still laminar due to their very small size.

This, of course, does not occur when the cyclone is used to separate out shredded paper from the air-classified waste stream. The paper particles, which tend to have dimensions in the centimeter range instead of in micrometers, are most certainly behaving in the turbulent flow regime.

Second, the use of the Stokes equation assumes that each particle is behaving as if it were not affected by other surrounding particles. This assumption is clearly not true for air-classified paper in a cyclone where the feed is a high concentration of paper and other light materials.

This is not to say that the above analysis is untrue. It simply must be applied with great care and understanding. Research into the processing of light material such as shredded paper is greatly needed.

Other Float/Sink Devices

Three additional options using float/sink technology have been used for MSW processing. The *heavy-liquid separators* substitute a denser liquid for water. For example, a mixture of tetrabromoethane and acetone has been used for the separation of aluminum from heavier materials. This liquid has a specific gravity of about 2.4, and the sink fraction of shredded, air-classified, and screened refuse, with the ferrous fraction removed, can produce a fairly high concentration of aluminum. Two major problems with the use of such heavy liquids are the inability to readily vary the specific gravity as needed and the cost of the chemicals. A significant fraction can be lost during the operation as a result of adsorption to the waste material. Another commercially successful heavy liquid is pentachlorethane (C_2HCl_5: specific gravity 1.67 at 20°C),[17] which has been used for coal cleaning.

Heavy-media separators differ from heavy liquids in that the specific gravity is varied by adding colloidal solids. For example, a mixture of ferrosilicon and water (85:15), with a surface-active agent, can be used to attain specific gravities over 3.0. In one study[23] aluminum was removed by first sinking it in a fluid with a specific gravity of 1.4 and then floating it in a liquid of specific gravity at 3.0. This method also suffers from the problem of capture and retention of the fluid, resulting in higher operating costs for replacement, as well as potential wastewater treatment problems. But it has advantages. The principal advantage of heavy-media separation can be demonstrated by the following argument.[24] Suppose that a mixture is composed of two solids, **a** and **b**, **a** being denser than **b**. Because of size differences, the larger **b** particles may have higher settling velocities than the small **a** particles, and hence separation by settling cannot be complete. The range of sizes that can be separated can be calculated by recognizing that, at equal settling velocities (assuming laminar flow and Stokes's law),

$$v_a = \frac{d_a^2 g\left(\rho_{s_a} - \rho\right)}{18\mu} \quad \text{and} \quad v_b = \frac{d_b^2 g\left(\rho_{s_b} - \rho\right)}{18\mu}$$

If the two velocities are equal,

$$\left(\frac{d_a}{d_b}\right)^2 = \frac{\rho_{s_b} - \rho}{\rho_{s_a} - \rho}$$

and separation is possible if

$$\frac{d_a}{d_b} > \left(\frac{\rho_{s_b} - \rho}{\rho_{s_a} - \rho} \right)^{1/2}$$

If it is possible to choose a fluid with a density very nearly equal to one of the solid materials, the classification can be made more efficient. For example, if the fluid density approaches that of particle b, the ratio d_a/d_b approaches zero, meaning that very large particles can be separated from even the very small **a** particles. If the fluid density is greater than the density of **b** and less than that of **a**, complete separation is possible.

The third method of achieving float/sink separation is to use an *upflow separator* with water as the fluid. Effective specific gravities of between 1.1 and 2.0 have been used in commercial devices. Such upflow devices have been used to separate heavy organics (leather, plastics, textiles, etc.) from metals and glass in the heavy fraction of air-classified refuse.[17]

A process that, strictly speaking, is a float/sink process, but has a special twist, is *flotation*. In this process the selected solids are selectively floated to the surface of the slurry by means of attached gas bubbles. The key to successful flotation is the selective adhesion of air bubbles to the material that is to be floated. The actual separation, after the material has been made lighter by air-bubble attachment, can be by frothing or by a simple gravity separator such as a shaking table. The usual separation method in resource recovery operations is froth flotation, and the common application is the removal of glass from ceramics and other contaminants.

MAGNETS AND ELECTROMECHANICAL SEPARATORS

Magnets

Magnets are used to separate ferrous materials from the rest of refuse. The code obviously is the magnetic property of ferrous materials such as steel. Typical magnet arrangements are shown in Figure 6-27. The usual process uses a belt magnet installed above a conveyor belt, with the belt on the magnet moving across the conveyor carrying the refuse. Steel cans and other ferrous materials are pulled from the refuse and adhere to the underside of the belt covering the electromagnet. The belt moves these materials off to the side where they cease to be under the influence of the magnetic field and drop off the conveyor. The effectiveness of magnets depends on several variables, such as the height of the magnet above the conveyor belt carrying the refuse. The closer the magnet is to the refuse, the better the effectiveness of ferrous removal (Figure 6-28). The greater the magnetic force employed, the greater will be the recovery of ferrous material.[25] The speed of the conveyor also affects the recovery, with higher speeds showing reduced recovery as would be expected. The magnetic field simply does not have enough time to act on the ferrous material and pull it out of the refuse. Finally, the greater the burden depth on the conveyor belt, the lower the recovery, as shown in Figure 6-29. This relationship is exponential because the buried ferrous material has an exponentially

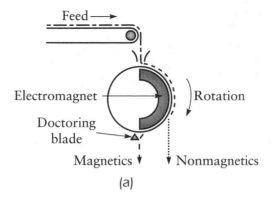

(a)

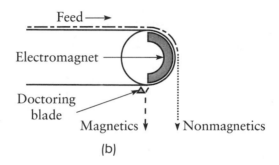

(b)

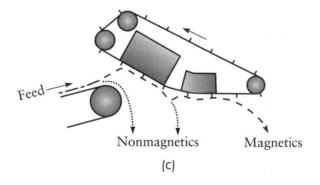

(c)

Figure 6-27 *Typical magnets used in removal of ferrous materials: (a) drum holding magnet, (b) belt holding magnet, (c) suspended type magnetic separator.*

increasing force on it due to the material above it. Thus burden depth, and the ability to maintain this at some shallow and even depth before it runs under the magnet, is the main concern in designing magnetic separation operations.[25]

A problem faced with the use of long conveyors prior to magnetic separation is that the jiggling of the material on the conveyor is not unlike a gold-miner's pan, and separation by density occurs, with the steel cans migrating to the bottom of the

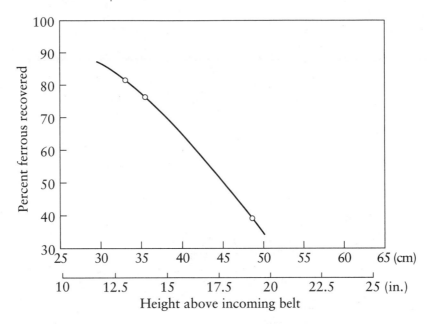

Figure 6-28 Height of a magnet above incoming refuse affects ferrous recovery. Source: (25)

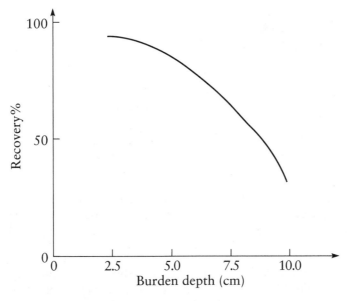

Figure 6-29 Greater burden depth reduces the recovery of ferrous from a conveyor belt. Source: (26)

charge on the conveyor belt. When the refuse is then passed under the magnet, the ferrous material must be able to move through the refuse piled on top to reach the underside of the magnet, an often impossible task. Good engineering requires either a one-particle thickness on the conveyor belt or a very short conveyor to prevent such separation from occurring.

Eddy Current Separators

Eddy current separators respond to the problem of separating nonferrous metals from the remainder of refuse, and depend on the ability of metals to conduct electrical current. If the magnetic induction in a material changes with time, a voltage is generated in that material, and the induced voltage will produce a current, called an eddy current. The feed to an eddy current separator might be the reject component from air classifiers from which the ferromagnetic components (steel cans mostly) has been removed.

Many eddy current separators are inclined tables. Underneath the table are several large magnets that produce an electrical field. If a particle that conducts electricity slides down the inclined table, the electrostatic forces push it in a direction perpendicular to its path. Only those particles that conduct electricity, as the particles move down the inclined table and come under the influence of the charge field, are laterally displaced. Nonconductors are not affected by the charge field and drop straight down.

A typical eddy current separator is shown in Figure 6-30. This device is a modification of a linear induction motor in that it generates a sine wave of magnetic intensity, which travels down the length of the motor with alternating north- and south-pole components. As the metal-rich concentrate passes over the linear induction motor, eddy currents are induced in an electrical conductor that appears on the surface of the table. The induced magnetic fields associated with the eddy currents in the metals interact with the moving field generated by the motor, which pushes the conductors (nonferrous metals) along the linear motor. All that is necessary to achieve removal is to orient the motor transverse to the direction of the feed, so as to repel the metal away from the main direction of travel. The mixed material is fed to one end of the nonmagnetic belt, which travels over the linear induction motors positioned on the underside of the belt. Recovery of the metal concentrate is on the top side of the belt, where the material to be removed is ejected by the linear induction motor against the retaining wall and into the extract area. The rejects are not affected by the eddy currents and therefore flow along the lower portion of the belt area. Table 6-4 shows the performance of a typical linear induction motor in separating aluminum from a shredded refuse from which ferrous material has been removed.[27]

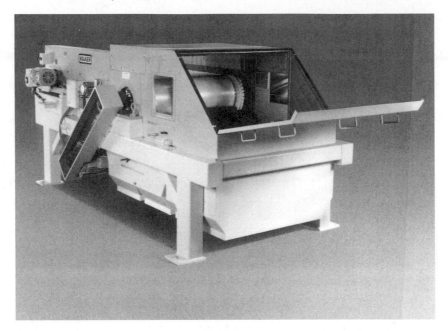

Figure 6-30   Eddy current separator. (Courtesy O. S. Walker Co.)

Table 6-4   Eddy Current Separation of Shredded Refuse from Which Ferrous
Has Been Removed

| | | Percent by weight | | | |
	Weight (kg)	Aluminum cans and other aluminum	Aluminum foil	Other inorganics	Organics
Extract	14.5	88.8	0.2	5.5	5.5
Middlings	7.4	60.2	2.1	11.6	25.1
Reject	85.7	85.7	7.0	3.0	86.6
Feed	107.5	21.9	2.6	3.9	71.5

Source: (27)

Using the numbers in Table 6-4, the eddy current separator was able to achieve a recovery of only little more than 50% with a purity of only 89%—not generally acceptable to secondary materials dealers. Thus a hand removal of contaminants is required after the eddy current separator.[28, 29]

Electrostatic Separation Processes

Charged particles under the influence of electrostatic forces obey laws of attraction and repulsion similar to those for magnets. Certain materials can be coded by being electrically charged, then separated by being attracted to the opposite-charged electrode, or by being repelled from a like-charged electrode.

Separating plastics has been a problem because the particle properties (the codes) are so similar for various types of plastics. One method of achieving separation is to allow the shredded or granulated plastic particles to acquire an electrical charge by friction as the particles rub against each other. Different plastics have a markedly different ability to take on electrostatic charge, as seen by the charging progression shown in Figure 6-31. The charge effectively codes the particles and they can then be separated by allowing them to fall freely through a high-voltage field and into separate compartments, as shown in Figure 6-32. Such a process is called a *triboelectric charging* process since the charge is placed on the particles due to friction. Each type of plastic has a different potential for acquiring a charge and thus separation can be made.[30]

OTHER DEVICES FOR MATERIALS SEPARATION

Listed here are some devices that may someday prove useful in MSW processing.

Stoners, also called pneumatic tables, differ from jigs only in that air is substituted for water as the pulsating fluid. The general principles of operation are the same as for jigs using water. Most commercial models, in addition to pulsating air, use shaking tables, thus providing two forms of the energy input. These devices have been applied for the recovery of aluminum from shredded and screened waste, although operating data have not been made public. Stoners have been used for many years in the agricultural industry for removing stones and other impurities from peanuts,

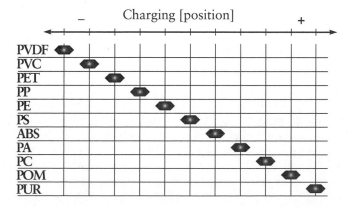

Figure 6-31 Triboelectric charging progression. (Courtesy Steinert)

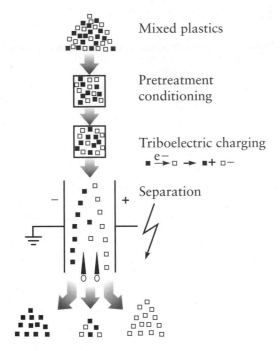

Mixed plastics

Pretreatment
conditioning

Triboelectric charging

Separation

Figure 6-32 *Separation of triboelectrically charged plastic. (Courtesy Steinert)*

beans, and so on. They are most frequently employed where the two-part separation into light and heavy fractions involves a minor fraction of heavies, with a density difference between the two of at least 1.5:1.[31]

Inclined tables can be used to separate particles of various densities and sizes. For example, coal can be washed by separating slate and other heavy materials from the raw ore by washing the mixture down an inclined table and removing the heavy contaminants through ports located along the incline.

Shaking tables differ from simple inclined tables only in that the table is shaken with a differential movement in the direction perpendicular to fluid flow. In addition all modern shaking tables, such as the *Wilfrey Table*, are equipped with riffles, which are long slots in the table, also perpendicular to the flow.

Optical sorting attempts to respond to a major problem with the recovery of glass. Waste glass is of many colors and such mixtures have low market value. Clear glass alone has substantial value, while contamination by even 5% amber glass makes it essentially unmarketable, since it cannot be used to produce clear glass products. At this time, the only technique, other than hand sorting, for color-sorting glass seems to be by the use of the wavelength of transmitted or reflected light from the glass.

The most popular optical sorter, which has in the past been used for the separation of diamonds in diamond mines and rotten corn and peanuts in agriculture, is shown in Figure 6-33, which illustrates the basic principle. The individual particles

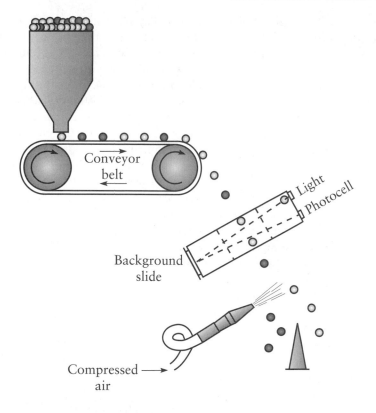

Conveyor
belt

Light

Photocell

Background
slide

Compressed ⟶
air

Figure 6-33 Optical sorting.

are moved by means of a high-speed pulley from a bin and flung through a light-detection box or sensor. Within this detector are three lights and three photocells. Each light is bounced off a background reference slide into a photocell, and the current produced is sensed electronically. Should the particle falling through the sensor be of the same color (some average of reflected and transmitted light) as the background slide, the photocell will not detect any difference and nothing happens. These particles are thus passed through the device as *extract*. Should a particle be lighter or darker than the background slide, however, the current change will trigger a compressed-air ejector located immediately below the sensor box and the *reject* particle is blown into a different bin. Dividing a feed of glass particles into three categories—flint (clear), green, and other—requires two passes, because the color sorters on the market are inherently binary devices. In such a case the first sorting would be based on light intensity reaching the photocells, passing the flint particles, and ejecting the colored particles. In the second sorting of the colored particles, green filters might be placed over the photocells and the electronics tuned so that green particles are perceived as *light* and are passed; amber (and other colors on the red end of the spectrum) particles are perceived as *dark* and are ejected.

This device works well when the particles are uniform in size and the electronics can be so set as to make the device reject even suspicious particles. For example, of 100 peanuts entering the sensor, two might be rotten. The device can be adjusted to sense these two, but since it is so finely tuned, may also eject two other peanuts, which may be perfectly good but which may have been so oriented during their trajectory as to produce a shadow and thus be rejected. The peanut farmer accepts this margin of safety, and gladly discards two good peanuts as long as he is reasonably sure that he has removed all rotten ones. With glass separation, however, the mix is not 2 colored pieces and 98 clear pieces out of 100. Instead, a 50/50 mix is common, and this presents problems of efficiency of operation. To date, it has not been possible to achieve the required purity of product that would make this device a practical alternative for glass sorting. In addition a color sorter requires a relatively narrow size distribution in the feed, and this requirement places considerable constraints on size-reduction processes.

Bounce and adherence separators have been used in Europe and in one facility in the United States. The bounce and adherence separator is a short conveyor on which the angle and speed can be adjusted. The material falls onto the middle of an angled conveyor, and as the belt moves the material upward, some types of material—such as paper, cardboard and film plastic—lies on (adheres to) the belt and travels up the conveyor and off the upper end. Other materials—such as bottles and cans—roll down the conveyor and fall off the lower end. The optimum efficiency of this separation must be achieved by trial and error as the angle and speed of the conveyor belt are adjusted.

The adherence material can be transported to a paper picking area, and the bounce material would proceed to a ferrous magnet, eddy current separator, or glass picking area. The advantage of this separation is that by using a first (if crude) cut, the process loading on each subsequent unit operation is reduced, enhancing the effectiveness of separation.

MATERIALS SEPARATION SYSTEMS

Each of the above unit operations accomplishes one separation, and its performance thus far has been analyzed in isolation. Obviously, the placement of the unit operations in series will affect the final products. While there is no "best" series for any given feed, experience over the years has shown that certain combinations seem to perform well.

For example, Figure 6-34 shows a system for processing mixed MSW. In this process the MSW is stored on a floor where unacceptable items—such as large appliances, old lawn mowers, etc.—are removed and sent to the landfill. The first separation step is a trommel screen, which also acts as a bag opener. Both the reject and extract from the trommel screen are sent to magnets for removal of ferrous products (mostly cans), which are then sent to a can flattener and then to storage. The extract from the trommel (undersize) is sent to the landfill. The reject (oversize) goes to a hand-sorting operation where corrugated cardboard, PETE plastic bottles, HDPE plastic, and aluminum cans are removed. The first three go to balers and then storage, while the aluminum cans are run through a magnet to remove any stray steel cans, and the product is then sent to storage. Following hand sorting, the remaining material is shredded and air-classified. Providing hand sorting ahead of

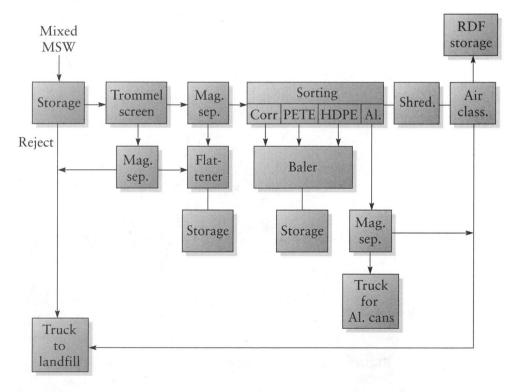

Figure 6-34 *A typical dirty materials recovery facility for mixed waste.*

the shredder is an excellent idea because it minimizes the chance of explosive materials finding their way to the shredder. The shredded material is then air-classified to produce a refuse-derived fuel, which is stored for transportation to a power plant, and the reject is again sent to the landfill.

Depending on the level of separation performed at the household level, source-separated material collected at curbside or at materials recycling centers can be processed in various combinations. Consider, for example, a system where the householder is asked to sort out all papers, regardless of type. Since mixed paper has little market value, a simple separating process consisting of hand sorting would be appropriate, as shown in Figure 6-35. The mixed paper is received at the plant, and a conveyor belt presents the papers to the pickers, who make decisions based on types of paper to be separated, such as office paper, newspaper, and magazines. Unwanted material, such as plastics, would be rejected.

If the separated material consisted of mixed papers and aluminum cans, the processing facility might look like Figure 6-36. Here the aluminum cans are removed from the collection vehicle and run through a magnet for the removal of stray steel cans. Unless the separation has been done by a person on a truck, an operator has to be assigned to the operation to prevent glass bottles and plastics from entering the magnet.

The role of the solid waste engineer, when employed by a municipality or waste authority, is to understand the overall performance of such facilities, and to relate

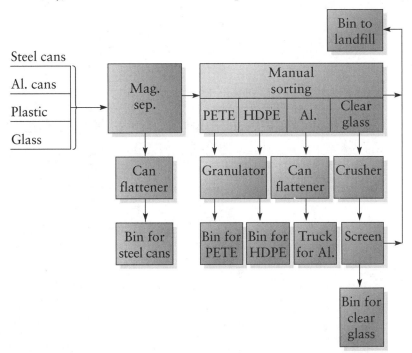

Figure 6-35 A typical clean materials recovery facility for previously separated waste.

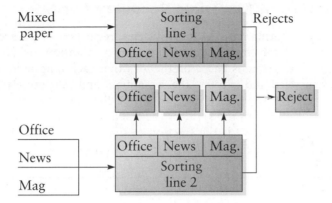

Figure 6-36 *An alternative materials recovery facility for previously separated waste.*

the performance to the available markets. A knowledge of the quality and quantity, as well as dependability, of the waste supply is critical. These engineers must also understand the operational costs and the details of the financing, including sale of the products.

Detailed design of a materials recovery facility is often done by consulting firms, and engineers practicing as consultants have to have a workable knowledge of the individual unit operations. These engineers have to understand the separation process as well as such ancillary requirements as traffic flow, power needs, and the structural aspects of the building in which the equipment is to function.

Finally, engineers employed by the manufacturers of the equipment must understand how their equipment will fit into a materials recovery system and what it will be expected to do. These engineers are as critical to the success of the facility as the other two types of engineers. They have a clear conflict, however, since they are employed by the manufacturer to sell equipment, but they have to make sure that this equipment works. They cannot, for example, say that the capacity of a trommel screen is 5 tons/hour, when they know full well that other facilities have found that it can function well only at less than 3 tons/hour. The problem is that there are many trommel screen manufacturers, and the tendency is to oversell the product's performance to make the sale.

In summary, there are three levels of engineering responsibility:

- engineers who must understand the overall nature of the system, including the nature of the feedstock and the markets for the products.
- engineers who must understand the system and how each unit operation is to perform within the materials recovery system.
- engineers who must understand each unit operation and who must be careful to apply such equipment appropriately.

Performance of Materials Recovery Facilities

The potential performance of all such systems, regardless of the arrangement of the unit operations, can be summarized using a notation originally devised by Hasselriis.[15] The system is based on the idea that each unit operation rejects some fraction of the feed and extracts the remaining, and that these fractions of reject and extract are the same regardless of where the unit operation is placed in the process train. Define f as the *split*, or the fraction of material rejected by any unit operation, and thus $1 - f$ is the fraction of material extracted by the unit operation. The use of this idea is illustrated by the example below.

EXAMPLE
6-4   Consider a processing operation consisting of an air classifier, a trommel screen, a magnet, and a disc screen, placed in various sequences. The feed is assumed to contain four materials: paper, glass, ferrous, and aluminum. The f-value (fraction of material rejected) for each of the unit operations for each of the materials is shown below:

	Air classifier	Trommel screen	Magnet	Disc screen
Paper	0.1	0.9	1.0	0.9
Glass	0.8	0.1	1.0	0
Ferrous	1.0	1.0	0	1.0
Aluminum	0.9	0.9	0.9	0.9

The assumption is that the air classifier will reject 10% of the paper and accept 90% of the paper ($f = 0.1$ and $1 - f = 0.9$). The trommel screen will, however, reject 90% of the paper.

Assume further that the feed rates of the mixed materials are as follows:

Material	Feed, lb/h
Paper	70
Glass	10
Ferrous	10
Aluminum	10

Consider now option 1, the placement of the unit operations as

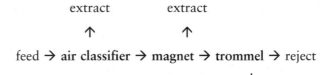

Begin by placing the feed quantities in the form of a vertical array (**A**) followed by an array of the f-values (**f**).

A   f

$$
\begin{bmatrix} 70 \\ 10 \\ 10 \\ 10 \end{bmatrix}
\times
\begin{bmatrix} 0.1 & 1.0 & 0.9 \\ 0.8 & 1.0 & 0.1 \\ 1.0 & 0 & 1.0 \\ 0.9 & 0.9 & 0.9 \end{bmatrix}
$$

Then calculate the reject from the air classifier by multiplying the *f*-value times the feed, or 70 × 0.1 for the paper, 10 × 0.8 for the glass, and so on. The resulting reject (**B**), or that material that moves on to the next unit operation, is thus 7 for paper, 8 for glass, and so on. The entire array looks like the following:

A   f   B

$$
\begin{bmatrix} 70 \\ 10 \\ 10 \\ 10 \end{bmatrix}
\times
\begin{bmatrix} 0.1 & 1.0 & 0.9 \\ 0.8 & 1.0 & 0.1 \\ 1.0 & 0 & 1.0 \\ 0.9 & 0.9 & 0.9 \end{bmatrix}
=
\begin{bmatrix} 7 & 7 & 6.3 \\ 8 & 8 & 0.8 \\ 10 & 0 & 0 \\ 9 & 8.1 & 7.3 \end{bmatrix}
$$

For example, the fraction of paper rejected by the air classifier is 0.1, so that 70(0.1) = 7. This reject then moves to the second unit operation, the magnet, where all of it is rejected, or 1.0(7) = 7. Moving on to the trommel screen, 0.9 is rejected, or 7(0.9) = 0.63, the reject from the trommel. This result can be used to calculate the recovery and purity of various products. For example, the recovery of paper, or that fraction of paper entering the system that is actually recovered as the extract from the air classifier, is calculated by first noting that the paper in the extract from the air classifier is 70 − 7 = 63. The recovery is then

$$
R_{\text{paper}_{\text{air classifier}}} = \frac{63}{70} 100 = 90\%
$$

The purity of the paper is calculated by noting than in addition to the 63 of paper, the air classifier extract includes 10 − 8 = 2 of glass, no ferrous, and 10 − 9 = 1 of aluminum, for a total of 66. The purity is then

$$
R_{\text{paper}_{\text{air classifier}}} = \frac{63}{66} 100 = 95\%
$$

Similarly, the recovery of glass (as the trommel extract) is

$$
R_{\text{glass}_{\text{trommel}}} = \frac{8 - 0.8}{10} 100 = 72\%
$$

and its purity is

$$
R_{\text{glass}_{\text{trommel}}} = \frac{7.2}{0.7 + 0.8 + 7.2} 100 = 83\%
$$

Suppose the unit operations are now arranged as follows:

extract   extract

↑   ↑

feed → **trommel screen** → **magnet** → **air classifier** → reject

↓

extract

The array would look like this:

A   f   B

$$\begin{bmatrix} 70 \\ 10 \\ 10 \\ 10 \end{bmatrix} \times \begin{bmatrix} 0.9 & 1.0 & 1.0 \\ 0.1 & 1.0 & 0.8 \\ 1.0 & 0 & 1.0 \\ 0.9 & 0.9 & 0.9 \end{bmatrix} = \begin{bmatrix} 63 & 63 & 6.3 \\ 1 & 1 & 0.8 \\ 10 & 0 & 0 \\ 9 & 8.1 & 7.3 \end{bmatrix}$$

Using the product array, we can calculate the recovery of paper by first noting that the extract from the air classifier is $63 - 6.3$ or about 56.7. Thus the recovery is

$$R_{paper_{air\ classifier}} = \frac{56.7}{70} 100 = 81\%$$

but now the purity of the paper is calculated by noting that the extract includes, in addition to the paper, $1 - 0.8 = 0.2$ of glass, 0 of ferrous, and $8.1 - 7.3 = 0.8$ aluminum, for a total of 57.7. The purity is then $(56.7/57.7)100 = 98.2\%$, which is far better than with the first option. Clearly, if the primary objective of the materials recovery facility is to produce a paper product of high purity (say, greater than 98% purity), then the second option would be the better arrangement of the unit operations (even though only 81% of the paper would be recovered).

A caution that ought to be obvious: We assume in the above analysis that the fraction of material extracted or rejected is the same regardless of what unit operation precedes the separation step, so that the quality of the feed does not matter in the recovery or purity of the product. This is not true, of course, but as a means of arriving at rough calculations and analyses for separation, the process is useful.

From the above discussion it is clear that there are a multitude of ways we can construct materials recovery facilities. No attempt has been made in this chapter to present an exhaustive list of the possibilities. Handbooks such as Taggart's *Handbook of Mineral Dressing*[32] present an impressive compilation of machines and processes and include some techniques that may eventually find use in future resource recovery facilities. An emerging field such as resource recovery has to borrow from many fields and select those devices that show most promise. With thousands of facilities throughout the United States and around the world processing solid waste, recyclables, and yard waste, the different designs and combinations of processing equipment are almost endless. The "perfect" system has yet to be designed; and given the variable nature of solid waste, it never will be designed. Each facility has a unique set of requirements, and the engineer must factor all of these into the design.

FINAL THOUGHTS

In the United States and many other countries, environmental decision making has become increasingly public and consultative. A decision to construct a power station, a prison, or a waste incinerator affects many people. These people rightly have a say in the planning process through formal and informal participation, including consultation, lobbying, and public hearings. The opportunity for class action suits and the relative ease with which United States law grants standing to concerned individuals in environmental cases provide additional access for public involvement. Despite all this representation, one large group of people receives no hearing. These are the people yet to be born.

The effects of today's management decisions will be felt for years to come. Indeed, many decisions have no effects until decades later. For example, it may take generations for waste containers to corrode, for their contents to leach, for the leachate to migrate and pollute groundwater, and for toxic effects to occur. The only persons to be adversely affected by a management decision may be the only persons who had no say in the decision. They cannot speak up for themselves, and unlike children, they have no one legally responsible for representing their interests. In fact, we often deliberately ignore the interests of future generations. For decades we have produced (and continue to produce) long-lived radioactive wastes, without any agreement about how to keep them safe in the future or even whether they can be kept safe at all. If we were serious about protecting future generations, we wouldn't generate the waste until we have figured out how to manage it safely.

Instead, there is no clear long-term responsibility for hazardous waste management. United States legislation is much more comprehensive and, in principle, tougher than that of most countries. The main statute, the Resource Conservation and Recovery Act of 1976, requires waste facility operators to provide "perpetual care" for depositories. "Perpetual care," however, is defined as a period of only 30 years. In terms of the interests of future generations, when we say "perpetual" we should *mean* perpetual, and not just a few decades.

REFERENCES

1. Rietema K. 1957. "On the Efficiency in Separating Mixtures of Two Components." *Chemical Engineering Science* 7: 89.
2. Worrell, W. A., and P. A. Vesilind. 1980. "Evaluation of Air Classifier Performance." *Resource Recovery and Conservation* 4.
3. Stessel, R. 1982. *Pulsed Flow Air Classification.* MS Thesis, Department of Civil and Environmental Engineering. Durham, N. C.: Duke University.
4. Hering, R., and S. A. Greely. 1921. *Collection and Disposal of Municipal Refuse.* New York: McGraw-Hill Book Company.
5. Engdahl, R. B. 1969. *Solid Waste Processing.* EPA OSWMP. Washington, D.C.
6. Henstock, M. 1978. Discussion at Engineering Foundation Conference, Rindge, N. H. (July).
7. Trezek, G. J., and G. Savage. 1976. "MSW Component Size Distribution Obtained from the Cal Resource Recovery System." *Resource Recovery and Conservation* 2: 67.
8. Matthews, C. W. 1972. "Screening." *Chemical Engineering* (July 10).
9. Rose, H. E., and R. M. E. Sullivan. 1957. *A Treatise on the Internal Mechanics of Ball, Tube and Rod Mills.* London: Constable and Co.
10. Nelson, W., A. Kruglack, and M. Overton. 1975. *Trommel Screening.* Duke Environmental Center, Duke University.
11. Makar, H. V., and R. S. DeCesare. 1975. "Unit Operations for Nonferrous Metal Recovery." *Resource Recovery and Utilization.* ASTM STP 592, pp. 71–88.
12. Beardsley, J. B. 1937. *From Wheat to Flour*; pp. 28–29. Chicago: Wheat Flour Institute.
13. Worrell, W. A. 1978. *Testing and Evaluation of Three Air Classifier Throat Designs.* Duke Environmental Center, Duke University.
14. Richardson, J. F., and W. N. Zaki. 1954. "Sedimentation and Fluidisation." *Transactions* Institute of Chemical Engineering 32: 35.
15. Hasselriis, F. 1984. *Refuse-Derived Fuel Processing.* Boston: Butterworths.
16. Boettcher, R. A. 1972. *Air Classification of Solid Waste.* EPA OSWMP, SW-30c. Washington, D.C.
17. Salton, K., I. Nagano, and S. Izumin. 1976. "New Separation Technique for Waste Plastics." *Resource Recovery and Conservation* 2: 127.
18. Murray, D. L., and C. L. Liddell. 1977. "The Dynamics, Operation and Evaluation of an Air Classifier." *Waste Age* (March).
19. Arthur D. Little, Inc. Personal communication.
20. Midwest Research Institute. 1979. *Study of Processing Equipment for Resource Recovery Systems.* EPA Contract 68-03-2387, Final Report, Washington, D.C.
21. Sweeney, P. J. 1977. *An Investigation of the Effects of Density, Size and Shape Upon the Air Classification of Municipal Type Solid Waste.* Report CEEDO-TR-77-25, Tyndall Air Force Base, Fla. Civil and Environmental Engineering Development Office.
22. Senden, M. M. G., and M. Tels. 1978. "Mathematical Model of Air Classifiers." *Resource Recovery and Conservation* 2: 129.
23. Michaels, E. L., K. L. Woodruff, W. L. Fryberger, and H. Alter. 1975. "Heavy Media Separation of Aluminum from Municipal Solid Waste." *Transactions* Society of Mining Engineering 258: 34.
24. Foust, A. S., et al. 1960. *Principles of Unit Operations.* New York: John Wiley & Sons, Inc.
25. Parker, B. L. 1983. *Magnetic Separation of Ferrous Materials from Shredded Refuse.* MS Thesis. Durham, N. C.: Duke University.

26. Spence, P. 1977. *Magnetic Separation—A Laboratory Study*. Duke Environmental Center (April). Durham, N. C.: Duke University.

27. Abert, J. G. 1977. "Aluminum Recovery—A Status Report." NCRR *Bulletin* 7, n. 2. Washington, D.C.: National Center for Resource Recovery.

28. Zavestski, S. 1992. "Reynolds Aluminum." *Waste Age* (October): p. 67.

29. Malloy, M. G. 1996. "Pouring It on in Ohio." *Waste Age* (March): p. 61.

30. Steiner. "Separating Plastics with the Same Density by Means of a Free Fall Separating Plant." Promotional literature. Köln, Germany.

31. Drobney, N. L., H. E. Hull, and R. F. Testin. 1971. *Recovery and Utilization of Municipal Solid Waste*. EPA OSWMT SW-10c. Washington, D.C.

32. Taggart, A. F. 1974. *Handbook of Mineral Dressing*. New York: John Wiley & Sons.

ABBREVIATIONS USED IN THIS CHAPTER

HDPE = high-density polyethylene
PETE = polyethylene terephthalate

MRF = materials recovery facility
RDF = refuse-derived fuel

PROBLEMS

6-1. Suppose that a proposal is made to hand-sort all the refuse from a town of 100,000. Estimate the manpower and cost required, and describe the problems involved in implementing such a program.

6-2. Estimate the critical speed and horsepower requirements for a trommel screen, 3 m in diameter, processing refuse so as to run 25% full.

6-3. If the angle of a trommel screen were increased (steeper), how would this affect the critical speed, horsepower, and efficiency?

6-4. Aluminum chips of uniform diameter 1.0 in. are to be separated from glass of 100% less than 0.5 in. by use of a single-deck reciprocating screen. The total feed rate is 1 ton/h, with 0.1 ton/h of aluminum and 0.9 ton/h of glass. It is required to produce an aluminum fraction that is 99% pure (by weight). Find the area and size of screen required.

6-5. Using the aluminum/glass mixture specified in Problem 6-4 (and assuming that the smallest glass particle of significance is 0.05 in. in diameter), estimate the air-classifier size and throat velocity required to separate the glass from the aluminum.

6-6. If the pressure drop through a cyclone is assumed to be 1 in. of water, estimate the diameter of cyclone required for various air-flow rates to remove 1-in.-diameter aluminum chips.

6-7. Estimate the air velocity required in an air classifier to just suspend a piece of glass of 0.5 in. in diameter. Suppose that a 0.4 volume fraction of the throat section were occupied by glass. What would the required air velocity be to suspend this glass?

6-8. What density of fluids is needed to separate glass, aluminum, and plastics into the three individual components? What specific fluids might be used for this purpose?

6-9. Two air-classifier manufacturers report the following performance for their units:

Manufacturer A:

Recovery of organics—80%
Recovery of inorganics—80%

Manufacturer B:

Overall recovery—60%
Purity of light fraction—95%

Assume a feed that consists of 80% organics. Compare the performance of the two units.

6-10. A shredded refuse is to be screened to recover the glass fraction. For the purposes of this analysis, the waste to be screened is divided into "glass" and "nonglass" fractions. The material to be screened has 25% by weight of "glass" and 75% by weight "nonglass." The size distribution of each fraction is shown below:

| Particle size (in.) | Percent finer than (%) | |
	Glass	Non-glass
2	100	20
1	100	10
0.5	80	8
0.25	20	5
0.125	5	0

a. Would it be possible to select a screen size such that all (100%) of the glass can be recovered (as screen extract)?

b. If a glass product of minimum 90% pure is to be obtained from the screening in Problem 6-10, what size holes much be used, and what is the recovery at this purity?

6-11. A 2.4-m-diameter trommel screen, rotating at 4 rpm, is 7.6 m long and set at an angle (β) of 15°. The feed rate is 10 tonnes per hour of raw MSW.
a. What is the critical speed?
b. At 4 rpm would you expect it to be cascading, cataracting, or centrifuging?

6-12. A drop test was conducted for paper and plastics. The results are shown in Figure 6-37. In this test the pieces were dropped from a height and their terminal settling velocities calculated. The paper is to be separated from the plastic using an air classifier, with at least 95% purity of the paper. What air velocity should be used in the air classifier to obtain the maximum recovery of paper at the required purity?

| Terminal settling velocity (m/sec) | Percent of particles with terminal settling velocities less than | |
	Paper	Plastic
0.5	0	0
1.0	15	0
1.5	80	25
2.0	100	68
2.5	100	95
3.0	100	100

6-13. A center-fed sink/float apparatus is 2 meters deep and has a fluid with a density of 1.2 g/cm^3 and a viscosity of 0.015 poise. Three plastics are to be separated, with the following results:

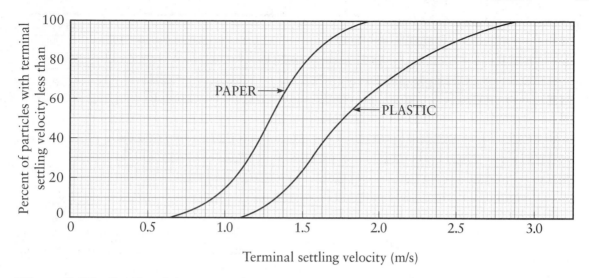

Figure 6-37 Results of drop test with paper and plastic pieces. See Problem 6-12.

Plastic	Feed (kg/h)	Density (g/cm³)	Overflow (kg/h)
A	42	1.4	5
B	38	1.1	35
C	20	0.9	18

a. What is the recovery of plastic B in the overflow?
b. What is the purity of plastic B in the overflow?
c. If plastic B is made of spheres 0.5 mm in diameter, how long would the spheres take to reach the top?

6-14. Two materials, A and B, are to be separated using two unit operations, 1 and 2. The feed has 10 tons/h A and 4 tons/h B.

The split (fraction of material rejected by each operation) is

Material	Split (fraction rejected) by unit operation	
	1	2
A	0.2	0.5
B	0.4	0.8

a. Which sequence, (1→ 2) or (2 → 1), will yield the greatest recovery of material A?
b. What will be purity of material A using that sequence?
c. What will be the efficiency of separation for the entire process train with regard to material A using the Worrell-Stessel equation?

6-15. Plot a curve showing the fraction of feed exiting as extract in a trommel screen versus the speed of the trommel. (Extract is what falls through the holes.)

6-16. The specific gravity of aluminum is about 1.6 and the viscosity of air is about 5×10^{-7} lb-s /ft^2. If a velocity of 1500 feet per minute is required to suspend shredded pieces of an aluminum can, what would be the aerodynamic diameter?

6-17. What feed characteristics and operational variables will make the eddy current separator less effective than it might theoretically be?

6-18. An air classifier receives a feed of organics and inorganics, and the two are to be separated. A preliminary test is conducted with the following results:

Organics in the feed (lb/h)	Inorganics in the feed (lb/h)	Air velocity (ft/sec)	Organics in extract (lb/h)	Inorganics in extract (lb/h)
18	20	20	5	0.5
18	20	40	8	1
18	20	50	12	2
18	20	60	17	3.5
18	20	80	19	6

a. If an extract of minimum 85% is required, what is the highest recovery possible?

b. Using the Worrell-Stessel efficiency definition, what air velocity yields the highest efficiency?

6-19. Three separation unit operations—trommel screen, magnet, and air classifier—are available for use in separating a feed made up of paper, ferrous, and glass. The splits (fraction of each material being rejected in each operation) are as follows:

Material	Feed in tons/ hour	Fraction rejected Trommel screen	Magnet	Air classifier
Paper	40	0.9	0.9	0.1
Ferrous	10	0.5	0.05	0.95
Glass	10	0.1	1.0	0.5

Using the array notation (see pp. 275–276), determine the order and combination of any three of the unit operations if the objective is to produce the highest recovery and purity of ferrous material. You can use one, two, or three unit operations.

6-20. A community collects mixed cans, aluminum and steel. Design a materials separation facility for separating aluminum cans from steel cans. Show the floor plan, including all conveying equipment, storage bins, etc. Analyze the robustness of your facility by assuming that each of the unit operations (including conveyors) goes down, one at a time. Can your facility still function?

6-21. A shredded MSW is to be air-classified in a straight wall air classifier with throat dimension of 18 inches × 18 inches. What should be the maximum capacity of the fan? Assume the air is incompressible and make all other necessary assumptions.

6-22. Explain why a pulsed flow air classifier would have better performance characteristics than a straight vertical tube air classifier. Start your explanation by first defining what is meant by "better performance"?

6-23. Figure 6-38 is a sketch of a clean material recovery facility. The flow rates of the various materials streams are indicated by the numbers, all having units of tons/day.

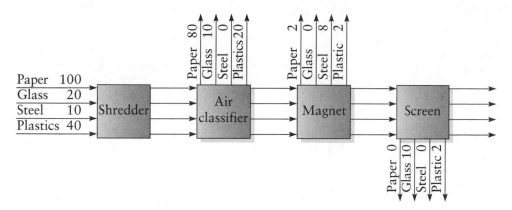

Figure 6-38 Flow of four materials through a materials recovery facility. See Problem 6-23.

a. What percent of steel entering the plant is recovered by the magnet?
b. What is purity of the paper produced by the air classifier?
c. What is the fraction of incoming MSW that ends up in the landfill?
d. Which of the unit operations is a polynary separator?

6-24. Design a materials recovery facility for a mixed municipal solid waste that will have as the end product the following materials: clear class, colored glass, aluminum, ferrous material, and corrugated cardboard. Draw a plan for such a facility, with unit operations clearly labeled. Include on this plan the receiving station and all methods of material transport between unit operations, and indicate the fate of material not collected.

6-25. Construct a matrix array using appropriate splits for the materials separation system consisting of a trommel screen, a magnet, and an air classifier. If the objective is to produce clean ferrous, expressed as both the recovery and purity, which configuration would be chosen? Assume three materials make up the mixed feed: paper, ferrous, and glass. The following are the feed tonnages and the splits (a split is defined as the fraction continuing on, or rejected by the unit operation):

| Material | Tons per hour | Split | | |
		Trommel screen	Magnet	Air classifier
Paper	40	0.90	0.90	0.10
Ferrous	10	0.5	0.05	0.95
Glass	10	0.10	1.00	0.50

6-26. Shredded solid waste was run through a set of sieves with the following results:

Sieve size (inches)	Fraction of feed retained on sieve
10	0.05
8	0.20
4	0.32
2	0.18
1	0.20

What is the characteristic size of this shredded material? (You may wish to use the Rosin-Rammler paper.)

6-27. What is an *eddy current separator* and how does it work?

6-28. Explain in your own words how a jig works.

6-29. The specific gravity of aluminum is about 1.6 and viscosity of air is about 5×10^{-7} lb-sec per ft^2. If a velocity of 1500 feet per minute is required to suspend a piece of aluminum, what would be its aerodynamic diameter?

6-30. Using your knowledge of trommel screen behavior, plot a typical curve showing the percent recovery (extract) versus trommel (a) speed, (b) slope, and (c) feed rate. Start by thinking of the extremes. For example, if the speed is zero rpm as one extreme.

6-31. Design a facility for separating aluminum cans from steel cans. Show all the storage points, unit operations, and means of conveyance. Submit a floor plan and an elevation.

6-32. An industrial materials recovery facility is to separate out two materials, A and B. Air classification is considered as one option for such separation. A drop test is performed on the materials, with the following results:

Terminal settling velocity, m/sec	Percent of particles with falling velocities less than the given velocity	
	A	B
0.25	10	0
0.50	30	5
0.75	85	10
1.00	100	50
1.25	100	80
1.50	100	100

a. Will it be possible to achieve 100% recovery of material A as the extract from the air classifier?

b. What would be the highest purity of material A that can be achieved?

c. At 50% recovery of material A as the extract, what will be the purity of the extract with respect to material A?

CHAPTER

7

Combustion and Energy Recovery

Much of municipal solid waste is combustible, and the destruction of this fraction, coupled with the recovery of energy, is an option in solid waste management. Refuse can be burned as is, or it can be processed to improve its heat value and to make it easier to handle in a combustor. Processed refuse can also be combined with other fuels, such as coal, and co-fired in a heat recovery combustor. About 15% of the total solid waste produced in the United States is processed in waste-to-energy facilities, or a total of over 34 million tons per year. Theoretically, the combustion of refuse produced by a community is sufficient to provide about 20% of the electrical power needs for that community. Refuse therefore is not an insignificant source of power.

HEAT VALUE OF REFUSE

One of the earliest measures of heat energy, still widely used by American engineers, is the *British thermal unit* (Btu), defined as that amount of energy necessary to heat one pound of water one degree Fahrenheit. The internationally accepted unit of energy is the *Joule*. Other common units for energy are the *calorie* and *kilowatt-hour* (kWh)—the former used in natural sciences, the latter in engineering. Table 7-1 shows the conversion factors for all of these units, emphasizing the fact that all are measures of energy, and are interchangeable.

The amount of energy or heat value in an unknown fuel can be estimated by ultimate analysis, compositional analysis, proximate analysis, and calorimetry.

Ultimate Analysis

Ultimate analysis uses the chemical makeup of the fuel to approximate its heat value. The most popular method using ultimate analysis is the DuLong equation, originally developed for estimating the heat value of coal:

$$\text{Btu/lb} = 145\,\text{C} + 620\left(\text{H} - \frac{1}{8}\text{O}\right) + 40\,\text{S}$$

283

where C, H, O, and S are the weight percentages (dry basis) of carbon, hydrogen, oxygen, and sulfur, respectively. The DuLong formula is cumbersome to use in practice, and it does not give acceptable estimates of heat value for materials other than coal.[1]

Table 7-1   Energy Conversions

To convert	to	multiply by
Btu	Calories	252
	Joules	1054
	kWh	2.93×10^{-4}
Calories	Btu	3.97×10^{-3}
	Joules	4.18
	kWh	1.16×10^{-6}
Joules	Btu	9.49×10^{-4}
	Calories	0.239
	kWh	2.78×10^{-7}
Kilowatt-hours	Btu	3413
	Calories	8.62×10^{5}
	Joules	3.6×10^{6}

Another equation for estimating the heat value of refuse, using ultimate analysis, is[2]

$$\text{Btu/lb} = 144\ C + 672\ H + 6.2\ O + 41.4\ S - 10.8\ N$$

where C, H, O, S, and N are again the weight percentages (dry basis) of carbon, hydrogen, oxygen, sulfur, and nitrogen, respectively, in the combustible fraction of the fuel. That is, the sum of all of these percentages has to add to 100%.

The elemental analysis can be conducted using standard methods published by the American Society for Testing and Materials (ASTM).[3] Carbon and hydrogen are measured by burning the sample in a tube furnace and capturing and analyzing the water and the CO_2 produced. Table 7-2 lists some typical values found in mixed municipal solid waste.

Table 7-2   Typical Ultimate Analyses of MSW

Constituent	Average of three different samples of MSW
Carbon, C	51.9
Hydrogen, H	7.0
Oxygen, O	39.6
Sulfur, S	0.37
Nitrogen, N	1.1

Source: (4)
(*Note:* The numbers in this table have been adjusted for zero inerts and zero moisture.)

Compositional Analysis

Formulas based on compositional analyses are an improvement over formulas based on ultimate analyses. One such formula is[5]

$$Btu/lb = 49R + 22.5(G + P) - 3.3W$$

where R = plastics, percent by weight of total MSW, on dry basis
G = food waste, percent by weight of total MSW, on dry basis
P = paper, percent by weight of total MSW, on dry basis
W = water, percent by weight on dry basis (NB!)

Using regression analysis and comparing the results to actual measurements of heat value, an improved form of a compositional model is suggested:[6]

$$Btu/lb = 1238 + 15.6R + 4.4P + 2.7G - 20.7W$$

where R = plastics, percent by weight on dry basis
P = paper, percent by weight on dry basis
G = food wastes, percent by weight on dry basis
W = water, percent by weight on dry basis (NB!)

An even more exact compositional analysis is possible if the composition of the MSW or processed fuel is known. The heat value of the complex fuel can be calculated by using typical heat values of its components, as listed in Table 7-3. The calculation is shown in Example 7-1.

EXAMPLE 7-1

A processed refuse-derived fuel has the following composition:

Component	Fraction by weight, dry basis
Paper	0.50
Food waste	0.10
Plastics	0.30
Glass	0.10

Estimate the heat value based on the typical values in Table 7-3.

$$0.50 (7200) + 0.10 (2000) + 0.30 (14,000) + 0.10 (60) = 8006 \text{ Btu/lb}$$

Table 7-3 Typical Heat Values of MSW Components

Component	Heat value, Btu/lb dry weight
Food waste	2000
Paper	7200
Cardboard	7000
Plastics	14,000
Textiles	7500
Rubber	10,000
Leather	7500
Garden trimmings	2800
Wood	8000
Glass	60
Nonferrous metals	300
Ferrous metals	300
Dirt, ashes, other fines	3000

Source: Modified from (7)

Some of the components of refuse may not always be tuned finely enough in such a compositional analysis. Paper, for example, comes in many categories, and each of these categories has its own heat value, as shown in Table 7-4.

Table 7-4 Heat Value of Various Types of Paper

Type of paper	Heat value, Btu/lb dry weight
Newspaper	7520
Cardboard	6900
Kraft	6900
Beverage and milk boxes	6800
Boxboard	6700
Tissue	6500
Colored office paper	6360
White office paper	6230
Envelopes	6150
Treated paper	6000
Glossy paper	5380

Source: (8)

If the waste composition is known in terms of the wet weight, the fraction of moisture has to be subtracted from the composition fractions. Typical moisture concentrations of MSW components are shown in Table 7-5. The calculation of heat value for a wet sample is shown in Example 7-2.

Table 7-5 Typical Moisture Contents of MSW

Component	Typical moisture, percent
Food waste	70
Paper	6
Cardboard	5
Plastics	2
Textiles	10
Rubber	2
Leather	10
Garden trimmings	60
Wood	60
Glass	2
Nonferrous metals	2
Ferrous metals	3
Dirt, ashes, other fines	8

Source: Adapted from (9)

EXAMPLE 7-2

Using the refuse-derived fuel composition shown in Example 7-1 but on a *wet* weight basis, calculate the heat value of this mixture.

As shown in Table 7-5, 6% of the wet weight of paper is water that does not contribute to the heat value. Hence the first term from the calculation in Example 7-1 is multiplied by 1.0 − 0.06 = 0.94, and so on.

0.50 (7200) 0.94 + 0.10 (2000) 0.30 + 0.30 (14,000) 0.98 + 0.10 (60) 0.98 = 7570 Btu/lb

Proximate Analysis

In proximate analysis it is assumed that the fuel is composed of two types of materials, volatiles and fixed carbon. The amount of volatiles can be estimated by loss of weight when the fuel sample is burned at some elevated temperature such as 600 or 800°C, and the fixed carbon is estimated by the weight loss when the sample is combusted at 950°C. A commonly used proximate analysis equation for estimating the heat value of refuse is

$$\text{Btu/lb} = 8000A + 14{,}500B$$

where A = volatiles, fraction of all dry matter lost at 600°C
B = fixed carbon, fraction of all dry matter lost between 600°C and 950°C

Another form of a proximate analysis equation is

$$\text{Btu/lb} = 2500D - 330W$$

where D = fraction volatile material, dry basis, defined as weight loss at 800°C
W = fraction water, dry basis

Depending on whether the moisture is included or not, the results are presented as Btu/lb as received or Btu/lb dry basis. Table 7-6 lists some representative moisture, volatiles, and fixed carbon fractions for several refuse components.[10]

Table 7-6   Typical Proximate Analysis of MSW Components

| Component | Moisture | Fraction by weight | | |
		Volatile	Fixed	Ash
Mixed paper	0.102	0.759	0.084	0.054
Yard waste	0.752	0.186	0.045	0.016
Food waste	0.783	0.170	0.036	0.010
Polyethylene	0.002	0.985	0.001	0.012
Wood	0.200	0.697	0.113	0.008

Source: (10)
(*Note:* If the values in Table 7-6 are to be used in the proximate analysis equation, the fractions have to be recalculated on the basis of *dry* matter.)

Calorimetry

Calorimetry is the referee method of determining heat value of mixed fuels. Figure 7-1 shows a schematic sketch of a *bomb calorimeter*. The bomb is a stainless steel ball that screws apart. The ball has an empty space inside into which the sample to be combusted is placed. A sample of known weight, such as a small piece of coal, is placed into the bomb, and the two halves are screwed shut. Oxygen under high pressure is then injected into the bomb, and the bomb is placed in an adiabatic water bath, with wires leading from the bomb to a source of electrical current. By means of a spark from the wires, the material in the steel ball combusts, heats the bomb, which in turn heats the water. The temperature rise in the water is measured with a thermometer and recorded as a function of time. Figure 7-2 shows the trace of a typical calorimeter curve.

More accurately, Figure 7-2, a plot of temperature (T) versus time (t), is called a *thermogram*. Because the initial (preperiod) and final (postperiod) slopes of the thermogram may be different, and because the temperature rise due to a chemical reaction is usually nonlinear, it is necessary to decide on a procedure to estimate the true net temperature rise (ΔT_c) in the experiment. A common method is to extrapolate the postperiod slope backward in time and the preperiod slope forward in time (the dotted lines in the figure). The net temperature rise (ΔT_c) can be measured from the difference of these two extrapolated lines at some intermediate time in the

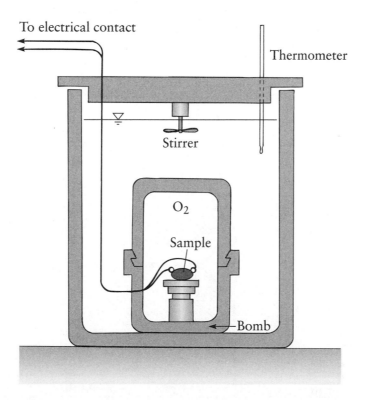

Figure 7-1   *Bomb calorimeter used to measure heat value of a fuel.*

reaction period. The exact location of this intermediate time depends on the calorimeter, and on the reaction being studied. A frequently used criterion for estimating this time is to measure ΔT at the point where the shaded areas in the figure are equal. It can be shown that these shaded areas are equal when the temperature increase is approximately $0.63\Delta T$. From the thermogram, the difference between the initial temperature (T_i) and the final temperature (T_f) is the temperature rise (ΔT) of the reaction.

The water container is well insulated, so no heat is assumed to escape from the system. All of the energy liberated during the combustion is used to heat the water and the stainless steel bomb. The heat energy is calculated as the temperature increase of the water times the mass of the water plus bomb. Since one calorie is defined as the amount of energy necessary to raise the temperature of one gram of water one degree Celsius, and knowing the grams of water in the calorimeter, it is possible to calculate the energy in calories. Knowing the weight of the sample, its heat value can be calculated.

Each calorimeter is different and must be standardized using a material for which the heat of combustion is known precisely. Plus, the heat generated by the combustion of the ignition wire must be taken into account if accurate analyses are required. Typically, benzoic acid is used as the standard. A benzoic acid pellet specially manufactured for this purpose is combusted in a bomb calorimeter, and the

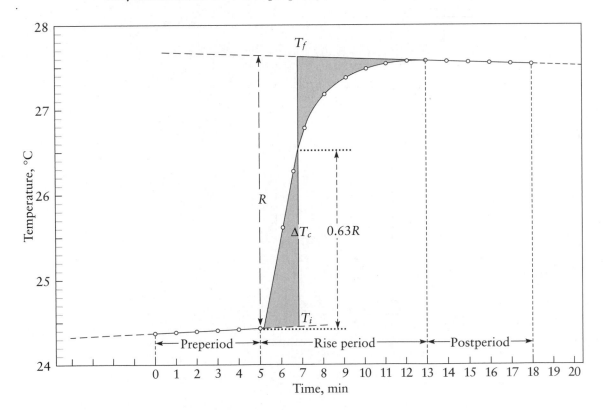

Figure 7-2 Temperature/time trace from a bomb calorimeter.

temperature rise (ΔT) is determined from the thermogram and used to calculate the heat capacity of the calorimeter C_v. That is,

$$C_v = \frac{UM_b}{\Delta T}$$

where C_v = heat capacity of the calorimeter, cal/°C
 U = heat of combustion of benzoic acid, cal/g
 M_b = mass of benzoic acid pellet, g
 ΔT = rise in temperature from thermogram, °C

For very accurate determinations, the heat generated by the combustion of the fuse wire must be included in the calculations.

$$C_v = \frac{\left[6318\,M_b + 1643\,M_w\right]}{\Delta T}$$

where C_v = heat capacity of the calorimeter, cal/°C
 M_b = mass of benzoic acid tablet used, g
 M_w = mass of ignition wire used, g
 ΔT = temperature rise, °C

Note that 6318 is the heat of combustion of benzoic acid expressed as cal/g, and 1643 is the heat of combustion of nickel chromium wire, cal/g.

E X A M P L E
7 - 3

A benzoic acid pellet weighing 5.00 g is placed in a bomb calorimeter, along with 0.20 g fuse wire. The benzoic acid is ignited, and the temperature rise is 3.56 °C. What is the heat capacity of this calorimeter?

$$C_v = \frac{\left[6318(5.00) + 1643(0.2)\right]}{3.56} = 8966\,\text{cal/°C}$$

Knowing the heat capacity of the calorimeter (C_v), the heat generated by the combustion of a material of unknown heat value may be determined by measurement of the temperature rise that occurs upon the combustion of the material:

$$U = \frac{C_v\,\Delta T}{M}$$

where U = heat value of unknown material, cal/g
 ΔT = rise in temperature from thermogram, °C
 M = mass of the unknown material, g
 C_v = heat capacity of the calorimeter, cal/°C

EXAMPLE
7-4

A 10-g sample of a refuse-derived fuel (RDF) is combusted in a calorimeter that has a heat capacity of 8966 cal/°C. The detected temperature rise is 4.72 °C. What is the heat value of this sample?

$$U = \frac{C_v \Delta T}{M} = \frac{8966 \times 4.72}{10.00} = 4231 \text{ cal/g}$$

The conversion from cal/g to Btu/lb is 1.78 × cal/g = Btu/lb, so that this fuel has a heat value of 7531 Btu/lb.

An important aspect of calorimetric heat values is the distinction between *higher heating value* and *lower heating value*. The higher heating value, or HHV, is also called the *gross calorific energy*, while the lower heating value (LHV) is also known as the *net calorific energy*. The distinction is important in design of combustion units.

In a calorimeter, as organic matter combusts, the products of combustion are (ideally) carbon dioxide and water. The water produced is in vapor form. As the bomb cools, however, this water condenses, yielding heat that is measured as part of the temperature rise. The HHV is calculated by including the contribution due to this *latent heat of vaporization*, or the heat required to produce steam from water. But such condensation does not occur in a large furnace. The hot flue gases carry the water vapor outside the furnace, and condensation cannot take place. The LHV is then the HHV minus the latent heat of vaporization that has occurred in the bomb calorimeter. For design purposes, the LHV is a much more realistic number, but American engineers usually express heat values in terms of HHV. When heat values of refuse or other fuels are specified, therefore, it must be clear if the numbers are as HHV or LHV. In this text, as in almost all American engineering practice, heat value refers only to HHV.

Although calorimetry is the referee method of measuring heat value of a fuel, it does not actually simulate the behavior of that fuel in a full-scale combustor. There are at least two reasons why the HHV number overestimates the actual heat value in combustion: the presence of metals and the incomplete combustion of organics.

Some metals, most notably aluminum, will oxidize at sufficiently high temperatures to yield heat. The oxidation of aluminum can be described as

$$4 \text{ Al} + 3O_2 \rightarrow 2 \ (Al_2O_3) + \text{heat}$$

This reaction is highly exothermic, yielding 13,359 Btu/lb (31,070 kJ/kg). If a RDF sample contains a significant amount of aluminum, the measured heat value in the calorimeter will reflect this exothermic reaction because the temperature of combustion in the calorimeter is so high. In a typical full-scale combustor, however, the temperature is not sufficiently high to force the oxidation of aluminum.[11]

The second problem with the use of calorimetric heat values is that while all organic material will oxidize in a calorimeter, this will not occur in a full-scale combustor. The amount of unburned organics can vary from 2% to 25%, depending on the effectiveness of the operation. In one laboratory study, samples of RDF were

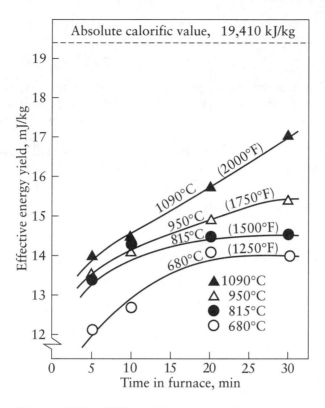

Figure 7-3 MSW combustion requires time to achieve full energy recovery. Source: (25)

combusted for different times in various furnace temperatures and the resulting ash analyzed for calorific value. The results, pictured in Figure 7-3, show that even at reasonable residence times and temperatures, a significant fraction of the fuel is unburned.

Because MSW is such a heterogeneous and unpredictable fuel, engineers often need to have "rules of thumb" for estimating the heat values. For MSW one rule of thumb is that one ton of MSW produces 5000 lb of steam, and this steam produces 500 kilowatts of electricity.

MATERIALS AND THERMAL BALANCES

Combustion Air

The energy from the sun is stored, using the process of photosynthesis, in organic molecules, and this energy is slowly released as the organic materials decompose. Animals, of course, receive their energy from plants by eating them, extracting energy for their maintenance needs as well as cell growth. In simplest of terms, the photosynthesis process is

$$CO_2 + \text{sunlight} + \text{nutrients} + H_2O \rightarrow (HC)_x + O_2$$

The $(HC)_x$ represents an infinite variety of hydrocarbons. The degradation of the high-energy organics is then

$$(HC)_x + O_2 \rightarrow CO_2 + H_2O + \text{nutrients} + \text{heat energy}$$

Combustion of the organic fraction of refuse is simply a very rapid decomposition process. The end products from the combustion of the hydrocarbons are essentially the same as in slower biochemical decomposition that reaches the final low-energy products.

Of interest to engineers is the amount of heat produced in the combustion operation. The heat value of pure materials can be estimated from thermodynamics. For example, the combustion of pure carbon is a two-stage reaction:

$$C + O \rightarrow CO + 10,100 \text{ J/g}$$

and

$$CO + O \rightarrow CO_2 + 22,700 \text{ J/g}$$

although it is usually expressed as a single reaction:

$$C + O_2 \rightarrow CO_2 + 32,800 \text{ J/g}$$

The amount of oxygen necessary to oxidize some hydrocarbon is known as *stoichiometric oxygen*. Consider the simple combustion of carbon:

$$C + O_2 \rightarrow CO_2 + \text{heat}$$

That is, one mole of carbon combines with one mole of molecular oxygen. The molecular weight of carbon is 12 and of oxygen 16, so it takes two oxygens or 32 grams of oxygen to react with 12 grams of carbon. The stoichiometric oxygen is then 2.67 g O_2/ g C.

EXAMPLE 7-5

Calculate the stoichiometric oxygen required for the combustion of methane gas.

The equation is

$$CH_4 + 2O_2 \rightarrow CO_2 + 2H_2O$$

That is, it takes 16 grams of methane (12 + 4) to react with $2 \times 2 \times 16 = 64$ moles of oxygen. (The molecular weight is 16, there are two oxygens, and there are two moles.) Thus the stoichiometric oxygen required for the combustion of methane is 64/16 = 4 g O_2/g CH_4.

Normally, refuse is not burned using pure oxygen, however, and air is used as the source of oxygen. A correction factor is therefore needed to recognize that the air is 23.15% oxygen by weight. Dividing the stoichiometric oxygen by 0.2315 yields *stoichiometric air*.

E X A M P L E
7 - 6

Calculate the stoichiometric air required for the combustion of methane gas.

Since the stoichiometric oxygen, from Example 7-5, is 4 g O_2/g CH_4, the stoichiometric air requirement is 4/0.2315 = 17.3 g air/g methane.

Efficiency

Figure 7-4 illustrates how a power plant operates. Water is heated to steam in a boiler, and the steam is used to turn a turbine, which drives a generator. This diagram can be simplified to a simple energy balance where *energy in* has to equal *energy out* (energy wasted in the conversion + useful energy) plus the energy accumulated in the box. Expressed as an equation:

$$
\begin{bmatrix} \text{Rate of} \\ \text{energy} \\ \text{ACCUMULATED} \end{bmatrix} = \begin{bmatrix} \text{Rate of} \\ \text{energy} \\ \text{IN} \end{bmatrix} - \begin{bmatrix} \text{Rate of} \\ \text{energy} \\ \text{OUT} \end{bmatrix} + \begin{bmatrix} \text{Rate of} \\ \text{energy} \\ \text{PRODUCED} \end{bmatrix} - \begin{bmatrix} \text{Rate of} \\ \text{energy} \\ \text{CONSUMED} \end{bmatrix}
$$

Of course, energy is never produced or consumed in the strict sense; it is simply changed in form.

Just as processes involving materials can be studied in their *steady-state* condition, defined as no change occurring over time, energy systems can also be in steady state. If there is no change over time, there cannot be a continuous accumulation of energy, and the equation must read

[Rate of energy IN] = [Rate of energy OUT]

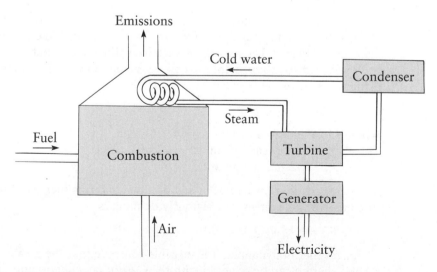

Figure 7-4   Operation of a fossil fuel power plant.

or if some of the energy out is useful and the rest is wasted,

$$\begin{bmatrix} \text{Rate of} \\ \text{energy} \\ \text{IN} \end{bmatrix} = \begin{bmatrix} \text{Rate of} \\ \text{energy} \\ \text{USED} \end{bmatrix} + \begin{bmatrix} \text{Rate of} \\ \text{energy} \\ \text{WASTED} \end{bmatrix}$$

If the input and useful output from a black box are known, the efficiency of the process can be calculated as

$$E = \frac{\text{Energy USED}}{\text{Energy IN}} \times 100$$

where E = efficiency (%).

EXAMPLE 7-7

A coal-fired power plant uses 1000 Mg (megagrams, or 1000 kg, commonly called a metric tonne) of coal per day. The energy value of the coal is 28,000 kJ/kg (kiloJoules/kilogram). The plant produces 2.8×10^6 kWh of electricity each day. What is the efficiency of the power plant?

Energy In = (28,000 kJ/kg $\times$ 1000 Mg/day) $\times$ (1×10^3 kg/Mg)

= 28×10^9 kJ/day

Useful Energy Out =

2.8×10^6 kWh/day $\times$ (3.6×10^6) J/kWh $\times$ 10^{-3} kJ/J

= 10.1×10^9 kJ/day

Efficiency(%) = $(10.1 \times 10^9)/(28 \times 10^9) \times 100 = 36\%$

One of the most distressing problems in the production of electricity from MSW—or from any fossil fuel for that matter—is that the power plants are less than 40% efficient. The reason for such low efficiency is that the waste steam must be condensed to water before it can again be converted to high-pressure steam. Using the previous energy balance:

$$0 = Q_0 - Q_U - Q_W$$

where Q_0 = energy flow in
Q_U = useful energy out
Q_W = wasted energy out

Note that energy flow can be in any number of terms such as kJ/sec or Btu/hr. The efficiency of this system, as previously defined, is

$$E(\%) = (Q_U)/(Q_0) \times 100$$

From thermodynamics, it is possible to prove that the greatest efficiency (least wasted energy) can be achieved with the *Carnot heat engine* and that this efficiency is determined by the absolute temperature of the surroundings. The efficiency of the Carnot heat engine is defined as

$$E_c\left(\%\right) = \frac{T_1 - T_0}{T_1} \times 100$$

where E_c = Carnot efficiency, %
  T_1 = absolute temperature of the boiler, °K = °C + 273
  T_0 = absolute temperature of the condenser (cooling water), K°

Since this is the best possible, any real-world system must be less efficient, or

$$E \le E_c$$

$$\frac{Q_U}{Q_0} \le \frac{T_1 - T_0}{T_1}$$

Modern boilers generate steam with temperatures as high as 600°C, while environmental restrictions limit condenser water temperature to about 20°C. Thus the best expected efficiency is

$$E_c\left(\%\right) = \frac{\left(600 + 273\right) - \left(20 + 273\right)}{\left(600 + 273\right)} \times 100 = 66\%$$

A power plant also has losses in energy due to hot stack gases, evaporation, friction losses, etc. The best plants so far have been hovering around 40% efficiency. When various energy losses are subtracted from nuclear powered plants, the efficiencies for these plants seem to be even lower than for fossil fuel plants.

The above analysis of combustion efficiency is applicable to situations where the fuel costs money, and therefore the efficiency is defined as the power produced relative to the cost of the fuel. When the cost of the fuel is a major component of the operating cost of a power plant, then a higher efficiency leads to higher profitability. As the prices of oil and natural gas continue to increase, this is even more important.

A solid waste–fueled plant is different, however. People actually pay the plant to take fuel (solid waste). The more inefficient a plant is, the more solid waste is needed, and the tipping fee revenue increases. The solid waste combustor has two jobs—it is to produce useful power, and it is to reduce the volume of the waste. Because people actually pay the power plant to take the fuel, the usual expressions of efficiency are not applicable to solid waste combustors.

Thermal Balance on a Waste-to-Energy Combustor

Energy (of whatever kind), when expressed in common units, can be pictured as a quantity that flows, and thus it is possible to analyze energy flows using the same concepts used for materials flows and balances. As before, a black box is any process or operation into which certain flows enter and others leave. If all of the flows can be correctly accounted for, then they must balance.

A thermal balance on a large combustion unit is difficult because much of the heat cannot be accurately measured. For instructional purposes, a black box can be used to describe the thermal balance, as shown in Figure 7-5. Assume that in this facility the heat is recovered as steam.

The heat input to this black box is from heat value in the fuel and the heat in the water entering the water-wall pipes. The output is the sensible heat in the stack gases, the latent heat of water, the heat in the ashes, the heat in the steam, and the heat lost due to radiation. If the black box is thought of in dynamic terms, the balance would be

$$
\begin{bmatrix} \text{Rate of} \\ \text{heat} \\ \text{accumulated} \end{bmatrix} = \begin{bmatrix} \text{Rate of} \\ \text{heat in} \\ \text{the fuel} \end{bmatrix} + \begin{bmatrix} \text{Rate of} \\ \text{heat in} \\ \text{the water} \end{bmatrix}
$$

$$
- \begin{bmatrix} \text{Rate of} \\ \text{heat out} \\ \text{in the} \\ \text{stack gases} \end{bmatrix} - \begin{bmatrix} \text{Rate of} \\ \text{heat out} \\ \text{in the} \\ \text{steam} \end{bmatrix}
$$

$$
- \begin{bmatrix} \text{Rate of heat out} \\ \text{as latent} \\ \text{heat of vaporization} \end{bmatrix} - \begin{bmatrix} \text{Rate of} \\ \text{heat out} \\ \text{in the ash} \end{bmatrix} - \begin{bmatrix} \text{Rate of} \\ \text{heat loss due} \\ \text{to radiation} \end{bmatrix}
$$

If the process is in steady state, the first term (accumulation) is zero, and the equation can be balanced. The best way to illustrate this is with an example.

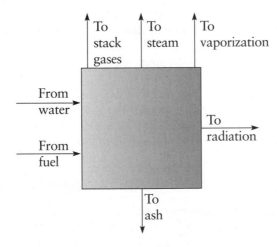

Figure 7-5   Black box showing energy flow in a combustor.

EXAMPLE
7-8

A combustion unit is burning RDF consisting of 85% organics, 10% water, and 5% inorganics (inerts) at a rate of 1000 kg/h. Assume the heat value of the fuel as 19,000 kJ/kg on a moisture-free basis. Assume the facility is refractory lined with no water wall and no heat recovery. Assume that the air flow is 10,000 kg/h and that the under- and overfire air contributes negligible heat. Assume further that 5% of the heat input is lost due to radiation and that 10% of the fuel remains uncombusted in the ash, which exits the combustion chamber at 800°C. The specific heat of ash is 0.837 kJ/kg/°C, and the specific heat of air is 1.0 kJ/kg/°C. What is the temperature of the stack gases?

(a) Heat from the combustion is

1000 kg/h × 0.85 × 19,000 kJ/kg = 16,150,000 kJ/h

(b) The heat due to water vaporization is

1000 kg/h × 0.10 × 2575 kJ/kg = 257,500 kJ/h

where 0.10 is the 10% water in the RDF and 2575 kJ/kg is the latent heat of vaporization for water.

(c) The heat loss due to radiation is

$0.05 \times 16.15 \times 10^6 = 807,500$ kJ/h

(d) The heat lost in the ashes is both the sensible heat in the ashes as well as the unburned fuel. The total amount of ash is the inerts plus the unburned organics or

[1000 kg/h × 0.05] + [1000 kg/h × 0.85 x 0.1] = 135 kg/h

If the temperature of the ash is 800°C, the heat loss due to the hot ash is

800°C × 135 kg/h × 0.837 kJ/kg/°C = 90,396 kJ/h

where 0.837 kJ/kg/°C is the specific heat of ash.

(e) The heat lost in the stack gases is then calculated by subtraction, or

$0 = 16,150,000 - 257,500 - 807,000 - 90,396 - X$

or $X \approx 15,000,000$ kJ/h. If the total air is 10,000 kg/h, and if the specific heat of air is 1.0 kJ/kg/°C, the temperature of the stack gases is

$$\frac{15,000,000 \text{ kJ/hr}}{10,000 \text{ kg/hr } 1.0 \text{ kJ/kg/°C}} = 1500°C$$

which is about 2700°F, or quite high.[31] These hot gases could be run through a heat exchanger to produce steam as a valuable product of the combustion process.

When purchasing a waste-to-energy plant, some municipalities have tried to have the manufacturer guarantee the electricity output per ton of waste. In such cases the manufacturer then requires the municipality to have a certain minimum heat value of its solid waste. No municipality can guarantee the heat value of its solid waste, and thus process efficiencies must be analyzed on the basis of a thermal balance. But the problem with this approach is that calculating the thermal balance is very difficult with such a variable fuel as solid waste, and the cost of monitoring and sampling the waste is significant. One alternative is to require the manufacturer to meet two criteria: (1) ash must not exceed a percent combustible level, and (2) exhaust gas in the stack must be within a predetermined temperature range. These two criteria ensure complete combustion of the solid waste and recovery of the heat, and both criteria can be easily monitored.

COMBUSTION HARDWARE USED FOR MSW

In years past when refuse was burned without recovering energy, the units were known as *incinerators*, a name no longer used by the industry because of the sorry record of these facilities. Poor design, inadequate engineering, and inept operation combined to produce an ash still high in organics, and smoke that even in the days of little industrial air pollution control caused many communities to shut down the incinerators.

Without energy recovery, the exhaust gas from these units was very hot. Typical air pollution control for particulates consisted of a dry cyclone. As the requirement for particulate control increased, electrostatic precipitators (ESP) were required for control of particulates, but the hot exhaust gas exceeded the acceptable inlet temperature for ESP. If a wet cyclone were used prior to the ESP to cool the gas, the moisture carryover would cause severe corrosion problems in the ESP. It became very difficult to upgrade old incinerators and most of them were shut down.

Waste-to-Energy Combustors

Modern combustors combine solid waste combustion with energy recovery (Figure 7-6). Such combustors have a *storage pit* for storing and sorting the incoming refuse (Figure 7-7), a *crane* for charging the combustion box, a *combustion chamber* consisting of bottom grates on which the combustion occurs (Figure 7-8), the *furnace* or *combustion chamber*, the *heat recovery system* of pipes in which water is turned to steam (Figure 7-9), the *ash-handling system* and the *air-pollution-control system*. The units take their name from the fact that the combustion chamber is lined with refractory bricks that are heat resistant, very much like the firebricks in the home fireplace. These units produce steam in a *boiler* located at the top of the combustion chamber.

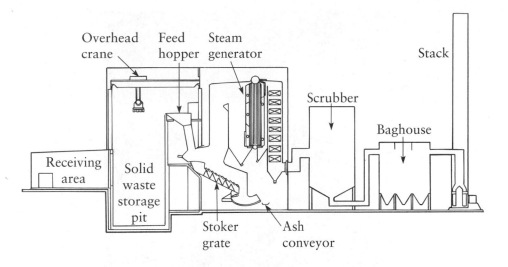

Figure 7-6 A typical municipal solid waste combustor.

The heart of the combustion process is the chamber in which the combustion occurs. In most units the refuse is moved through the combustion chamber on a moving grate, and the design and operation of these grates often determines the

Figure 7-7   *Storage pit and crane in a typical MSW combustor.*

success or failure of the entire process. The functions of the grates are to provide turbulence so that the MSW can be thoroughly burned, to move the refuse down and through the combustion chamber, and finally, to provide *underfire air* to the refuse through openings in the grates. The underfire air both assists in the combustion as well as cools the grates. The control of underfire air is also the most important variable in maintaining a desired operating temperature in the combustion chamber. Most refuse combustors operate in the range of 1800 to 2000°F (980 to 1090 °C), which ensures good combustion and elimination of odors, and is still sufficiently low to protect the refractory materials lining the combustion chamber.

The temperature within the combustion chamber is critical for successful operation. If it is too low, say below 1400°F (770 °C), then many of the plastics will not burn, resulting in poor combustion. Above 2000°F (1090 °C) the refractories in the furnace will not be able to stand the heat. Thus the window for effective operation is not large, and close control needs to be kept on the charge to the combustion chamber and the amount of overfire air and underfire air.

(a)

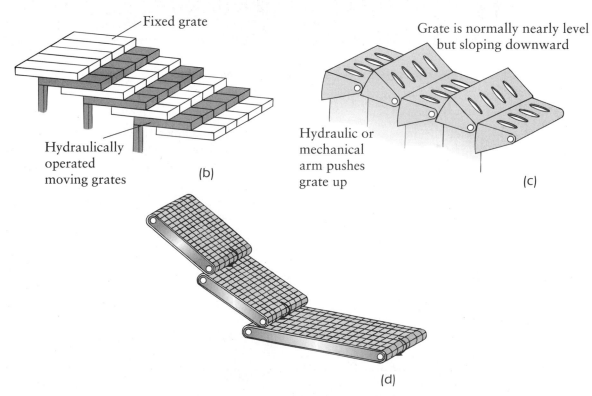

Fixed grate

Hydraulically
operated
moving grates

(b)

Grate is normally nearly level
but sloping downward

Hydraulic or
mechanical
arm pushes
grate up

(c)

(d)

Figure 7-8 *Grates in a MSW combustor. The underfire air is blown through the holes in the grates. The drawings show three types of grates: (b) reciprocating, (c) rocking, (d) traveling.*

Figure 7-9 Boiler tubes for steam generation.

As the amount of excess air is increased, the temperature drops. The relationship between the availability of air and the temperature in the combustion chamber is shown in Figure 7-10. Note that at stoichiometric air quantities the temperature

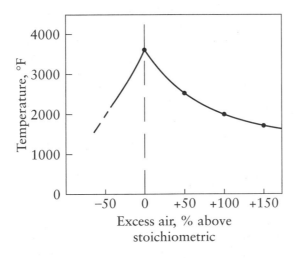

Figure 7-10 Excess underfire air and temperature relationship in MSW combustion.

increases to intolerable levels. As shown in Figure 7-10, maintaining a temperature of 2000°F (1090 °C) in the combustion chamber requires about 100% excess air. Also note that the temperature drops if the unit is operated in the starved air mode, providing less than stoichiometric air to the combustion unit. Such combustors are discussed below.

The air blown into the combustion chamber above the refuse is logically *over-fire air*, and its purpose is to provide the oxygen necessary for combustion as well as to enhance the turbulence in the combustion chamber.

A modification of the combustion chamber is the *rotary kiln*, shown in Figure 7-11. In this unit the refuse travels down an ignition grate by gravity and into a rotating kiln, where the combustion occurs. Rotary kilns provide the most turbulence of any grate system and thereby enhance the rate and completion of combustion.

The furnace walls of modern combustors are lined with metal tubing through which water is circulated. This *water wall* is then part of the boiler or heat recovery system. The water tubes protect the combustion chamber housing by transferring the heat into the water. Figure 7-12a shows how the water wall is placed in the furnace. Figure 7-12b shows a close-up photograph of a water wall in an MSW combustor.

The steam generating system has been a source of problems with solid waste fuels, in particular the generation of superheated steam, which is needed when generating electricity. The superheated steam, between 600 and 950°F (330 and 530°C), allows the use of multistage high efficiency turbine-generators. The exhaust gas from the combustion unit is still very hot when it encounters the superheater

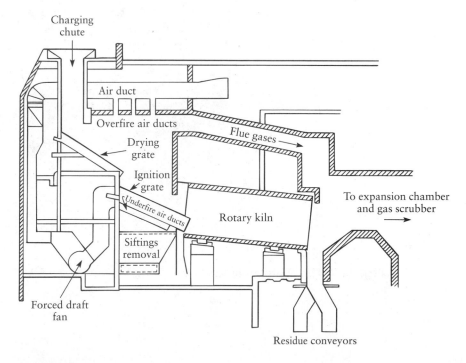

Figure 7-11   Rotary kiln.

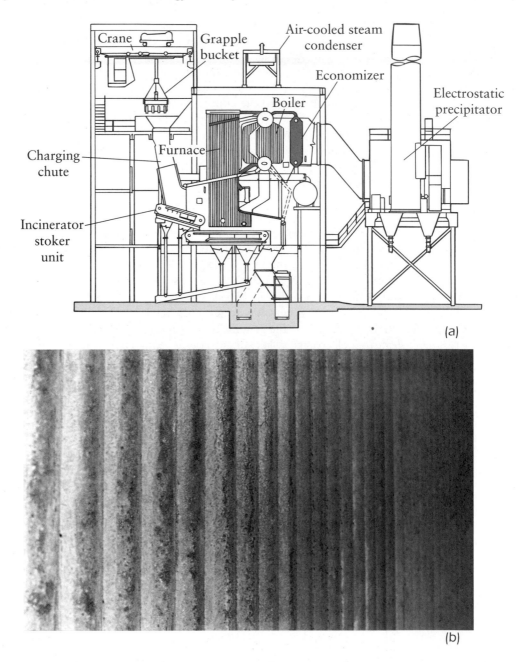

(a)

(b)

Figure 7-12 Water-wall tubes lining the furnace of an MSW combustor.

tubes, causing rapid corrosion of the tubes. These corrosion problems have required the retrofitting of refractory materials on the tubes, decreasing the efficiency of heat transfer.

The efficiency of steam production also depends on the quality of the fuel. As shown in Table 7-7, the production of steam drops dramatically with increases in moisture in the fuel as well as the fraction of noncombustibles.

Table 7-7   Steam Production as Related to Quality of MSW as a Fuel

	As-received heat value, Btu/lb				
	6500	6000	5000	4000	3000
Refuse					
Percent moisture	15	18	25	32	39
Percent noncombustible	14	16	20	24	28
Percent combustible	71	66	55	44	33
Steam generated, tons/ton refuse	4.3	3.9	3.2	2.3	1.5

Source: (1)

Modular Starved Air Combustors

A third type of combustion unit is the *modular starved air combustor*, shown in Figure 7-13. These units are characterized by a two-stage combustion system, with the first stage being operated in a starved air mode, producing a large quantity of suspended carbon, which is then burned using a fossil fuel in the second stage. These units are batch fed using a double-ram system shown in Figure 7-14, and range in size from 15 to 100 tons per day. Typically, such units do not incorporate heat recovery systems.

Modular combustors are useful in cases where the waste has to be combusted but the quantity is insufficient to warrant the construction of a large refractory-lined or water-wall combustor. Modular units are also flexible in that more units can be purchased as the need arises. For example, if the average production of solid waste is 100 tons per day, three 50-ton/day units can be purchased, with one remaining idle as a standby and for scheduled maintenance. One of the most widely used applications of these units is in the destruction of some hazardous materials, such as biohazards waste from hospitals.

Pyrolysis

Pyrolysis is destructive distillation, or combustion in the absence of oxygen. The products of pyrolysis include a solid, a liquid, and a gas. In true pyrolysis, heat is added to the complex organic feed. For example, if the feed material is pure cellulose, the unbalanced reaction might be

$$C_6H_{10}O_5 + \text{heat energy} \rightarrow CH_4 + H_2 + CO_2 + C_2H_4 \text{ (ethylene)} + C + H_2O$$

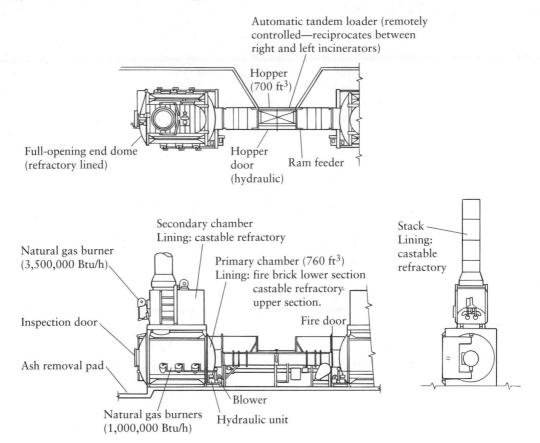

Figure 7-13 Modular combustor.

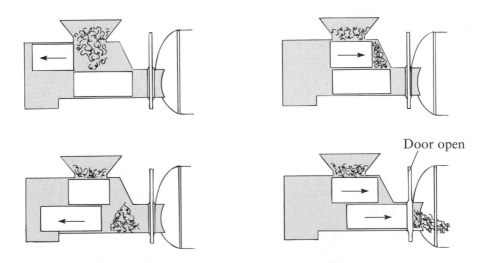

Figure 7-14 Ram feed system for a modular incinerator.

The solid is the carbon, the liquid is ethylene, and the gas is methane. All are fuels for subsequent use.

A modification of pyrolysis is *gasification*, in which a limited quantity of oxygen is introduced, as pure oxygen or as air, and the resulting oxidation produces enough heat to make the system self-sustaining. Thus the reaction can be exo- or endothermic, depending on the amount of heat and oxygen added.

The two important variables in a pyrolysis system are heating rates (how rapidly the fuel is brought to a high temperature) and the final temperature. The ranges of these are listed in Table 7-8.

Table 7-8   Primary Variables in Pyrolysis

Heating rate	°C/sec	Temperature	°C
Slow	<1	Low	500 to 750
Intermediate	5 to 100	Intermediate	750 to 1000
Rapid	500 to 10^6	High	1000 to 1200
Flash	>10^6	Very high	>1200

Source: (15)

The choices of these variables, heating rate and temperature, determine the products obtained from the pyrolysis system, as shown in Figure 7-15. At very high temperatures and slow heating, the product is mostly gas, while at very slow heating rates and low temperature, mostly solid product results.[12]

Pyrolysis (and gasification) has a lot to recommend it theoretically.[13] The process is environmentally excellent, producing little pollution, and it results in the production of various useful fuels. It would seem therefore that pyrolyzing MSW would be an ideal application. Unfortunately, pyrolysis has had a sorry history with MSW. Large facilities were constructed in the 1970s to produce both liquid fuel (oil from garbage!)[14] and solid fuels, and all of these facilities failed because of operating problems. The pyrolysis of predictable and homogeneous fuels, such as sugarcane bagasse, would seem reasonable and logical, and these units are successful. The heterogenous and unpredictable nature of MSW has resulted in numerous failures. Nevertheless, the process has its supporters, especially since gasification plants are able to produce a highly valuable and useful gaseous fuel.[15] An advantage of gasification systems is that they appear to be able to readily meet the air emission requirements for solid waste combustion, including the strict dioxin standards (see below). Nevertheless, not a single unit has yet to be successfully field tested in full scale, and it may well be that we will continue to see one failure after another as we learn that this is not the proper technology for processing municipal solid waste.

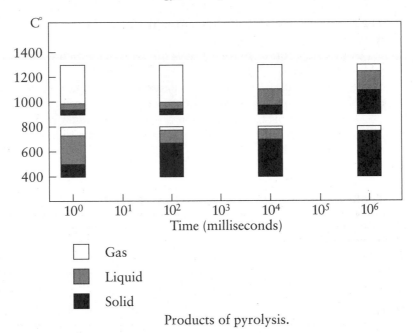

Products of pyrolysis.

Figure 7-15 Effect of temperature and time in the formation of pyrolysis products. Source: (12)

Mass Burn versus RDF

Combustion systems are characterized as either *mass burn* units or *refuse-derived fuel* (RDF) units. A mass burn unit has no preprocessing of the solid waste prior to being fed into the combustion unit. The solid waste is loaded into a feed chute, which leads to the grates, and the solid waste in the chute provides an air lock, which allows the amount of combustion air to be controlled. From the chute the solid waste moves directly onto the combustion grate.[16]

In an RDF system the solid waste is processed prior to combustion to remove noncombustible items and to reduce the size of the combustible fraction, thus producing a more uniform fuel at a higher heat value. The RDF is fed through a rotary feeder and injected into the combustion unit above the grate. Some combustion takes place above the grate with the remaining combustion occurring on the grate. This is referred to as *semi-suspension firing*. If no grate exists and all combustion occurs in the air, this would be called *suspension firing*. Suspension firing is common when burning pulverized coal, but is not generally applicable to RDF.

The advantage of an RDF plant is that the heat value of the fuel is more uniform and thus the amount of excess air required for combustion is reduced. The amount of combustion air used is important because if there is insufficient oxygen in the combustion chamber, a reducing atmosphere is created, which leads to corrosion problems. For RDF systems the excess air (above stoichiometric requirements) is about 50%, while in mass burn plants, because of the large variation in fuel value between items, about 100% excess air is needed. For the same amount of fuel a mass burn plant requires more air and larger air-pollution-control devices. In addition, by preprocessing the solid waste, some of the potential problem items—such as batteries that contain mercury—can be removed.[17]

While there appear to be several theoretical advantages of RDF over mass burn plants, they have had their share of operating problems. Processing of solid waste is not easy, and RDF plants have encountered corrosion and erosion problems. But with RDF the total amount of combustion air needed is sharply reduced, and the cost of emission controls (which could be as much as 100% of the operating cost of the combustion process) is considerably reduced. Whether there is an advantage to incurring the additional cost related to the production of RDF over mass burn is still unclear.

What is clear is that if the RDF is produced from mixed paper that has been sorted as part of the recycling effort, the fuel is a valuable resource. The characteristics of shredded mixed paper as a fuel are shown in Table 7-9. The data in this table are the average values obtained from 12 different communities.[18]

Table 7-9   Characteristics of Mixed Paper as a Refuse-Derived Fuel

Component	Average as percent of total weight
Moisture	7.3
Ash	9.9
Volatile	82.8
Fixed carbon	7.1
Carbon	41.1
Hydrogen	6.2
Nitrogen	0.1
Sulfur	0.1
Chloride	0.1
	Average concentration as parts per million
Arsenic	0.48
Barium	46.21
Beryllium	0.70
Cadmium	0.55
Cobalt	6.78
Chromium	6.48
Copper	18.12
Manganese	27.34
Mercury	0.05
Molybdenum	7.42
Nickel	6.92
Lead	7.51
Antimony	3.88
Selenium	0.06
Tin	7.92
Zinc	149.21

Source: (18)

The heat value of mixed paper is surprisingly high, considering the high ash content. Overall, the HHV of mixed paper is about 7200 Btu/lb (16,700 kJ/kg).[18] The amount of heat value contributed by various components of mixed paper varies, as shown previously in Table 7-4. In addition to mixed papers, some other secondary fuels can become valuable sources of energy, as shown in Table 7-10.

Table 7-10 Combustion Properties of Various Fuels

Fuel	Percent moisture	Heat value, Btu/lb	Percent ash
Bark and sawdust	55	3160	2.0
Cardboard boxes	5	7180	3.6
Carpet edges	1	11,600	22.1
Peat	35	5800	5.0
Tires	1	16,200	1.8
Wood chips	38	5230	0.2

To convert from Btu/lb to kJ/kg, multiply by 2.32.
Source: (19)

The American Society for Testing and Materials (ASTM) has developed designations of refuse-derived fuels, as shown in Table 7-11. RDF-1 is mixed refuse without any processing, while RDF-2 is refuse that has been shredded. This homogenizes the fuel and makes it much easier to handle in combustion. RDF-3 is shredded refuse from which most of the inorganic materials have been removed. Such a fuel would be produced in a typical materials recovery facility (MRF), which might process mostly preseparated solid waste. Further shredding into a *fluff* (RDF-4) (Figure 7-16) or pelletized into dog-food-sized pellets (RDF-5) further improves the usefulness of the fuel (Figure 7-17).[20] The last categories, RDF-6 and 7, have been tried on a pilot basis but have not been found to be successful at full-scale plants.

Table 7-11 ASTM Refuse-Derived Fuel Designations

Name	Description
RDF-1	Unprocessed MSW (the mass burn option)
RDF-2	MSW shredded but no separation of materials
RDF-3	Organic fraction of shredded MSW. This is usually produced in a materials recovery facility (MRF) or from source-separated organics such as newsprint.
RDF-4	Organic waste produced by a MRF that has been further shredded into a fine, almost powder, form, sometimes called *fluff*
RDF-5	Organic waste produced by a MRF that has been densified by a pelletizer or a similar device. These pellets can often be fired with coal in existing furnaces.
RDF-6	Organic fraction of the waste that has been further processed into a liquid fuel such as oil
RDF-7	Organic waste processed into a gaseous fuel

Figure 7-16   Fluff, or finely shredded RDF.

Figure 7-17   Pellets used as RDF.

One advantage of RDF over mass burn is that the fuel can be stored in large bulk storage containers such as shown in Figure 7-18 and burned as needed, unlike the raw MSW in mass burn facilities where the storage time is limited.

UNDESIRABLE EFFECTS OF COMBUSTION

The combustion of fuel such as RDF can have undesirable side effects. In this discussion we cover the waste heat generated by a power plant, the ash requiring disposal, the materials that can escape in the stack and cause air pollution, and the production of carbon dioxide.

Waste Heat

MSW is a low-grade fuel that can be used for the production of steam. This steam is sufficiently useful for driving turbines, but the remaining steam has little industrial use, unless it is located sufficiently close to buildings to use it for heating. Often the residual steam has to be condensed to water, and the water is either cooled and used again in the power plant or disposed of into the environment as hot water.

Figure 7-18   Storage bin for RDF at a combustion facility.

Usually, the boiler water is treated and reused because it is too expensive to be used only once. Small amounts—less than 10%—are blowdown (fresh water added to the recycle) to keep dissolved solids low. If the hot water is discharged into a watercourse, it can cause serious deleterious ecological effects in streams, rivers, and estuaries, so heat discharges are governed by environmental regulations. Typically, the limit on heat discharges is that the temperature of the receiving water cannot be raised by more than one degree Celsius.

The calculation of the temperature of a receiving body of water is straightforward. Heat energy is easy to analyze by energy balances since the quantity of heat energy in a material is simply its mass times its absolute temperature. This is true if the heat capacity is independent of temperature. In particular, if phase changes do not occur, as in the conversion of water to steam, an energy balance for heat energy in terms of the quantity of heat is

$$\begin{bmatrix} \text{Heat} \\ \text{energy} \end{bmatrix} = \begin{bmatrix} \text{Mass of} \\ \text{material} \end{bmatrix} \times \begin{bmatrix} \text{Absolute temperature} \\ \text{of the material} \end{bmatrix}$$

Energy flows are analogous to mass flows. When two heat energy flows are combined, for example, the temperature of the resulting flow at equilibrium is calculated using the energy balance,

$$0 = (\text{Heat energy IN}) - (\text{Heat energy OUT}) + 0 - 0$$

or stated another way,

$$0 = (T_1 Q_1 + T_2 Q_2) - (T_3 Q_3)$$

or

$$T_3 = \frac{T_1 Q_1 + T_2 Q_2}{Q_3}$$

where T = absolute temperature
Q = flow, mass/unit time (or volume constant density)
1 and 2 = input streams
3 = output stream

The mass/volume balance is

$$Q_3 = Q_1 + Q_2$$

Although strict thermodynamics requires that the temperature be expressed in *absolute* terms, in the conversion from Celsius (C) to Kelvin (K) the 273 simply cancels out, and T can be conveniently expressed in degrees C. Recall that $0°C = 273°K$.

E X A M P L E
7 - 9

A coal-fired power plant discharges 3 m³/sec of cooling water at 80°C into a river, which has a flow of 15 m³/sec and a temperature of 20°C. What will be the temperature in the river immediately below the discharge?

The confluence of the river and cooling water can be thought of as a black box, and

$$T_3 = \frac{T_1 Q_1 + T_2 Q_2}{Q_3}$$

$$T_3 = \frac{\left[(80 + 273)(3)\right] + \left[(20 + 273)(15)\right]}{(3 + 15)} = 303°\text{K}$$

or $303 - 273 = 30°\text{C}$. Note that the use of absolute temperatures is not necessary since the 273 cancels out (the right-hand side of the equation is really 30 + 273).

Since most states restrict thermal discharges to less than 1°C rise above ambient stream temperature, the heat in the cooling water must be dissipated into the atmosphere before the cooling water is discharged. Various means are used for dissipating this energy, including large shallow ponds and cooling towers. A cutaway drawing of a typical cooling tower is shown in Figure 7-19. Cooling towers represent a substantial additional cost to the generation of electricity.

Even with this expense, watercourses immediately below cooling water discharges are often significantly warmer than normal. This results in the absence of ice during hard winters, and the growth of immense fish. Stories about the size of fish caught in artificially warmed streams and lakes abound, and these places become not only favorite fishing sites for people, but winter roosting places for birds. Wild animals similarly use the unfrozen water during winter when other surface waters are frozen. The heat clearly changes the aquatic and terrestrial ecosystem, and some would claim that this change is for the better. Everyone seems to benefit from having the warm water. Yet changes in aquatic ecosystems are often unpredictable and potentially disastrous. Heat can increase the chances of various types of disease in fish, and heat will certainly restrict the types of fish that can exist in the warm water. Many cold-water fish, such as trout, cannot spawn in warmer water and they will die out, their place taken by fish that can survive, such as catfish and carp. It is unclear what the net environmental effect is due to governmental restrictions on thermal discharges such as "no more than 1°C rise in temperature."

Ash

Power plants produce both *bottom ash* and *fly ash*. Bottom ash is recovered from the combustion chamber and consists of the inorganic material as well as some unburned organics while fly ash is the particulates removed from the gaseous emis-

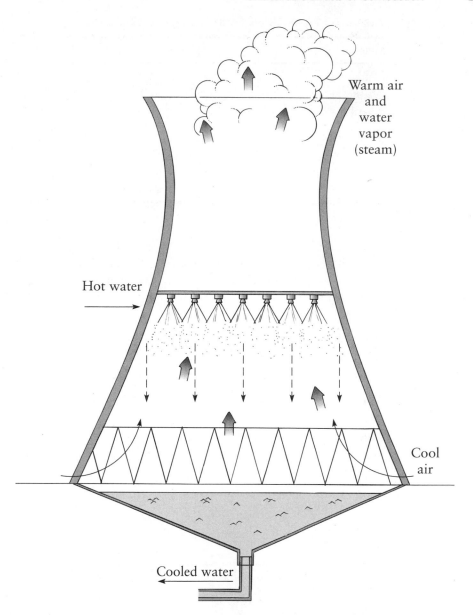

Figure 7-19 Cooling towers typically used in fossil fuel and solid waste combustion facilities.

sions. Considering both types of ash together, MSW combustion typically achieves 75% reduction in material by weight (and 95% reduction by volume). Thus 25% of the original mass is ash, with a high density, of about 1200 to 1800 lb/yd^3. The materials in typical MSW ash are shown in Table 7-12.

Table 7-12 Materials Found in Typical MSW Ash

Material	Percent by weight
Metals	16.1
Combustibles	4.0
Ferrous metal	18.3
Nonferrous metal	2.7
Glass	26.2
Ceramics	8.3
Mineral, ash, other	24.1

Source: (21)

The major problem with ash from MSW combustion is the presence of heavy metals. Table 7-13 shows a representative array of some heavy metals found in combined fly ash and bottom ash from a MSW waste-to-energy unit.

Table 7-13 Total Metal in Combined Ash

Metal	mg/kg of ash by weight
Aluminum	17,800
Calcium	33,600
Sodium	3,800
Iron	20,400
Lead	3,100
Cadmium	35
Zinc	4,100
Manganese	500
Mercury	less than 3

Source: Modified from (22)

Ash from MSW combustion comes perilously close to being classified as a hazardous waste by the EPA. The test usually used is an extraction procedure where the ash is shaken with a solvent and the amount of metal extracted is mea-sured. If this concentration exceeds 100 times the drinking water standard, the waste is classified as hazardous and requires special and very expensive disposal. If fly ash is measured by itself, its constituents often do not pass the test, and it is classified as hazardous. Combined with the bottom ash, however, the mixture most often meets the requirements for a nonhazardous waste.

The most important variable is the pH of the solution into which the metal is leached. If the solution is either at very low or at very high pH, then lead is quite soluble and can exceed the EPA limits.[23] Treatment of the emissions with lime or limestone usually produces an ash in the alkaline range, and this helps in maintaining the metals in the nonsoluble hydroxide form. Ash disposal is either in special landfills, or in regular municipal solid waste landfills. If the ash is compacted, the density increases from the noncompacted 1500 lb/yd^3 to as high as 3300 lb/yd^3. At this density the ash is highly impermeable, with a permeability as low as 1×10^{-9} cm/sec.

As more and more ash is being produced and landfill space becomes too valuable to use it for the disposal of such ash, alternative uses are being sought. In Europe, where a large fraction of MSW is incinerated and where landfill space is very expensive, ash processing has been quite successful.[24] Some of the uses of ash include

- Road base material
- Structural fill
- Gravel drainage ditches
- Capping strip mines
- Mixing with cement to make building blocks

In addition, ash from MSW combustion contains metals that can be reclaimed, especially steel and aluminum. This process is increasingly feasible because modern air quality controls are dry systems instead of wet so there is no quenching of the ash, making recovery difficult.

Air Pollutants

The air pollutants of concern in municipal waste incineration can be classified as *gases* or *particulates*. Another classification is *primary pollutants* and *secondary pollutants*. Primary pollutants are products of the combustion process that can be shown to be harmful in the form they are emitted. Secondary pollutants are those that are formed in the atmosphere as a direct result of the emission of primary pollutants. Because the combustion of MSW is a complex and dirty process, many undesirable reactions can occur that produce undesirable gases.

A particularly difficult problem is the sulfur oxides produced during combustion. In the presence of high temperature the sulfur in the fuel can combine with oxygen to produce sulfur dioxide.

$$S + O_2 \rightarrow SO_2$$

Sulfur dioxide is itself a primary pollutant and can cause respiratory distress and even property damage. But sulfur emissions can also be a secondary pollutant when sulfur dioxide, in the presence of water vapor and oxygen, can oxidize further to sulfur trioxide:

$$O_2 + 2SO_2 \rightarrow 2SO_3$$

Sulfur trioxide can then dissolve in water to form sulfuric acid,

$$SO_3 + H_2O \rightarrow H_2SO_4$$

The sulfuric acid, in combination with the hydrochloric acid produced in the reaction of chlorine compounds and the nitric acid produced from nitrogen oxides, produces acid rain. Normal, uncontaminated rain has a pH of about 5.6, but acid rain can be as low as pH 2 or even lower.

The effect of acid rain has been devastating. Hundreds of lakes in North America and Scandinavia have become so acidic that they no longer can support fish life. In a study of Norwegian lakes, more than 70% of the lakes having a pH of less than 4.5 contained no fish, while nearly all lakes with a pH of 5.5 and above contained fish. The low pH not only affects fish directly, but contributes to the release of potentially toxic metals such as aluminum, thus magnifying the problem. In North America acid rain has already wiped out all fish and many plants in 50% of the high mountain lakes in the Adirondacks. The pH in many of these lakes has reached such levels of acidity as to replace the trout and native plants with acid-tolerant mats of algae.

The deposition of atmospheric acid on freshwater aquatic systems prompted the EPA to suggest a limit of from 10 to 20 kg SO_4^{-2} per hectare per year. If "Newton's law of air pollution" is used (what goes up must come down), it is easy to see that the amount of sulfuric and nitric oxides emitted from power plants, municipal combustors, cars, and the many other sources is vastly greater than this limit. For example, just for the state of Ohio alone, the total annual emissions are 2.4×10^6 metric tons of SO_2 per year. If all of this were converted to SO_4^{-2} and deposited on the state of Ohio, the total would be 360 kg per hectare per year.

But not all of this sulfur falls on the folks in Ohio, and much of it is exported by the atmosphere to places far away. Similar calculations for the sulfur emissions for northeastern United States indicate that the rate of sulfur emissions is 4 to 5 times greater than the rate of deposition.

Another secondary pollutant of concern in municipal waste incineration is the formation of photochemical smog. The well-known and much-discussed Los Angeles smog is a case of secondary pollutant formation. Table 7-14 lists in simplified form some of the key reactions in the formation of photochemical smog.

The reaction sequence illustrates how nitrogen oxides formed in the combustion of fuels, such as gasoline and municipal waste, and emitted to the atmosphere are acted upon by sunlight to yield ozone (O_3), a compound not emitted as such from sources and hence considered a *secondary pollutant*. Ozone in turn reacts with hydrocarbons to form a series of compounds that includes aldehydes, organic acids, and epoxy compounds. The atmosphere can be viewed as a huge reaction vessel wherein new compounds are being formed while others are being destroyed.

The formation of photochemical smog is a dynamic process that begins with the production of nitrogen oxides from automobiles, industrial facilities, and MSW combustion. As the nitrogen oxides react with sunlight, O_3 and other oxidants are produced. The hydrocarbon level similarly increases at the beginning of the day and then drops off in the evening.

Table 7-14 Simplified Reaction Scheme for Photochemical Smog

NO_2 + Light		→	NO + O
O	+ O_2	→	O_3
O_3	+ NO	→	$NO_2 + O_2$
O	+ $(HC)_x$	→	$HCO°$
$HCO°$	+ O_2	→	$HCO_3°$
$HCO_3°$	+ HC	→	Aldehydes, ketones, etc.
$HCO_3°$	+ NO	→	$HCO_2° + NO_2$
$HCO_3°$	+ O_2	→	$O_3 + HCO_2°$
$HCO_x°$	+ NO_2	→	Peroxyacetyl nitrates

The escape of heavy metals with the emission gases has been another concern with combustion of MSW. Lead, cadmium, and mercury have been the most studied and represent the metals of most likely health concern and are presently regulated under the Clean Air Act. Mercury is especially difficult to control because it volatilizes so readily and escapes with the gaseous emissions. Difficulties in the control of mercury are exacerbated by the difficulties in controlling its inclusion in the waste stream. The best practice for controlling mercury emissions is source reduction. With household battery collection programs and the virtual elimination of mercury from batteries in the early 1990s, mercury emissions have been reduced. The mercury in the stack emissions can be highly variable, creating problems in setting emission standards for regulating mercury emission.[25] Figure 7-20

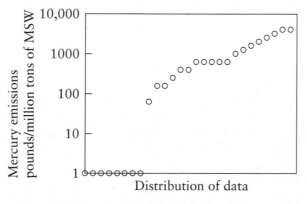

Figure 7-20 Mercury emissions during combustion of MSW. Source: (25)

shows the results of one study in which the highest emission recorded was 2000 pounds per million tons of MSW, while many readings were essentially zero!

Another concern with solid waste combustion is the formation of global warming gases. The earth acts as a reflector to the sun's rays, receiving the radiation from the sun, reflecting some of it into space (called *albedo*) and adsorbing the rest, only to reradiate this into space as heat. In effect the earth acts as a wave converter, receiving the high-energy, high-frequency radiation from the sun and converting most of it into low-energy, low-frequency heat to be radiated back into space. In this manner the earth maintains a balance of temperature, so that

$$\begin{bmatrix} \text{Energy from the sun} \\ \text{IN} \end{bmatrix} = \begin{bmatrix} \text{Energy radiated back to space} \\ \text{OUT} \end{bmatrix}$$

Unfortunately, some gases—such as methane (CH_4) and carbon dioxide (CO_2)—adsorb radiation at wavelengths approximately the same as the heat radiation trying to find its way back to space. Because the radiation is adsorbed in the atmosphere by these gases, the temperature of the atmosphere increases, heating the earth. The system works exactly like a greenhouse in that light energy (short-wave, high-frequency radiation) passes through the greenhouse glass, but the long-wavelength, low-frequency heat radiation is prevented from escaping. The gases that adsorb the heat energy radiation are properly referred to as *greenhouse gases* since they in effect cause the earth to heat up just like a greenhouse.

Not all gases have the same effect in global warming. Methane, for example, is 17 times more potent as a greenhouse gas than carbon dioxide. It can therefore be argued that the production of CO_2 in solid waste combustion is far better for the global environment than the production of methane in landfills. The requirement of flaring the gases in modern landfills is directly related to the reduction in the production of global warming gases.

While it is fairly easy to measure the concentrations of greenhouse gases such as CO_2, CH_4, and N_2O, and to show that the levels of these gases have been increasing during the past 50 years, it is another matter to argue from empirical evidence that these gases have caused an increase in the earth's temperature. The temperature of the earth undergoes continual change, with fluctuation frequencies ranging from a few years to thousands of years. Even if it is possible to measure accurately the earth's temperature, it is no proof that the change is being caused by the higher concentration of the greenhouse gases.

There seems to be growing consensus, however, that even though it is not possible to prove without doubt that global warming is occurring, the net effect can be so devastating to the earth that it would be imprudent to simply sit back and wait for the irrefutable proof. By then the change could be so immense that the effect may be irreversible.

Control of Particulates

The Clean Air Act emission standards allow states to set strict emissions limits for various sources. Municipal waste combustors are regulated under Subpart E. This regulation limits the emission of particulates to 24 mg/dry standard m^3, corrected to 7% oxygen. The oxygen is specified because the concentration of the emissions can be reduced by the simple expedient of sucking in more air to be emitted through the stack. Because normal air has about 20% oxygen, the requirement to maintain less than 7% oxygen (by volume) prevents operators from cheating. In addition the federal regulations limit the emissions to less than 10% opacity.

The simplest devices for controlling particulates are *settling chambers* consisting of nothing more than wide places in the exhaust flue where larger particles can settle out, usually with a baffle to slow the emission stream. Obviously, only very large particulates (> 100 μ) can be efficiently removed in settling chambers, and these are used, if ever, only as pretreatment in modern combustion units. Economizers that transfer heat from the exhaust gas to combustion air are examples of a settling chamber.

Possibly the most popular, economical, and effective means of controlling particulates from many industrial sources is the *cyclone*, similar to the cyclones used for removing shredded particles following air classification (see Chapter 6). Figure 7-21 shows a simple single-stage cyclone and a bank of high-efficiency cyclones. The dirty air coming to the cyclone is blasted into a conical cylinder, but off centerline. This creates a violent swirl within the cone, and the heavy solids migrate to the wall of the cylinder, where they slow down due to friction, slide down the cone, and finally exit at the bottom. The clean air is in the middle of the cylinder and exits out the top. Cyclones are not sufficiently effective for removing small particles and thus need to be backed up by other particulate-removal devices.

Bag (or *fabric*) *filters* used for controlling particulates (Figure 7-22) operate like the common vacuum cleaner. Fabric bags are used to collect the dust, which must be periodically shaken out of the bags. The fabric will remove nearly all particulates, including submicron sizes. Bag filters are widely used in many industrial applications, including MSW combustion. The basic mechanism of dust removal in fabric filters is thought to be similar to the action of sand filters in water quality management. The dust particles adhere to the fabric due to entrapment and surface forces. They are brought into contact by impingement and/or Brownian diffusion. Since fabric filters commonly have an air space-to-fiber ratio of 1:1, the removal mechanisms cannot be simple sieving.

The *scrubber* (Figure 7-23) is another method for removing large particulates. More efficient scrubbers promote the contact between air and water by violent action in a narrow throat section into which the water or a chemical slurry is introduced. Generally, the more violent the encounter, and hence the smaller the gas bubbles or water droplets, the more effective the scrubbing. Scrubbers are used in MSW combustion mostly for the removal of gaseous pollutants, but they also help in the removal of particulates. Wet scrubbers, which use water sprays, are efficient devices, but have three major drawbacks:

- They produce a visible plume, albeit only water vapor. The lay public seldom differentiates between a water vapor plume and any other visible plume, and hence public relations often dictate no visible plume.
- The waste is now in liquid form, and some manner of water treatment is necessary.
- The ash is wet, and recovery of metals is difficult.

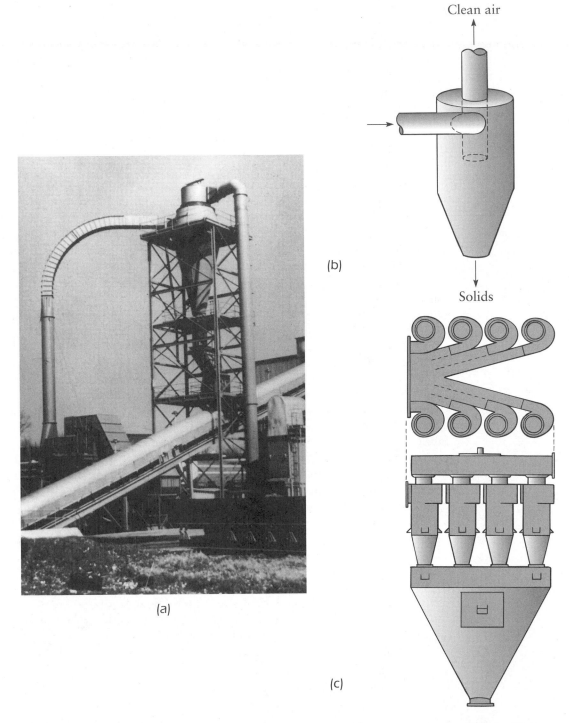

Figure 7-21 Cyclones. (a) simple cyclone, (b) simple cyclone, (c) bank of high-efficiency cyclones.

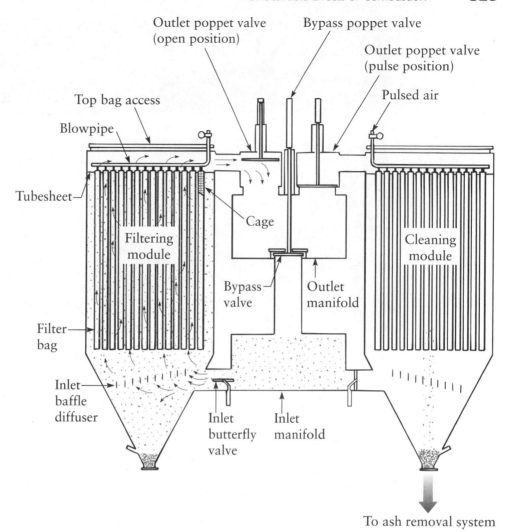

Outlet poppet valve
(open position)

Bypass poppet valve

Outlet poppet valve
(pulse position)

Pulsed air

Top bag access

Blowpipe

Cage

Tubesheet

Filtering
module

Cleaning
module

Bypass
valve

Outlet
manifold

Filter
bag

Inlet
baffle
diffuser

Inlet
butterfly
valve

Inlet
manifold

To ash removal system

Figure 7-22   Bag filters. (Courtesy Bundy Environmental Technology, Inc.)

Dry scrubbers inject a chemical slurry such as lime. This type of scrubber does not produce a visible plume, and the waste is a powder, not liquid.

Electrostatic precipitators (Figure 7-24) are widely used in power plants, mainly because power is readily available. The particulate matter is removed by first being charged by electrons jumping from one electrode to the other. The negatively charged particles then migrate to the positively charged collecting electrode. The type of electrostatic precipitator shown in Figure 7-24a consists of a negatively

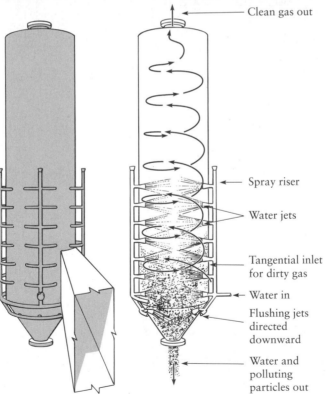

Clean gas out

Spray riser

Water jets

Tangential inlet
for dirty gas

Water in

Flushing jets
directed
downward

Water and
polluting
particles out

Figure 7-23 Scrubbers used for air pollution control.

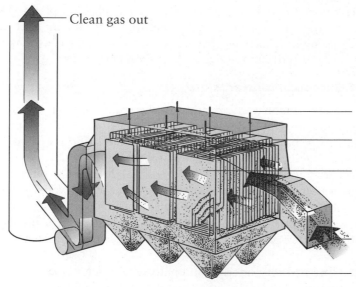

Clean gas out

Negative electrode connected to electric power source

Negatively charged wire

Grounded collecting plate with positive charge

Dirty gas in

Hopper to discharge

(a)

(b)

Figure 7-24 Electrostatic precipitator.

charged wire hanging down the middle of positively charged plates. The particulates collect on the plates and must be removed by banging with a hammer. Electrostatic precipitators have no moving parts, require only electricity to operate, and are extremely effective in removing submicron particulates.

Control of Gaseous Pollutants

The control of gases involves the removal of the pollutant from the gaseous emissions, a chemical change in the pollutant, or a change in the process producing the pollutant. In MSW combustion, where the fuel composition is seldom under control, the gaseous removal processes must be robust and effective.

Wet scrubbers, which can be used for partial particulate removal, can also remove gaseous pollutants by simply dissolving them in the water. Alternatively, a chemical may be injected into the scrubber water, which then reacts with the pollutants. This is the basis for most SO_2 removal techniques, as discussed below. Because of the moisture carryover, wet scrubbers are usually placed after the ESP or baghouse.

Dry scrubbers are very effective in controlling sulfur oxides. A lime slurry is injected into this unit, but the liquid evaporates due to the high temperature of the exhaust gas. The lime slurry also reduces hydrogen sulfide and hydrogen chloride in the exhaust gases.

In simple terms the reaction in the dry scrubber is

$$Ca(OH)_2 + heat \rightarrow CaO + H_2O$$

$$SO_2 + CaO \rightarrow CaSO_3$$

or if limestone is used,

$$SO_2 + CaCO_3 \rightarrow CaSO_4 + CO_2$$

Both the calcium sulfite ($CaSO_3$) and the calcium sulfate (or gypsum) ($CaSO_4$) are solids that represent a staggering disposal problem. It is possible to convert the sulfur to H_2S, H_2SO_4 or elemental sulfur and market these raw materials, but the total possible markets for these chemicals is far less than the anticipated production from desulfurization.

Waste-to-energy combustors also are a source of nitrogen oxides that can lead to the formation of photochemical smog and contribute to acid rain. Nitrogen oxides, designated by the general expression NO_x, are produced in two ways. *Thermal NO_x* results from the reaction of excess oxygen (from air) with nitrogen (from air) at high temperatures. *Fuel NO_x* are produced when the nitrogen in the fuel produces the oxide during combustion. Generally, the thermal NO_x represents only about 25% of the total nitrogen oxide production.[26]

Unlike sulfur, nitrogen oxides cannot be readily removed from fossil fuels, and the reduction of nitrogen is commonly accomplished using the *denox* system, a selective non-catalytic reaction that involves the injection of ammonia into the flue gases that pass through a catalysis bed. The ammonia and nitrogen oxides react to form nonpolluting nitrogen gas, N_2, and water vapor.

Dioxin

Of particular concern in waste combustion is the production of *dioxin*. Dioxin is actually a combination of many members of a family of organic compounds called polychlorinated dibenzodioxins (PCDD). Members of this family are characterized by a triple ring structure of two benzene rings connected by a pair of oxygen atoms (Figure 7-25). A related family of organic chemicals are the polychlorinated dibenzofurans (PCDF), which have a similar structure except that the two benzene rings are connected by only one oxygen. Since any of the carbon sites are able to attach either a hydrogen or a chlorine atom, the number of possibilities is great. The sites that are used for the attachment of chlorine atoms are identified by number, and this signature identifies the specific form of PCDD or PCDF. For example, 2,3,7,8-tetrachloro-dibenzo-p-dioxin (or 2,3,7,8-TCDD in shorthand) has four chlorine atoms at the four outside corners, as shown in Figure 7-26. This form of dioxin is especially toxic to laboratory animals, and is often identified as a primary constituent of contaminated pesticides and emissions from waste-to-energy plants.[27]

All of the PCDD and PCDF compounds have been found to be extremely toxic to animals, with the LD_{50} for guinea pigs being about 1 microgram/kg body weight. Neither PCDD or PCDF compounds have found any commercial use and are not manufactured. They do occur, however, as contaminants in other organic chemicals.[28] Various forms of dioxins have been found in pesticides (such as Agent Orange, widely used during the Viet Nam War) and in various chlorinated organic chemicals such as chlorophenols. Curiously, recent evidence has not borne out the same level of toxicity to humans, and it seems likely that dioxins are actually less harmful than they might appear from laboratory studies. A large chemical spill in Sevesco, Italy, was expected to result in a public health disaster, based on the extrapolations from animal experiments, but thus far this has not materialized. One estimate of the dose suffered by the children in Sevesco was 3,000,000 pg of dioxin/kg of body weight, compared to the EPA risk-specific level of 0.006 pg/kg body weight, or the Canadian tolerable daily intake of 10 pg/kg body weight. Although the children suffered from chloracne (a temporary skin condition), none have seemed to have the more serious cancers predicted. Nevertheless, dioxins are able, at very low concentrations, to disrupt normal metabolic processes, and this has caused the EPA to continue to place severe limitations on the emission of dioxins from incinerators.[29]

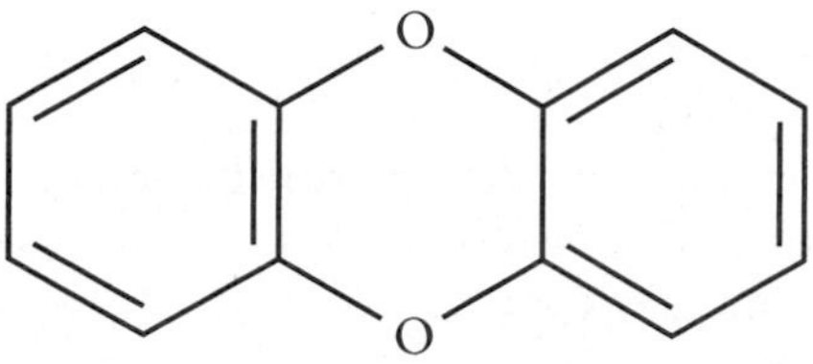

Figure 7-25 A dioxin molecule.

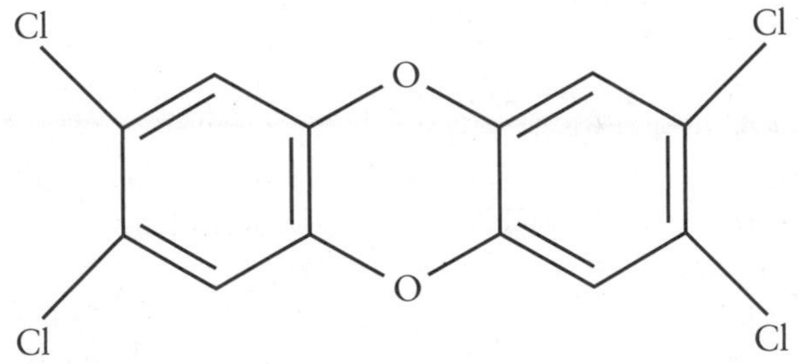

Figure 7-26   A particularly toxic form of dioxin: 2,3,7,8-TCDD.

Dioxins emitted from waste-to-energy facilities come from two sources: dioxins that are in the waste and are not combusted in the furnace, and *de novo* dioxins that are created during combustion.[30] Tests on waste-to-energy plants that have included slug loads of chlorinated plastics, thought to be the main precursors in dioxin formation, have produced negative results.[31] Nevertheless, the EPA estimates that a significant part of environmental dioxins (of all forms, including difurans) come from combustion, with the combustion of hospital wastes being the greatest single source, as shown in Table 7-15.

Table 7-15   Air Emissions of Dioxins and Difurans in the United States

Emission source	Percent of total
Hospital waste combustion	55.4
Municipal waste combustion	32.6
Wood burning	3.9
Cement kilns	3.8
Nonferrous metal refining	2.7
Diesel fuel combustion	0.9
Hazardous waste incineration	0.4
All other sources	0.3

Source: (25)

The measurement of dioxin in a combustion operation is difficult and expensive. On a gross level the emissions of dioxins can be tracked by measuring the emission of carbon monoxide, CO. Hasselriis[31] has proposed that the production of PCDDs is proportional to the CO concentration as

$$PCDDs = \left(\frac{CO}{A}\right)^2$$

where PCDDs = concentration of dioxins in the off-gases, ng/m^3

CO = concentration of carbon monoxide in the off-gases, as percent of total gas

A = a constant, function of operation system

As would be expected from this relationship, the emission of PCDDs increases with increasing CO, which is controlled by both the amount of excess air used and the combustion temperature. There seems to be a point of lowest dioxin production at a given amount of excess air and a given temperature. From empirical evidence it has been possible to develop relationships that are good predictors of dioxin formation.[27]

For modular combustors:

$$PCDDs + PCDFs = 2670.2 - 1.37\ T + 100.06\ CO$$

For water-wall combustors:

$$PCDDs + PCDFs = 4754.6 - 5.14\ T + 103.41\ CO$$

where T is the temperature in the secondary chamber for modular combustors, and the furnace temperature in water-wall combustors, respectively, in degrees C. The concentration of CO is in percent of gases.

There is little doubt that waste-to-energy plants emit trace amounts of dioxins, but nobody knows for sure how the dioxins originate. Some dioxin is in the waste, and this may escape in the bottom ash or the fly ash or the off-gases. On the other hand, it is also likely that *any* combustion process that has even trace amounts of chlorine produces dioxins, and that these are simply an end-product of the combustion process. Another theory is that the high temperatures in the combustion unit destroy the dioxins but that new dioxins form as the exhaust gas is cooled. The presence of trace quantities of dioxins in emissions from wood stoves and fireplaces seems to confirm this view. Whatever the mechanism, it seems clear that it is not a simple chemical equation and that many parallel reactions are occurring to produce the various forms of dioxins and difurans.[29]

It might be well to remember here that the two sources of risk, incinerators versus fireplaces, are clearly unequal. The effect of the latter on human health is greater than the effect of incinerator emissions. But the fireplace is a *voluntary risk*, whereas the incinerator is an *involuntary risk*. People are willing to accept voluntary risks 1000 times higher than involuntary risks, and they are therefore able to vehemently oppose incinerators while enjoying a romantic fire in the fireplace.[32]

FINAL THOUGHTS

Earle Phelps was the first to recognize that most environmental regulatory decisions are made using what he called the *principle of expediency*. A sanitary engineer known for his work with stream sanitation and the development of the Streeter-Phelps oxygen sag curve equation, Phelps described expediency as "the attempt to reduce the numerical measure of probable harm, or the logical measure of existing hazard, to the lowest level that is practicable and feasible within the limitations of financial resources and engineering skill." He recognized that "the optimal or ideal condition is seldom obtainable in practice, and that it is wasteful and therefore inexpedient to require a nearer approach to it than is readily obtainable under current engineering practices and at justifiable costs." Most importantly for today's standard setters, who often have difficulty defending their decisions, he advised that "the principle of expediency is the logical basis for administrative standards and should be frankly stated in their defense."

Phelps saw nothing wrong with the use of standards as a kind of speed limit on pollution affecting human health. He also understood the laws of diminishing returns and a lag time for technical feasibility. Yet he always pushed towards reducing environmental hazards to the lowest expedient levels. Just as utilitarianism is an ethical model that can resolve moral dilemmas, Phelps' expediency principle can be used to resolve the moral dilemmas of setting environmental regulations. The regulator must balance two primary moral values—do not deprive liberty, and do no harm. Setting strict regulations would result in unwarranted reduction in liberty, while the absence of adequate regulations can damage public health. By using the principle of expediency, the regulator can establish the proper balance and resolve a moral dilemma.

All ethical models, if they are to be useful, need adequate information. Using the utilitarian ethical model, for example, a just decision is possible only if the amount of happiness and unhappiness that results from decisions can be calculated. Similarly, the regulator must have scientific evidence on pollutant quantities, concentrations, vectors, and health effects to make a just environmental decision. Unfortunately, there will always be gaps between what we know and what we would like to know about environmental hazards, and the absence of adequate scientific knowledge makes the regulatory decision difficult.

A classical case is the setting of the dioxin standard for MSW combustion. The best we can do at this time, in the absence of adequate information, is to set the standard as low as we can without completely shutting down all waste burn combustors. It is not expedient to set the standard too high and allow the facilities not to worry about dioxin emissions. This chemical is exceedingly toxic, and we should be concerned; and as better information becomes available, we should refine our standard. But in the meantime we set the standard as low as is expedient.

REFERENCES

1. Wilson, D. L. 1972. "Prediction of Heat of Combustion of Solid Wastes from Ultimate Analyses." *Environmental Science and Technology* 13 (June).

2. Brunner, C. 1994. "Waste-to-energy." In *Handbook of Solid Waste Management*, F. Keith (ed.). New York: McGraw-Hill.

3. Test Method E1037-84, Standard Test Method for Measuring Particle Size Distribution of RDF-5, ASTM, Philadelphia, Pa., 1996.

4. Neissen, W. R. 1977. "Properties of Waste Materials." In *Handbook of Solid Waste Management*, D. G. Wilson (ed.). New York: Van Nostrand Reinhold.

5. AliKhan, M. Z. A., and Z. H. Abu-Ghararah. 1991. "New Approach for Estimating Energy Content of Municipal Solid Waste." *Journal of the Environmental Engineering Division* ASCE 117, n. 3: 376–380.

6. Liu, J. and R. D. Paode. 1996. "Modeling the Energy Content of MSW Using Multiple Regression Analysis." *Journal of the Air & Waste Management Association* 46, n. 7: 650–656.

7. Brunner, C., and S. Schwartz. 1983. *Energy and Resource Recovery from Wastes*. Park Ridge, N. J.: Noyes.

8. Erdincler, A. U., and P. A. Vesilind. 1993. "Energy Recovery from Mixed Paper." *Waste Management and Research* 11: 507–513.

9. Wilson, D. G. 1977. *Handbook of Solid Waste Engineering*. New York: Van Nostrand Reinhold.

10. Neissen, W. R. 1995. *Combustion and Incineration Processes: Applications in Environmental Engineering*, 2nd edition. New York: Marcel Decker.

11. Vesilind, P. A., W. P. Martello, and B. Gullett. 1981. "Calorimetry of Refuse Derived Fuels." *Conservation and Recycling* 4, n. 2: 89–97.

12. Wen, C. Y., and E. S. Stanley. 1979. *Coal Conversion Technology*. Reading, Mass.: Addison-Wesley.

13. Levy, S. J. 1974. "Pyrolysis of Municipal Solid Waste." *Waste Age* (October).

14. *Pyrolysis*. 1973. Washington, D.C.: National Center for Resource Recovery.

15. Neissen, W. R. 1996. "Evaluating Gasification and Advanced Thermal Technologies for Processing MSW." *Solid Waste Technologies*, n. 4: 28–36.

16. International Solid Waste Association. 1997. *Energy from Waste: State of the Art Report*. Copenhagen: ISWA.

17. Boley, G. L. 1991. "Refuse-derived fuel (RDF)—Quality Requirements for Firing in Utility, Industrial, or Dedicated Boilers." *Proceedings* International Joint Power Generation Conference, San Diego.

18. Kersletter, J. D., and J. K. Lyons. 1991. "Mixed Waste Paper as a Fuel." *Waste Age* (November): 41–43.

19. Philipson Industries. 1999. Private communication. Mississauga, Ontario, Canada.

20. Eimers, J. L., and P. A. Vesilind. 1984. "Physical Properties of Densified Refuse Derived Fuel." *Journal of Testing and Evaluation* 12, n. 4: 238–240.

21. Chesner, W. H., R. J. Collins, and T. Fung. 1994. "Assessment of the Potential Stability of Southwest Brooklyn Incinerator Residue in Asphaltic Concrete Mixes." Quoted in Hasselriis, F. "Ash disposal." In *Handbook of Solid Waste Management*, R. Keith (ed.). New York: McGraw-Hill.

22. Forrester, K. 1994. "State-of-the-art in Refuse-to-energy Facility Ash Residue Characteristics." Quoted in Hasselriis (35).

23. Sawell, S. E., T. R. Bridle, and T. W. Constable. 1988. "Heavy Metal Leachability from Solid Waste Incinerator Ashes." *Waste Management and Research* 6: 227–238.

24. "Ash Use on the Rise in United States." 1999. *World Wastes*, pp. 16–18.

25. Hasselriis, F. 1994. Data published in *Waste Age* (November), p. 96.

26. Harrison, K. W., R. D. Dumas, S. R. Nishtala, and M. A. Barlaz. 2000. "A Life Cycle Inventory Model of Municipal Solid Waste Combustion." *Journal of the Air and Waste Management Association* 50: 993–1003.

27. Chang, N-B. 1994. "The Impact of PCDD/PCDF Emissions on the Engineering Design of Municipal Incinerators." Reported in Hasselriis (31).

28. Chang, N-B, and S-H Huang. 1995. "Statistical Modeling for the Precision and Control of PCDDs and PCDFs Emissions from Municipal Solid Waste Incinerators." *Waste Management and Research* 13: 379–400.

29. McAdams, C. L., and J. T. Aquino. 1994. "Dioxin: Impact on Solid Waste Industry Uncertain." *Waste Age* (November): 103–106.

30. Johnke, B., and E. Stelzner. 1992. "Results of the German Dioxin Measurement Programme at MSW Incinerators." *Waste Management and Research* 10: 345–355.

31. Hasselriis, F. 1987. "Optimization of Combustion Conditions to Minimize Dioxin Emissions." *Waste Management and Research* 5: 311–325.

32. Elliott, S. J. 1998. "A Comparative Analysis of Public Concern over Solid Waste Incinerators." *Waste Management and Research* 16, n. 4: 351–364.

ABBREVIATIONS USED IN THIS CHAPTER

ASTM = American Society for Testing and Materials
ESP = electrostatic precipitator
HHV = higher heating value
LHV = lower heating value

MSW = municipal solid waste
NB = *nota bene* = take notice
RDF = refuse-derived fuel
PCDD = polychlorinated dibenzodioxins
PCDF = polychlorinated dibenzofurans

PROBLEMS

7-1. Describe the difference between pyrolysis and gasification.

7-2. The ideal equation for the combustion of cellulose is $C_6H_{10}O_5 + 6O_2 \rightarrow 6CO_2 + 5H_2O$. What is a similar idealized equation for pyrolysis? What end-products might you expect?

7-3. You serve as the president of XYZ Corporation, which is an energy-intensive manufacturing operation. You have four boilers to produce steam and electricity for your operations. One is a semi-suspension fired boiler with a capacity of 100 tons of coal per hour. The boiler can also burn supplemental fuel such as coconut or macadamia nut shells. The other three boilers are peaking units and burn fuel oil. The City of Podunk has approached you about the possibility of burning their solid waste as a supplemental fuel. As the president of the XYZ Corporation you need to develop answers to many political, social, and technical questions before you make a commitment to the city.

a. What specific questions would you ask your own power plant engineer about the possibility of accepting Podunk's MSW?

b. What specific additional data would you need from the city before you discuss this further?

7-4. A furnace dedicated to paper (assume pure cellulose, $C_6H_{10}O_5$) operates at 15 tons/hour, at 100% excess air. How much air is required?

7-5. Draw flow diagrams of unit operations for producing from raw MSW the following products:

a. RDF-1

b. RDF-2

c. RDF-3

d. RDF-4

No other materials are to be recovered, and the purity of the fuels is important.

7-6. Suppose your nonengineering roommate asked you to explain how a refuse waste-to-energy facility worked. In one 8 1/2 by 11 page, using words and sketches, help him/her understand how such facilities take in MSW and produce electric power.

7-7. Why are automobile tires troublesome in MSW combustion?

7-8. Two members of Greenpeace climb the stack of your waste-to-energy plant and chain themselves to it. You are the city public works director, and the waste-to-energy plant operator calls you in a panic. What actions do you take?

7-9. A *Citizens' Guide to the Care of the Environment* stated the following: "Avoid buying milk in white plastic containers. They are virtually indestructible, except by burning, which produces a poisonous gas." Comment on this assertion.

7-10. Cellulose is to be burned in a waste-to-energy facility. The chemical equation for cellulose is $C_6H_{10}O_5$. The atomic weights of C, H, and O are 12, 1, and 16, respectively.

a. Calculate the stoichiometric oxygen necessary for the combustion of cellulose.

b. Calculate stoichiometric air.

7-11. Draw the organic chemical formula for a polychlorinated dibenzodioxin, and describe how dioxins are controlled in MSW combustion.

7-12. A sample of refuse has a moisture content of 20%. A 1.2-g sample is placed in a calorimeter, and a gross calorific value of 6200 Btu/lb is measured based on the temperature rise. Further analysis shows that 0.3 g of ash remains in the bomb calorimeter after combustion. Calculate the HHV in terms of (a) moisture-free, and (b) moisture- and ash-free.

7-13. Using words and sketches, show how a bomb calorimeter is used to measure the heating value of a sample of RDF.

7-14. Explain the difference between the HHV and LHV. Why is the HHV always higher than the LHV?

7-15. What are the two primary reasons for using underfired air in a solid waste combustion system? Use a sketch to show how underfired air is used.

Biochemical Processes

Municipal refuse contains about 75% organic material, which can be converted to useful energy by simple combustion, as discussed in the previous chapter, or to useful products by biochemical processes—the topic of this chapter.

The three components of MSW of greatest interest in the bioconversion processes are garbage (food waste), paper products, and yard wastes. The last two are especially valuable in biochemical processes as a source of cellulose, a potentially useful industrial raw material.

The garbage fraction of refuse varies with geographical location and season. Dietary habits, of course, affect its composition and quantity, as does the standard of living. Kitchen garbage grinders in more-affluent communities transfer much of the putrescent waste from the refuse stream to the sewerage system, and the reduction of the garbage fraction is a continuing trend in the United States and in many other countries.[1]

The garbage fraction also has by far the highest moisture content of any constituent in MSW, but the moisture is rapidly transferred to absorbent materials such as newspapers as soon as contact is made. Garbage also tends to be well mixed in MSW, and therefore it is often difficult to find identifiable bits of garbage in mixed refuse other than the large pieces, such as orange peels or apple cores. Garbage is even better distributed in MSW if the waste is shredded.

The fraction of paper in MSW tends to remain fairly stable throughout the year, while yard waste in most locations is seasonal. The latter also varies from week to week, often reflecting the weather.

On the average the organic components of refuse can be described by an analysis as shown in Table 8-1. The largest single constituent is, of course, cellulose.

The five methods of biochemical energy conversion discussed here all make use of the organic fraction of refuse. The first three methods (two means of anaerobic digestion and composting) are broad-spectrum processes, where the specific organisms responsible for the bioconversion are not identified or isolated, and the processes are described by empirical data. The last two techniques, enzyme and acid hydrolysis, on the other hand, involve the use of known chemicals and specific chemical pathways to produce a known pure product (glucose). Because composting and anaerobic digestion do not begin with raw material composed of only one chemical, the specific biochemical reactions involved in these processes are numerous, and therefore it is not possible to approach them as one would describe the hydrolysis of cellulose.

Table 8-1   Organic Analysis of Refuse

	Percent by weight	
Organics	54.4	
Cellulose, sugar starch		46.6
Lipids (fats, oils, waxes)		4.5
Protein		2.1
Other organics (e.g., plastics)		1.2
Inorganics	24.9	
Moisture	20.7	
	100	

Source: (2)

METHANE GENERATION BY ANAEROBIC DIGESTION

When organic matter decays under anaerobic (without free or combined oxygen) conditions, the end-products include such gases as methane (CH_4), carbon dioxide (CO_2), small amounts of hydrogen sulfide (H_2S), ammonia (NH_3), and a few others. The recognition that methane is an excellent fuel long ago prompted wastewater treatment plant design engineers to digest (decompose) waste solids and capture this gas for use in heating buildings and running machinery in the treatment plant. While the quantity of methane generated in a wastewater treatment plant is not sufficient to consider its conversion to pipeline gas, the potential for producing pipeline gas from decomposing refuse has a lot of merit.

Ideally, the production of methane and carbon dioxide can be calculated using the following equation:

$$C_a H_b O_c N_d + \left(\frac{4a - b - 2c + 3d}{4} \right) H_2O \rightarrow$$

$$\left(\frac{4a + b - 2c - 3d}{8} \right) CH_4 + \left(\frac{4a - b + 2c + 3d}{8} \right) CO_2 + dNH_3$$

Example 8-1 illustrates how this equation can be used if the chemical composition of a material is known.

E X A M P L E

8 - 1

Estimate the production of CO_2 and CH_4 during the anaerobic decomposition of glucose.

The general formula for glucose is $C_6H_{12}O_6$, hence by the equation above, $a = 6$, $b = 12$, $c = 6$, and $d = 0$.

$$C_6H_{12}O_6 + \left(\frac{24 - 12 - 12}{4}\right)H_2O \rightarrow \left(\frac{24 + 12 - 12}{8}\right)CH_4 + \left(\frac{24 - 12 + 12}{8}\right)CO_2$$

$$C_6H_{12}O_6 \rightarrow 3CH_4 + 3CO_2$$

Note that the equation balances. The molecular weights are $180 \rightarrow 3(16) + 3(44)$, hence 1 kg of glucose produces 0.73 kg of CO_2 and 0.27 kg of CH_4. Recalling that 1 gram molecular weight of a gas at standard temperature and pressure occupies 22.4 liters, the production of CO_2 and CH_4 from one kg of glucose is 746 liters each of methane and carbon dioxide.

Unfortunately, the chemical composition of MSW is difficult, if not impossible, to determine, although some attempts have been made to do so. The best approximation is that the organic fraction of refuse can be described by the chemical formula $C_{99}H_{149}O_{59}N$. With this formula the above equation estimates that the production of methane from a landfill is 257 liters of methane per kg of wet refuse (total, organic plus inorganic, assuming wet refuse is 50% biodegradable organic). In using this equation note that the only carbon that can participate in the production of gas is from decomposable materials such as food waste and paper. Other organics, most importantly plastics, do not decompose to produce gas.

The two ways of generating methane are to capture the gases produced in landfills, or to pretreat the refuse and digest it in tanks similar to those used in wastewater treatment plants. Methane generation both in anaerobic digesters and in landfills is discussed in this section, although a more complete presentation of landfill gas production is found in Chapter 4. Much of the following anaerobic decomposition theory applies to both processes, however.

Anaerobic Decomposition in Mixed Digesters

The two basic metabolic pathways for the decomposition or degradation of wastes are aerobic (with oxygen) and anaerobic (in the absence of oxygen). While an aerobic system might be generally represented as

[complex organics] + O_2 → CO_2 + H_2O + NO_3^- + SO_4^{-2} + other stable products + heat

the anaerobic decomposition of organics can be described as

[complex organics] + heat → CO_2 + CH_4 + H_2S + NH_4^+

The end-products in aerobic decomposition are all stable, possessing no additional energy to be used by decomposing organisms (they are at their highest oxidation state). The products of anaerobic decomposition, on the other hand, still contain energy. Ammonia and hydrogen sulfide could be still further oxidized, and methane contains considerable energy.

The microorganisms responsible for anaerobic decomposition can be divided into two broad categories:

1. *Acid formers* that ferment the complex organic compounds to more simple organic forms such as acetic and propionic acids. These hardy organisms can be either facultative or strict anaerobes.
2. *Methane formers* that convert the organic acids to methane. These organisms are strict anaerobes and have very slow growth rates—two characteristics that cause considerable problems in anaerobic processes in wastewater treatment, and will similarly plague anaerobic decomposition of refuse. Methane formers are very sensitive to various environmental factors. They are strict anaerobes and quite sensitive to temperature changes. Two different groups of methane formers seem to exist, with one group (*mesophilic*) operating best around 30 to 38°C (85 to 100°F) and a second group (*thermophilic*) operating best around 50 to 58°C (120 to 135°F). The methane formers also require stable and neutral pH. Sufficient alkalinity (resistance to pH drop) should be present to prevent the pH from falling below 6.8. Finally, methane formers are sensitive to the presence of toxic materials, such as heavy metals and pesticides.

During the acid-forming stage, the first step in the process involves extracellular enzymes produced by acid formers, which break down the large complex organic molecules. For example, the enzymes cellobiase and cellulase break down cellulose to glucose, and lipase breaks fat to shorter-chained fatty acids. This process is energy consuming.

Other bacteria then metabolize the glucose and other products into organic acids, mostly acetic and propionic. These simple organic acids then serve as substrate for methanogens. This methane formation is performed by a number of organisms that have specific substrates and roles in the overall reaction. The two reactions

$$CH_3COOH \rightarrow CH_4 + CO_2$$

$$4CH_3CH_2COOH + 2H_2O \rightarrow 7CH_4 + 5CO_2$$

for acetic and propionic acids, respectively, are actually the net results of a large number of steps. The resulting gas varies in composition but averages about 60% methane, with a heating value between 500 and 700 Btu/ft^3 (4700 kJ/m^3 and 6500 kJ/m^3).[3]

The total amount of gas theoretically available from the anaerobic digestion of MSW is considerably more than has been captured to date in pilot plant facilities. About 54% of the volatile solids have been found to pass through the digester[4] and have not been converted to CO_2 and CH_4.

The digestion of refuse involves hardware and a flow diagram not unlike the anaerobic digestion used in wastewater treatment. In the case of MSW the organics are first separated from the refuse and are slurried with sewage sludge or some other suitable liquid. The resulting mixture is digested in a heated and enclosed tank. The gas is captured either under a floating cover or in a separate tank. The residual of the digestion process is a dark, odoriferous slurry that must be disposed of.

Pilot plant studies have shown that the total gas produced is strongly influenced by detention time in the digester and the digester temperature, as shown in Figure 8-1. Note that the 45°C result is lower than 40°C, suggesting the existence of both a mesophilic (40°C) and a thermophilic (60°C) operating range.

In addition to temperature and residence time, other variables are important in this process, such as the maintenance of total anaerobiosis. It is also necessary to maintain a neutral pH level—never below 6.2, at which point methane production ceases. Since this is a biological process, the provision of adequate nutrients, such as nitrogen, is required. Usually, the C/N ratio of shredded waste is not sufficient for full decomposition, and another source of N is needed, such as sewage sludge rich in nitrogen. The C/N ratio for typical MSW has been reported as 24:1, with some values as high as 40:1,[5] while a ratio of 20:1 as a minimum is required for active anaerobic decomposition. Raw primary sludge[6] has a C/N ratio of about 16:1.

Finally, toxic materials can be detrimental to anaerobic digestion, and these must be controlled by removing potential toxins before they get to the digester, or if there, precipitating them out. The latter method has been used in the removal of metals by precipitation with sulfide in wastewater treatment plants.[7]

MSW digestion might be described as the decay or reduction in volatile (organic) matter as

$$\frac{dS}{dt} = -K_d S$$

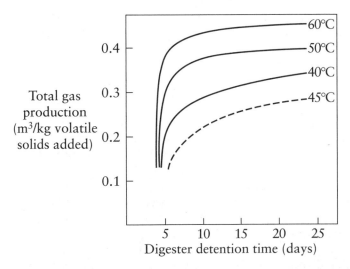

Figure 8-1   Gas production from anaerobic digestion of MSW. Source: (4)

where S = concentration of the biodegradable material (measured as volatile suspended solids, or a specific material if the system feed is controlled), mg/liter at time t

K_d = decay constant, days^{-1}

t = time, days

This is simply a first-order decay equation, stating that the rate of decay is proportional to the organics remaining, a reasonable assumption if the process rate is not time dependent. After integration,

$$\frac{S}{S_0} = e^{-K_d t}$$

where S_0 is the original organic solids concentration (t = 0), mg/liter.

The materials balance within a completely mixed continuous digester would be

[rate of input] − [rate of output] − or + [rate of accumulation] = [rate of net change]

If the digester is operating at steady state, the net change is zero, and

$$\frac{QS_0}{V} - \frac{QS}{V} - K_d S = 0$$

where Q = flow rate through digester, m^3/day

V = volume of digester, m^3

The accumulation term is negative because the organic material is being destroyed.

The hydraulic residence time $\bar{t}$ = Q/V, or

$$\bar{t} = \frac{S_0 - S}{K_d S}$$

Hence if K_d is known, the required residence time for any reduction in solids can be calculated. Batch laboratory experiments can be used to obtain values of K_d by plotting the values of log S/S_0 versus time and measuring the slope as ($K_d/2.303$). Values of K_d for refuse slurries have not been reported.

The process kinetics may also be described in terms of the gas produced instead of the volatile matter destroyed. Using a similar mass balance Pfeffer[8] found that it was possible to describe the reactor performance by the model

$$\frac{G_0 - G}{G} = K_g \bar{t}$$

where G_0 = maximum gas production attainable, estimated at 0.547 liter gas per gram volatile solids in the reactor

G = daily gas production, liters/g volatile solids

$\bar{t}$ = hydraulic residence time, days

K_g = rate constant, days^{-1}

K_g seems to have two distinct values. The initial rate is rapid and lasts between 5 and 10 days, followed by a significantly lower rate. Table 8-2 is a listing of the K_g values. At 45°C there is a substantial drop in K_g from 40°C, indicating again the existence of mesophilic and thermophilic regimes in anaerobic digestion.

Table 8-2   Rate Constants, K_g, for Gas Production in Anaerobic Digesters

Temperature, °C	Rate constant (day^{-1})	
	Initial	Final
35	0.055	0.003
40	0.084	0.043
45	0.052	0.007
50	0.117	0.030
55	0.623	0.042
60	0.990	0.040

Source: (8)

The quantity of gas generated can be estimated by entering a plot such as Figure 8-1 at the calculated $\bar{t}$ (hydraulic residence time) and reading off the gas production. Because of the heterogeneous nature of the waste and the fact that not all the organics decompose, any theoretical calculations would probably be fruitless. Therefore laboratory studies to determine kinetic constants for a particular waste are necessary.

Potential for the Application of Anaerobic Digesters

Since the anaerobic digestion of refuse in digesters has never been successfully operated on a prototype scale in the United States, design guidelines are not available. Bench-scale data indicate that one 1000-tons/day (900-tonnes/day) plant could produce about 3.6 million ft^3 (16,000 m^3) of methane per day.[9] There are about 65 metropolitan areas in the United States that could support such a plant, and thus the total potential production is 1,000,000 m^3/day.[9] Since the total use of natural gas in the United States is about 100×10^9 m^3/day, the impact of methane from waste would be a substitution of only 0.001% of the national need—not a very significant fraction.

In addition to the minuscule impact, the process is plagued by potential problems. There is no way to ensure the removal of toxic materials before the waste goes into the digesters, and "sour" digesters, such as those encountered in wastewater treatment plants, are a definite possibility.

The problem of mixing a paper slurry has not yet been solved. Even pilot plant scale mixing with fairly dilute slurries has been found to be a problem. The desired solids concentration for these digesters is 10%, which is a highly viscous and thixotropic slurry. In wastewater treatment practice, where solids concentrations normally range from 3 to 5%, mixing has always been a problem. Tracer studies have shown that typical primary digesters seem to have only 25% of their volume mixed, the remaining being dead space.[10] Such problems will surely plague refuse-digestion facilities as well. A demonstration project in Florida continued to break shafts on their mixers because of the high fibrous content of the waste.

Large land areas are required by the digesters, a minimum 12 acres (5 ha) for a 1000-ton/day (900-tonne/day) plant.[9] This can be a problem where transport costs prohibit long-range refuse movement, and the treatment facility must be located on expensive urban land. And finally, the problem of what to do with the residue has not been solved. The sludge does seem to dewater readily, as it should, with all the fiber in it, but its ultimate disposal is an additional problem in the application of this process.

Methane Extraction from Landfills

The extraction of gas from landfills is discussed in detail in Chapter 4. Gas is extracted by placing gas collectors into the landfill when it is being constructed, or by drilling wells into completed landfills. The design of methane extraction from landfills involves the proper spacing of wells, the type of wells, and the gas cleaning or processing facility. Wells can be either vertical or horizontal, depending on the need. Typically, a 1- to 3-foot (0.3- to 1.0-m) diameter auger is used to drill the well and a 4- or 6-inch (10- to 15-cm) PVC (polyvinyl chloride) pipe is placed inside the well with the remaining space filled in with gravel. The screen is typically on the bottom portion of the well. At the header of each well is a valve to control the suction on the well and a sampling port. Oxygen readings are taken, and if oxygen is detected, the suction is reduced. Each well is connected to a HDPE (high-density polyethylene) flexible pipe collection system, and the collection piping connects to the blower and the energy recovery system or flare. Because landfills are continuously settling, gas extraction wells may have to be replaced on a regular basis. Wells can be installed as the landfill is constructed or drilled afterwards. Care is required during the drilling operation, since methane gas is escaping and mixing with oxygen.

Landfill gas is an excellent fuel, but its transport is expensive if the distance is great, thus production of electricity on-site is advantageous. The gas cannot be fed directly into a natural pipeline because of its low heating value and presence of contaminants such as water. Upgrading the gas can be accomplished by molecular sieves, and the product has a heating value of 950 kJ/m^3 (1000 Btu/ft^3). Molecular sieve process uses hydrated metal aluminum silicates characterized by a structure such that it allows only certain molecules to enter, depending on their size, shape, and polarity, and excludes all others. The gas is circulated through a vessel that contains the absorbent until the capacity of the silicate to absorb the impurities, in this case CO_2, is reached. As the gas flow is switched to a second vessel, the first one is regenerated by depressurizing. Eventually, simple depressurization no longer completely cleans the screen, and the silicate has to be heated to be regenerated. The cleaning is cyclical but wholly automated, resulting in a continuous flow of gas.

Potential for the Application of Methane Extraction from Landfills

Estimates show that it is unlikely that landfill gas will ever contribute more than about 0.5% of our national gas use[11] so it is not a major source of pipeline gas. It is, however, a useful fuel as it is extracted from the landfill, and the direct use of this 500 Btu/ft^3 (4.4 $\times$ 10^6 cal/m^3) gas should increase in the future.

Landfill gas is currently being used in several different ways. Some landfills use a gas turbine (1 to 10 megawatt) to generate electricity. This is the most efficient method of generation, but also takes a high-quality fuel. As the gas is pressurized prior to injection into the turbine, water vapor is condensed and removed. The condensate is likely a hazardous waste and must be treated as such. Common practice used to be to drain the condensate back into the landfill. One option is to inject the condensate into the gas flare for destruction.

A second option is to burn the methane in an internal combustion engine (up to 5 megawatts) and produce electricity. The landfill gas for this operation does not have to be as clean as for a gas turbine. The third option is to burn the gas in a gas-fired boiler. Some units are as large as 49 megawatts. Finally, the gas can be mixed with existing natural gas if a source is near the landfill. While all of the above options are technically feasible, one constraint is economics. If local utilities are willing to pay only 2 or 3 cents per kilowatt, electrical generation may not be economically feasible.

COMPOSTING

Composting differs from the previous two processes in that it is an aerobic process, and the end-product is the partially decomposed organic fraction. Composting is often promoted as a "natural" process of solid waste treatment. One reason for this reputation is that compost piles can be readily constructed in the backyard, and the product is a useful soil conditioner. It is little wonder, therefore, that municipal engineers and city councils are besieged by citizens groups urging that composting be initiated in their community in place of alternative solid waste disposal schemes such as landfilling and combustion, which many people view as a waste of money and natural resources.

Fundamentals of Composting

Aerobic microorganisms extract energy from the organic matter through a series of exothermic reactions that break the material down to simpler materials. The basic aerobic decay equation holds:

$$[\text{complex organics}] + O_2 \xrightarrow[\text{microorganisms}]{\text{aerobic}} >$$

$$CO_2 + H_2O + NO_3^- + SO_4^{-2} + [\text{other less complex organics}] + [\text{heat}]$$

During this decomposition the temperature increases to about 70°C (160°F) in most well-operated composting operations. As the reaction develops, the early decomposers are mesophilic bacteria followed after about a week by thermophilic bacteria, actinomycetes, and thermophilic fungi.[12] Above 70°C, spore-forming bacteria predominate. As the decomposition slows, the temperature drops and mesophilic bacteria and fungi reappear. Protozoa, nematodes, millipedes, and worms are also present during the later stages. The concentration of dead and living organisms in compost can be as high as 25%.[12]

The elevated temperatures destroy most of the pathogenic bacteria, eggs, and cysts. Some of the more common pathogens and their survival at elevated temperatures are shown in Table 8-3. The product of thermophilic composting is essentially free of pathogens. All potential pathogens, including resistant parasites such as Ascaris eggs and cysts of *Entamoeba histolytica*, are destroyed.[13]

Table 8-3   Destruction of Some Common Pathogens and Parasites During Composting

Salmonella typhosa	No growth beyond 46°C, death within 30 min at 55–60°C and within 20 minutes at 60°C; destroyed in a short time in compost environment
Salmonelia sp.	Death within 1 h at 55°C and within 15–20 min at 60°C
Escherichia coli	Death for most within 1 h at 55°C and within 15–20 min at 60°C
Shigella sp.	Death in 1 h at 55°C
Entamoeba histolytica cysts	Death within a few minutes at 45°C
Trichinella spiralis larvae	Quickly killed at 55°C
Brucella abortus or *Br. suis*	Death within 3 min at 62°C and within 1 hr at 55°C
Streptococcus pyogenes	Death within 10 min at 50°C
Mycobacterium tuberculosis var. *hominis*	Death within 15–20 minutes at 66°C or after momentary heating at 67°C
Corynebacterium diphtheriae	Death within 45 min at 55°C
Ascaris lumbricoides eggs	Death in less than 1 h at 50°C

Source: (12)

A critical variable in composting is the moisture content. If the mixture is too dry the microorganisms cannot survive, and composting stops. If there is too much water, the oxygen from the air is not able to penetrate to where the microorganisms are, and the mixture becomes anaerobic. The right amount of moisture, whether wastewater sludge or other sources of water, that needs to be added to the solids to achieve just the right moisture content can be calculated from a simple mass balance:

$$M_p = \frac{M_a X_a + 100 X_s}{X_s + X_a}$$

where M_p = moisture in the mixed pile ready to begin composting, as percent moisture

M_a = moisture in the solids such as the shredded and screened refuse, as percent moisture

X_a = mass of solids, wet tons

X_s = mass of sludge or other source of water, tons (This assumes that the solids content of the sludge is very low, a good assumption if waste activated sludge is used, which is commonly less than 1% solids.)

EXAMPLE 8-2

Ten tons of a mixture of paper and other compostable materials has a moisture content of 7%. The intent is to make a mixture for composting of 50% moisture. How many tons of water or sludge must be added to the solids to achieve this moisture concentration in the compost pile?

$$M_p = \frac{M_a X_a + 100 X_s}{X_s + X_a} = \frac{(10 \times 7) + (100 \times X_s)}{10 + X_s} = 50$$

Solving for X_s yields 8.6 tons of water or sludge.

If the water to be added is expressed in gallons instead of tons, the water balance equation reads

$$M_p = \frac{M_a X_a + 0.416 W_s}{X_a + 0.00416 W_s}$$

where W_s = water or sludge to be added in gallons. The other variables are as defined previously.

All biochemical conversion processes such as composting are in essence two-step operations. The first step is the decomposition of complex molecules of waste materials into simpler entities. If there is no nitrogen available, this is the full extent of the process. If nitrogen is available, however, the second step is the synthesis of the breakdown products into new cells. These new microorganisms contribute to the process, and the system operates in balance.

Because of the high rate of microbial activity, a large supply of nitrogen is required by the bacteria. If the reaction were slower, the nitrogen could be recycled, but since many reactions are occurring concurrently, a sufficient nitrogen supply is necessary. The requirement for nitrogen can, as before, be expressed as the C/N ratio.

A C/N of 20:1 is the ratio at which nitrogen is not limiting the rate of decomposition. Above a C/N of 80:1, thermophilic composting cannot occur because the nitrogen severely limits the rate of decomposition. Most systems operate between these extremes. Some researchers recommend an optimal C/N ratio of 25:1.[14] A higher C/N ratio than this can increase the time to maturity. Nitrogen can become limiting at a C/N ratio greater than about 40:1.[15] At higher pH levels the nitrogen will be lost into the atmosphere as ammonia gas if the C/N ratio exceeds 35:1.

The C/N ratio generally decreases during the composting process as the organic carbon is converted to carbon dioxide. Unless there is a concurrent loss of ammonia (which is possible as the pH becomes more basic), the nitrogen content remains fairly constant, resulting in a decrease in the C/N ratio.[15]

The C/N ratios for various materials used in composting are shown in Table 8-4.

Table 8-4   Carbon/Nitrogen Ratios for Various Materials

	C/N
Food waste	
Raleigh, NC	15.4
Louisville, KY	14.9
MSW (including garbage)	
Berkeley, CA	33.8
Savannah, GA	38.5
Johnson City, TN	80
Raleigh, NC	57.5
Chandler, AZ	65.8
Sewage sludge	
Waste activated	6.3
Mixed digested	15.7
Wood (pine)	723
Paper	173
Grass	20
Leaves	40–80
Sawdust	511

Source: (15)

The calculation of carbon and nitrogen levels and the C/N ratio is straightforward and based on mass balances. If two materials, such as shredded refuse and sewage sludge, are mixed, the carbon of the mixture is calculated as

$$C_p = \frac{C_r X_r + C_s X_s}{X_r + X_s}$$

where C_p = carbon concentration in the mixture prior to composting, as percent of total wet mass of mixture

C_r = carbon concentration in the refuse, as percent of total wet refuse mass

C_s = carbon concentration in the sludge, as percent of total wet sludge mass

X_s = total mass of sludge, wet tons per day

X_r = total mass of refuse, wet tons per day

The pH of the compost pile varies with time, showing an initial drop and then increasing to between 8.0 and 9.0, finally leveling off between 7.0 and 8.9.[16] If the compost heap becomes anaerobic, however, the pH continues to drop due to the action of the anaerobic acid formers. As long as the pile stays aerobic, there is sufficient buffering within the compost to allow the pH to stabilize at an alkaline level. For educational purposes the progression of pH and temperature in a compost pile can be readily demonstrated by laboratory-scale apparatus.[17]

The pH also affects nitrogen loss, because ammonia escapes as ammonium hydroxide above a pH of 7.0. Thus efficient compost operations, which operate around a pH of 8.0, cannot retain nitrogen at a greater concentration than C/N of about 35:1.[12]

The time required for a compost pile to mature depends on such factors as the putrescence of the feed, the insulation and aeration provided, the C/N ratio, the particle size, and other conditions. Usually, two weeks is considered the minimum time for the adequate composting of shredded municipal refuse in windrows. Mechanical composting plants, using inoculation of previously composted materials, can accomplish decomposition in 2 or 3 days. This material is still quite active, however, and usually requires further stabilization.[12]

The completion of composting is judged primarily on the basis of a slight drop in temperature and a dark brown color. A more accurate measure is the determination of starch concentration in the compost. Starch is readily decomposable, and thus its disappearance is a good indicator of mature compost. A simple laboratory method for measuring starch in compost is available, although the technique yields only qualitative information.[18, 19] This technique can also be applied to a composting demonstration project for the classroom.[17] A more rigorous measure of the end point is the drop in the C/N ratio to perhaps 12:1. Higher C/N ratios will result in continued decomposition of the compost after it is applied, and the subsequent robbing of nitrogen from the ground.

Recognizing the difficulties involved in the processing of a heterogeneous material such as municipal solid waste, Golueke[15] suggests that the viability of any biochemical process be judged on the basis of the organisms employed. Regardless of what biochemical process is used, the organisms should have the following characteristics:

Not fastidious—they will work under adverse conditions (e.g., wide temperature range) and be tolerant of environmental change.

Ubiquitous—they should exist in nature, since pure stock cultures degenerate with time and rarely stay pure.

Persistent—they must grow in the environment without special assistance.

Not picky—they should be able to use a broad spectrum of substrates.

If these criteria are used to judge the efficacy of composting, the process would pass with flying colors. Composting is able to handle many organic wastes and seems to be insensitive to changes in flow rates and feed characteristics. From a purely fundamental and biochemical perspective, composting of MSW makes a great deal of sense. Actual experiences with MSW composting, however, have been for the most part negative.

Composting Municipal Solid Waste

Composting on a municipal scale is an uncomplicated process. At its simplest, shredded and/or screened refuse or source-separated organic waste is placed in long parallel piles, called *windrows,* and the moisture content is maintained near 50%. The piles are periodically aerated by fluffing and moving the material around. After several weeks, such accelerated aerobic decomposition results in a dark brown earthy-smelling material that has low nutrient value but is an excellent soil conditioner. The ranges of nitrogen, phosphorus, and potassium in finished compost from MSW are shown in Table 8-5:

Table 8-5   Fertilizer Value of Compost from MSW

Nutrient	Range, as fraction of total solids
Nitrogen	0.4 to 1.6
Phosphorus	0.1 to 0.4
Potassium	0.2 to 0.6

Source: (12)

Typically, windrows are constructed of sorted and shredded MSW, mixed with wastewater sludge, and laid in long rows of about 4 to 6 ft (1.2 to 2 m) high (Figure 8-2). Because the reaction is aerobic, oxygen must be made available to the microorganisms, and this is done either by turning the pile with a specially constructed agitator (Figure 8-3) or by placing the pile on PVC pipes so that air can be pulled through the pile (Figure 8-4). This is called the *aerated static pile* composting method. Both of these operations offer the advantage of low capital cost and the simultaneous use of wastewater sludge, but the operating cost can be high, and odor problems can be serious if the piles are not sufficiently well aerated.

More sophisticated are the *in-vessel* composting plants. In one in-vessel process the shredded and sorted refuse is mixed in an aerobic digester with air being injected through hollow augers (Figure 8-5). The residence time within such a unit is short—often as little as 24 hours, but usually more like 5 days—and the rate of decomposition is quite rapid. Typically, the composted material is removed from the aerobic digester and allowed to cure in windrows for several weeks prior to sale or disposal. A different mechanical technique uses a long rotary drum (Figure 8-6) in which the MSW is slowly turned in a long barrel during a 24-hour detention time.

Figure 8-2 Windrow composting system.

Some of these units use refuse in the unshredded and unsorted state, with perhaps only a bag opener preceding the composting barrel, and screens following digestion. As before, the screened fines must then be stored in windrows for several weeks prior to use. The rejects from the screens are sent to a landfill.

Figure 8-3 Mobile aerator for windrows. (Courtesy Scat Engineering)

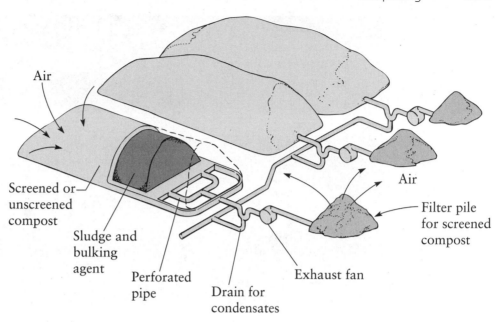

Figure 8-4 Aerated static pile composting system.

Potential for Composting Municipal Solid Waste

Composting is an old process and quite well understood. The practical application of this process to MSW, however, is limited by three serious problems:

1. Lack of markets for the finished product.
2. Small reduction in the total refuse volume requiring disposal.
3. Environmental factors of composting plants, specifically odor.

The first is the most serious of the three problems. The majority of the municipal composting plants in the United States have closed because they could not sell (or even give away) the product. When composting plants are designed, however, the economic analysis invariably includes a profit from the sale of compost. The closing of some plants—the one in Portland, Oregon, and an 800-ton/day plant in Dade County, Florida—had serious financial consequences on the letter of credit bank, making future financing difficult.

The second problem, the limited reduction in the volume of waste, usually has technical solutions because the process can be changed to actually capture such items as glass, metals, and so on. In fact, composting may be considered as simply another process within a complete materials recovery plant. One pilot-scale process has been run to produce a putrescent material from the screening (7-mesh) of the light fraction of air-classified refuse.[20] This material consisted of 6% of the raw refuse, and contained 74% volatile solids. Although in this study the material was digested for methane production, it could also be readily composted.

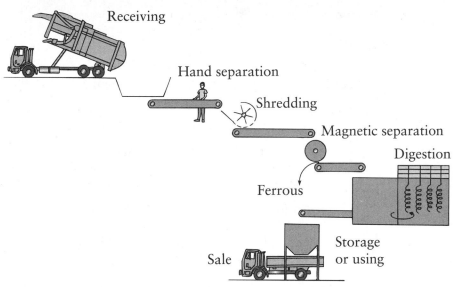

Figure 8-5 Mechanical composting system using hollow augers in an aerated digester.

The fraction of putrescent material in refuse has been steadily declining over the years, and will no doubt continue to do so. As the organic fraction is reduced, the compostable fraction similarly drops, and it is unlikely that more than 50% of refuse could ever be recovered as compost. The remaining 50% must be disposed of by normal means, and this cost must be added to the total cost of the composting facility.

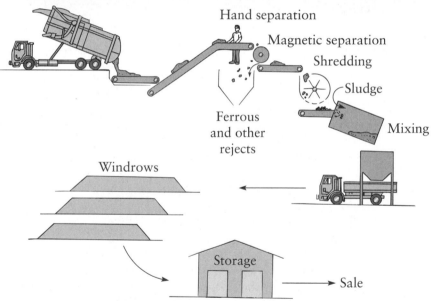

Figure 8-6 *Rotating composting system.*

While composting is often touted as the "natural" means of solid waste disposal, it is not without adverse environmental impact. In fact, if a risk analysis is conducted on the effect of composting and the potential pathways of dioxins and pathogenic organisms causing health problems, the effect of composting seems to be much more adverse than the effect of MSW combustion.[21] While data to substantiate such conclusions are impossible to obtain, it is nevertheless true that composting is not without its health risks.

The last problem is mostly one of odor production. Although some people insist that the odor from compost piles is pleasant, these people are in the distinct minority. Compost plants do smell, there is no doubt of that, and thus the plants must be located fairly far from residential areas. This requirement adds another cost—transportation.

MSW composting plants are paid a tipping fee to accept MSW. If the plant cannot sell the compost, it must continue to accept incoming MSW to maintain a cash flow. This results in an ever-increasing pile of compost on-site. As the pile builds, it eventually is too big to be turned and it becomes anaerobic, causing odor problems.

In conclusion, it seems unlikely that municipal refuse will be composted on a large scale, at least in more developed countries. Whenever composting of refuse on a municipal scale is suggested, the decision makers should always bear in mind the dismal record of past composting efforts. All of these plants were built under an aura of optimism and enthusiasm, and all of the economic analyses promised success. The engineer and decision maker should ask what is so different in the proposed new compost plant that would allow it to succeed while all the others have failed.

Composting Wastes Other Than Refuse

While the composting of MSW is problematical, many other wastes can be composted at a much lower cost, and the products from such facilities may have significant environmental and cash value.

Yard wastes are now composted in thousands of communities in the United States. A typical yard waste operation consists of first grinding the materials in tub grinders, wetting thoroughly, and placing the mixture in windrows about 8 ft (2.5 m) wide. Periodic aeration is required, and since leaves are low in nitrogen, a nutrient such as sodium nitrate, at a rate of 1 lb/yd^3 (0.6 kg/m^3) of stacked leaves, is added.[22] After about 6 months, the composted yard waste can be again ground in a simple compost grinder and sold to the public. The entire operation is inexpensive and produces a high-quality soil conditioner.

The level of undesirable contaminants in compost made of yard waste and leaves is surprisingly low,[23] so the compost produced from this raw material can be used in gardens without threat of contamination from pesticides or heavy metals. Metals in compost manufactured from MSW, by contrast, can be at levels that present significant public health concern.[24] Lead, for example, is typically at a level of 800 ppm in compost made from MSW.[25] The largest single problem is the plastic bags many homeowners use to package their leaves and other yard waste.

Composting of organic materials such as leaves and grass can also be used as a potent method for the destruction of other undesirable wastes. For example, such toxic organic compounds as chlorinated pesticides are often found to be amenable to degradation in compost piles.[26, 27] Compost placed on contaminated soils such as are found in many of the *brownfields* in the rust belt of the United States can also be highly beneficial. The advantages of composting in cleaning up the contamination seems well established.[28]

Various types of industrial and agricultural wastes can also be readily composted. A combination such as sawdust and chicken manure, for example, produces a superior compost, high in nitrogen.

Sludges from sewage treatment plants have for many years been used as sources of nutrients in composting.[29] The problem with composting sludge has always been that the solids compact too tightly, leaving no spaces for air to enter the pile. This problem can be solved by mixing the sludge with wood chips and composting the mixture. The wood chips are composed of poorly decomposable cellulose and lignin and can be readily removed from the stabilized material by simple screening and can then be reused. Raw sludges can be composted to a passable soil conditioner or a high-grade topsoil in about two weeks by this method.

A number of special composting operations have been successful. In Altoona, Pennsylvania, compost is used for growing mushrooms, while Disney World in Florida uses static pile composting to treat the food waste and sewage sludge from its complex of theme parks and hotels.

OTHER BIOCHEMICAL PROCESSES

Glucose Production by Acid and Enzymatic Hydrolysis

Cellulose is abundant in nature, comprising at least one-third of all the material in the plant world. The approximate empirical formula of cellulose is $C_6H_{10}O_5$, with a molecular weight of about 500,000. Pure cellulose has a low solubility in water, and the molecules are tightly bound together by hydrogen bridges to form chains.

Cellulose breaks down to form glucose by either acid or enzymatic hydrolysis. One kilogram of cellulose is capable of being converted to 0.5 kg of glucose. This process is used by the ruminants (cows, deer, sheep, etc.) for the production of glucose, whereas other animals cannot digest cellulose and must ingest sugars directly.

Although cellulose is a renewable resource, its energy coming from the sun, its supply is not unlimited. One constraint to the production of cellulose is available land, since land that could be used for cellulose production may instead have to be used for other purposes, such as food production. On the other hand, the cellulose found in waste is a "free" source, as no additional land is needed for its production and no additional resources must be allocated. The reclamation of cellulose from wastes thus makes a great deal of sense.

Foremost among cellulosic wastes other than MSW are agricultural wastes, which are usually free of contaminants and often occur in large quantities. The ideal use of agricultural wastes occurs when the normal procedure is to collect the residues as part of the harvesting or processing operation. One example of this is bagasse (sugar cane after the sugar has been squeezed out). This is a clean, shredded, homogeneous material that still contains considerable energy, as illustrated by the fact that it is used as a primary electric fuel in power production in several sugar plantations in Hawaii. It could just as readily be used as a source of cellulose for the production of glucose. Wastes from the wood-processing industries also yield clean products that might be used as raw materials for cellulose production.[30]

In MSW cellulose is found mostly in paper and paper products, which constitute a major fraction of refuse. Wood and cotton are also sources of cellulose, but their fraction within the waste stream is small. Wood is mostly lignaceous in composition and, relative to paper, is not a good source of cellulose.

Cellulose is first converted to glucose, which can then be used to produce various products including ethanol, acetone, and other useful organic chemicals. The various ways by which glucose is further processed, such as fermentation and microbial and chemical conversion of glucose, are not covered here and may be found in standard biochemistry or modern chemical engineering texts. The process of interest here is the conversion of cellulose to glucose.

Cellulose belongs to a family of organic compounds called polysaccharides, which are polymers of simpler compounds called monosaccharides, such as glucose. The structure of cellulose is shown in Figure 8-7. The number of monomers making up the chain is very large, in the range 2000 to 3000. The cellulose molecules are bound together tightly by a large number of hydrogen bridges along the chain.

The conversion of cellulose to glucose is through a process known as hydrolysis, which in organic chemistry denotes any reaction in which water is involved. The empirical equation for this process is

$$C_6H_{10}O_5 + H_2O \rightarrow C_6H_{12}O_6$$

where the product is glucose. In structural terms hydrolysis involves the severing of the bonds (β - 1:4 glucosidic) between the cellulose monomers, and the reaction of these with water.

This reaction can be achieved by several processes, particularly by the use of either acids or enzymes. Under carefully controlled conditions, mild aqueous acids will break down the cellulose into glucose. In practice acid hydrolysis requires corrosion-proof equipment, and the high temperature and the low pH can cause decomposition of the resulting sugars. The process must therefore be so balanced as to produce more sugar than that which is destroyed during the process.[31]

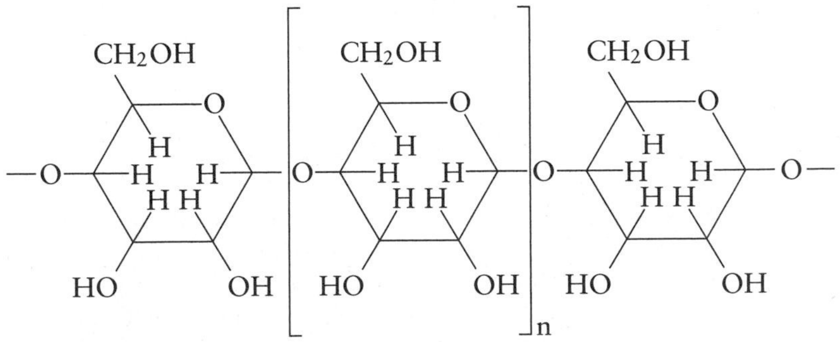

Figure 8-7 *Simplified structure of cellulose.*

Glucose yields of about 50% per weight of cellulose have been obtained, but this invariably contains unwanted by-products.[32]

Enzymatic hydrolysis of cellulose, on the other hand, seems to perform well with high glucose yields obtained without unwanted by-products.[31] The process is based on the use of cellulose enzymes derived from mutant strains of the fungus *Trichoderma viride* or other organisms that are capable of hydrolyzing insoluble cellulose.

The first step in the process is growing the fungus in a culture medium containing pure cellulose and various nutrients. The fungus is then filtered out, and the enzymes are captured in the filtrate. The waste-containing cellulose is then mixed with the enzyme broth in a reactor at 50°C and at pH adjusted to 4.8. The glucose is produced in the reactor and filtered out.

The enzymes that hydrolyze glucose are called cellulases. In strict chemical terms cellulase is known as β - 1:4 glucanase, since these enzymes hydrolyze the β - 1:4 glucosidic linkages of cellulose. The two main cellulases are called C_1 and C_x. The C_x component attacks amorphous cellulose, whereas C_1 together with C_x is required for hydrolyzing insoluble crystalline cellulose.

The action of the cellulase enzyme complex (C_1, C_x) on cellulose can be approximated as

(native cellulose) – C_1 → (hydrated polyanhydroglucose chains) – C_2 →

(cellobiose) – β-glucosidase → (glucose)

The actual reactions are quite complex, with multicomponent enzymes and many reaction pathways. The susceptibility of native cellulose to enzymatic hydrolysis is affected by a number of variables, the most important being the degree of crystallinity, the presence of lignin and its association with cellulose, and the degree of polymerization of the native cellulose.[33] X-ray diffraction analysis has shown the cellulose has four stable crystal lattices. Any of these crystalline highly ordered forms of cellulose are more resistant to enzymatic hydrolysis than is the amorphous or less-ordered variety. The crystalline lattice structures can be broken down physically by crushing or grinding the cellulose.

The presence of lignin can interfere with the enzymatic hydrolysis of cellulose. The exact nature of the association between lignin and cellulose is still unknown, and the possibilities of a covalent chemical bond, hydrogen bonding, and incrustation of the cellulose in a lignin network, thus preventing the access of the enzyme molecules, all have some scientific evidence.[33] Whatever the association, however, it is possible to break down the complex by various chemical means as well as by mechanical grinding.

The third important variable affecting the rate of enzymatic hydrolysis is the degree of polymerization. Again, the degree of polymerization can be significantly reduced by mechanical milling of the cellulose.

Other Bacterial Fermentation Processes

Single-cell protein can be produced by the bacterial fermentation of bagasse.[34] The term *fermentation* is used in biochemical engineering to mean any process wherein CO_2 is produced. Fungus can be used to convert cellulose to protein.[35] This process can be speeded up by the treatment of cellulose with inorganic salts and ultraviolet light, but the long required irradiation time makes the practical application of the process questionable.[36] The production of ethanol by the fermentation of glucose has also received considerable interest.[37] Ethanol can be used as a substrate for the growth of single-cell protein, with up to 93% yield.[38]

FINAL THOUGHTS

Engineers working for municipalities, either directly or indirectly, have to have some degree of autonomy. Professionals correctly believe that professional autonomy is beneficial to the welfare of the public. If the government starts telling physicians how to treat people, or telling preachers what to preach, or telling engineers how to build things, then the public loses. Accordingly, the professions have jealously guarded their autonomy in the name of the public good. The engineering profession recognizes that if engineering is to maintain its professional autonomy, the public has to trust engineers and that it is very much to the advantage of engineering and the public at large to maintain this trust.

Such autonomy can, of course, be taken away by the state, as witnessed in nations having totalitarian governments such as the former Soviet Union, in which once-proud and independent engineers became tools of the communist government and had little say in technical decisions. The inability of the engineers to voice their concerns about projects that were counterproductive, wasteful in resources, and harmful to the public was in great part responsible for the eventual downfall of the Soviet empire.[39]

Engineers, as all other professionals, must work to maintain public trust. To promote trust in the engineering profession, professional engineering societies have all drafted statements that express the values and aspirations of the profession, statements commonly referred to as a "Code of Ethics."

One of the earliest codes of ethics in the United States was adopted in 1914 by the American Society of Civil Engineers (ASCE). Based in spirit on the original Code of Hammurabi,[40] the 1914 ASCE Code addressed the interactions between engineers and their clients, and among engineers themselves. Only in the 1963 revisions did the ASCE Code include statements about the engineer's responsibility to the general public, stating as a fundamental canon the engineer's responsibility for the health, safety, and welfare of the public.

In 1997 ASCE modified its code to include the commitment of engineers to sustainable development. The term *sustainable development* was first popularized by the World Commission on Environment and Development (also known as the Brundtland Commission), sponsored by the United Nations. Within this report sustainable development is defined as "development that meets the needs of the present without compromising the ability of future generations to meet their own needs."[41] Sustainable development can be defined in a number of ways—and indeed the Brundtland Report itself includes ten different definitions—while a report for the United Kingdom Department of the Environment contains thirteen pages of definitions.[42]

Although the original purpose of introducing the idea of sustainable development was to recognize the rights of the developing nations in using their resources, sustainable development has gained a wider meaning and now includes educational needs and cultural activities, as well as health, justice, peace, and security.[43] All these are possible if the global ecosystem is to continue to support the human species. We owe it to future generations, therefore, not to destroy the earth they will occupy. Using biochemical processes to produce useful products such as ethanol may still be far into the future, but we should keep our eye on the ball and recognize that the use of fossil fuels for our energy use is not in keeping with the principles of sustainable development, and eventually we have to find ways of producing useful products from other raw materials, such as MSW.

REFERENCES

1. Alter, H. 1989. "The Origins of Municipal Solid Waste: The Relationship Between Residues from Packaging Materials and Food." *Waste Management and Research* 7: 103–114.

2. Bell, J. M. 1964. "Characteristics of Municipal Refuse." *Proceedings* National Conference on Solid Waste Research, American Public Works Association, February.

3. Pfeffer, J. T. 1974. *Reclamation of Energy from Organic Refuse.* 670/2-74-016. Cincinnati, Ohio.

4. Pfeffer, J. T., and J. C. Liebman. 1976. "Energy from Refuse by Bioconversion Fermentation and Residue Disposal Processes." *Resource Recovery and Conservation* 1: 295.

5. Hagerty, D. J., J. L. Pavoni, and J. E. Heer. 1973. *Solid Waste Management.* New York: Van Nostrand Reinhold Company.

6. Vesilind, P. A. 1979. *Treatment and Disposal of Wastewater Sludge.* Ann Arbor, Mich.: Ann Arbor Science Publishers.

7. Lawrence, A. W., and P. L. McCarty. 1965. "The Role of Sulfide in Preventing Heavy Metal Toxicity in Anaerobic Treatment." *Journal of the Water Pollution Control Federation* 37: 113.

8. Pfeffer, J. T. 1974. "Temperature Effects on Anaerobic Fermentation of Domestic Refuse." *Biotechnology and Bioengineering* 16: 77.

9. Hille, S. J. 1975. *Anaerobic Digestion of Solid Waste and Sewage Sludge to Methane.* EPA OSWMP SW-15g, Washington, D.C.

10. Monteith, H. D., and J. P. Stephenson. 1997. "Mixing Efficiencies in Full-scale Anaerobic Digesters by Tracer Methods." *Proceedings* Symposium on Sludge Treatment. Wastewater Technology Centre, Burlington, Ontario, Canada.

11. Bowerman, F. 1979. "Methane Generation from Deep Landfills." *Proceedings* Engineering Foundation Conference on Resource Recovery, Henniker, N. H. (July).

12. Golueke, C. G. 1972. *Composting.* Emmaus, Pa.: Rodale Press, Inc.

13. Wilby, J. S. 1962. "Pathogen Survival in Composting Municipal Wastes." *Journal* Water Pollution Control Federation 34: 80.

14. Barktoll, A. W., and R. A. Nordstedt. 1991. "Strategies for Yard Waste Composting." *BioCycle* 32, n. 5: 60–65.

15. Golueke, C. 1978. "State of the Art of Bioconversion Processes." *Proceedings* Engineering Foundation Conference on Resource Recovery, Rindge, N. H. (July).

16. *An Analysis of Composting as an Environmental Remediation Technology.* 1998. EPA 530-R-98-008.

17. Vesilind, P. A. 1973. "A Laboratory Exercise in Composting." *Compost Science* (Sept.–Oct.).

18. Lossin, R. D. 1971. "Compost Studies." *Compost Science* (Mar.–Apr.).

19. Vesilind, P. A. 1973. *Solid Waste Engineering Laboratory Manual.* Durham, N. C.: Department of Civil Engineering, Duke University.

20. Diaz, L. F., F. Kurz, and G. J. Trezek. 1976. "Methane Gas Production as Part of a Refuse Recycling System." *Compost Science.*

21. Jones, K. H. 1991. "Risk Assessment: Comparing Composting and Incineration Alternatives." *MSW Management* 1, n. 3: 29–32, 36–39.

22. Walter, R. 1971. "How to Compost Leaves." *American City* (June): p. 116.

23. Roderique, J. O., and D. S. Roderique. 1995. Quoted in *Decision Maker's Guide to Solid Waste Management.* EPA 530-r-95-023. Washington, D.C.

24. Oosthnoek, J., and J. P. N. Smit. 1987. "Future of Composting in the Netherlands." *BioCycle* (July).

25. Richard, T. L., and P. Woodbury. 1993. *Strategies for Separating Contaminants from Municipal Solid Waste*. Cornell University Waste Management Institute.

26. Cole, M. A., L. Zhang, and X. Liu. 1995. "Remediation of Pesticide Contaminated Soil by Planting and Compost Addition." *Compost Science and Unitization* 3: 20–30.

27. Savage, G., L. Diaz, and C. G. Golueke. 1985. "Disposing of Hazardous Wastes by Composting." *BioCycle* 26: 31–34.

28. McKinley, V. 1984. *Microbial Activity in Composting Sewage Sludge*. Ph.D. Dissertation, U. of Cincinnati.

29. Wilson, G. B., and J. M. Walker. 1973. "Composting Sewage Sludge: How?" *Compost Science* (Sept.-Oct.).

30. Trezek, G. J., and C. G. Golueke. 1978. "Availability of Cellulosic Wastes for Chemical or Biochemical Processing." *Proceedings* Energy, Renewable Resources and New Foods, AIChE Symposium Series N. 72, p. 158.

31. Span, L. A., J. Medeiros, and M. Mandel. 1976. "Enzymatic Hydrolysis of Cellulosic Wastes to Glucose." *Resource Recovery and Conservation* 1.

32. Goldstein, I. S. 1974. *The Potential for Converting Wood into Plastics and Polymers or into Chemicals for the Production of These Materials*. Raleigh, N. C.: School of Forest Resources, North Carolina State University.

33. Hajny, G. J., and E. T. Reese (Eds.). 1969. *Cellulases and Their Applications*, Advances in Chemistry Series 95. Washington, D.C.: American Chemical Society.

34. Callahan, C., and C. E. Dunlop. 1974. *Construction of a Chemical-Microbial Plant for Production of Single Cell Protein from Cellulosic Wastes*. EPA, SW 24C. Washington, D.C.

35. Rogers, C. J., E. Coleman, D. F. Sino, T. C. Purcell, and P. V. Scoparius. 1972. "Production of Fungal Proteins from Cellulose and Waste Celluloids." *Environmental Science and Technology* 8: 715.

36. Cookson, A., and G. Froeisdorph. 1973. *The Nitrite Accelerated Photochemical Degradation of Cellulose as a Pre-treatment for Microbiological Conversion to Protein*. EPA. 670-2-73-052. Washington, D.C.

37. Porteus, A. 1972. "WP Disposal Process Turns Cellulose Material into Alcohol." *Paper Trade Journal* (Feb. 7).

38. McAbee, M. K. 1974. "Japan Pushes Alcohol for Protein Process." *Chemical Engineering News* (Dec. 9): p. 11.

39. Graham, L. 1996. *The Ghost of the Executed Engineer*. Cambridge, Mass.: Harvard University Press.

40. Hope, R. F. (Ed) 1994. "Hammurabi, King of Babylonia." *The Code of Hammurabi, King of Babylonia*, about 2250 B.C. Holmes Beach, Fla.: Wm. W. Gaunt.

41. Herkert, J. R., A. Farrell, and J. Winebrake. 1996. "Technology Choice for Sustainable Development." *Technology and Society*, IEEE 15, n.2.

42. Pearce, D., A. Markanya, and E. B. Barber. 1989. *Blueprint for a Green Economy*. Report for the UK Department of the Environment. London: Earthscan Publication.

43. World Commission on Environment and Development. 1987. *Our Common Future*. Oxford: Oxford University Press.

ABBREVIATIONS USED IN THIS CHAPTER

ASCE = American Society of Civil Engineers
HDPE = high-density polyethylene

MSW = municipal solid waste
PVC = polyvinyl chloride

PROBLEMS

8-1.   Assume that refuse has a C/N ratio of 24:1. If raw sludge, with a C/N of 16:1, is to be added to refuse to reach a required C/N of 20:1, how much of each, refuse and sludge, is needed? (*Note:* Use a wet weight basis.)

8-2.   Assume that a city of 50,000 people uses, on the average, 20 million m^3 of natural gas per year. What fraction of this demand could be met by digesting the refuse from this community?

8-3.   Using the approximate chemical analyses of refuse, $C_{99}H_{149}O_{59}N$, develop an empirical formula for the organic fraction of refuse, and estimate the theoretical production of methane from this hypothetical compound.

8-4.   Estimate the methane gas production from a local landfill. Show all assumptions made.

8-5.   Suppose you are a town engineer. An advertisement appears in a trade journal for a new in-vessel composting system that produces methane gas as a valuable end-product. Your town manager, who has no technical training, asks you to find out more about this system. You decide to write a letter to the company, Bioscam Inc., requesting more information. What questions would you ask? Write such a letter.

8-6.   Assume a rough estimate of MSW composition is as follows:

	% by weight
Paper (assume all cellulose, $C_6H_{10}O_5$)	50
Glass	10
Steel	10
Aluminum	10
Food waste (assume sugar $C_6H_{12}O_6$)	20

If 10,000 metric tonnes of this material is placed in a landfill, how much methane gas would theoretically be produced by anaerobic decomposition?

8-7.   Suppose a wet mixture of old bread, leaves, and soil is placed in a pile 1 m tall, and a long thermometer is stuck in the middle of the pile. Every other day the pile is aerated by throwing the mixture into a new pile. Draw a graph showing the temperature in the pile during the first two weeks and explain why this occurs.

8-8.   An industry operates a landfill into which an organic waste having the formula $C_4H_8O_4$ decays anaerobically.
   a.   How much methane gas will be generated if the decomposition is complete? Express your answer as "lb gas/lb waste."
   b.   Realistically, given the nature of the anaerobic reaction, how much methane gas might actually be produced (approximately), and why only this much?

8-9.   Estimate the production of CO_2 and CH_4 during the anaerobic decomposition of ethanol, C_2H_6O.

8-10.   Describe how acid or enzymatic hydrolysis occurs, using simple chemical equations.

Current Issues in Solid Waste Management

While most of this book deals with the technical aspects of integrated solid waste management, many issues related to solid waste engineering are managerial, financial, regulatory, and even political. This chapter includes a number of current issues that influence solid waste engineering today and may have an increasing effect in the future.

LIFE CYCLE ANALYSIS AND MANAGEMENT

One means of understanding questions of material and product use and waste production is to conduct what has become known as a *life cycle assessment*. Such an assessment is a holistic approach to pollution prevention by analyzing the entire life of a product, process, or activity—encompassing raw materials, manufacturing, transportation, distribution, use, maintenance, recycling, and final disposal. In other words, assessing its *life cycle* should yield a complete picture of the environmental impact of a product.

The first step in a life cycle assessment is to gather data on the flow of a material through an identifiable society. Figure 9-1, for example, shows the flow of paper through Switzerland. Once the quantities of various components of such a flow are known, the environmental effect of each step in the production, manufacture, use, and recovery/disposal is estimated.

Life Cycle Analysis

Life cycle analyses are performed for several reasons, including the comparison of products for purchasing and a comparison of products by industry. In the former case the total environmental effect of glass returnable bottles, for example, could be compared to the environmental effect of nonrecyclable plastic bottles. If all of the factors going into the manufacture, distribution, and disposal of both types of bottles are considered, one container might be shown to be clearly superior. In the case of comparing the products of an industry, we might determine if the use of phosphate builders in detergents is more detrimental than the use of substitutes that have their own problems in treatment and disposal.

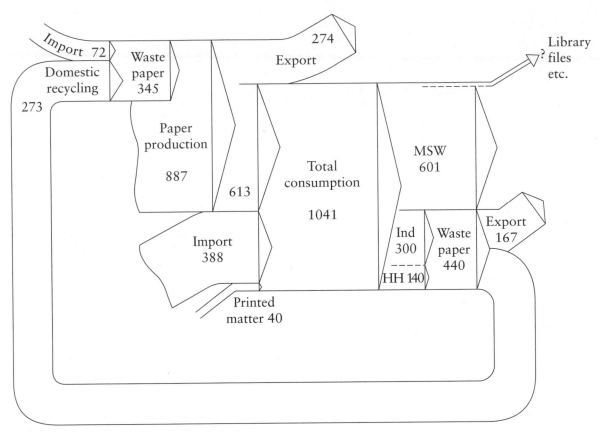

Figure 9-1   Flow of paper through Swiss society. All figures in tonnes/yr. Source: (1)

One problem with such studies is that they are often conducted by industry groups or individual corporations, and (surprise!) the results often promote their own product. For example, Proctor & Gamble, the manufacturer of a popular brand of disposable baby diapers, found in a study conducted for them that the cloth diapers consume three times more energy than the disposable kind. But a study by the National Association of Diaper Services found disposable diapers consume 70% more energy than cloth diapers. The difference was in the accounting procedure. If one uses the energy contained in the disposable diaper as recoverable in a waste-to-energy facility, then the disposable diaper is more energy efficient.[2]

Life cycle analyses also suffer from a dearth of data. Some of the information critical to the calculations is virtually impossible to obtain. For example, something as simple as the tonnage of solid waste collected in the United States is not readily calculable or measurable. And even if the data *were* there, the procedure suffers from the unavailability of a single accounting system. Is there an optimal level of pollution, or must all pollutants be removed 100% (a virtual impossibility)? If there are both air pollution and water pollution, how must these be compared?

A recent study supported by the EPA developed complex models using principles of life cycle analysis to estimate the cost of materials recycling.[3] The models were able to calculate the dollar cost, as well as the cost in environmental damage caused at various levels of recycling. Contrary to intuition, and the stated public policy of the EPA, it seems that there is a breakpoint at about 25% diversion. That is, as shown in Figure 9-2, the cost in dollars and adverse environmental impact starts to increase at an exponential rate at about 25% diversion. Should we therefore even strive for greater diversion rates, if this results in unreasonable cost in dollars and actually does harm to the environment?

A simple example of the difficulties in life cycle analysis would be in finding the solution to the great coffee cup debate—whether to use paper coffee cups or polystyrene coffee cups. The answer most people would give is not to use either, but instead to rely on the permanent mug. But there nevertheless are times when disposable cups are necessary (e.g., in hospitals), and a decision must be made as to which type to choose.[4] So let's use life cycle analysis to make a decision.

The paper cup comes from trees, but the act of cutting trees results in environmental degradation. The foam cup comes from hydrocarbons such as oil and gas, and this also results in adverse environmental impact, including the use of nonrenewable resources. The production of the paper cup results in significant water pollution, while the production of the foam cup contributes essentially no water pollution. The production of the paper cup results in the emission of chlorine, chlorine dioxide, reduced sulfides, and particulates, while the production of the foam cup results in none of these. The paper cup does not require chlorofluorocarbons (CFCs), but neither do the newer foam cups ever since the CFCs in polystyrene were phased out. The foam cups, however, result in the emission of pentane, while the paper cup contributes none. From a materials separation perspective, the recyclability of the foam cup is much higher than the paper cup since the latter is made

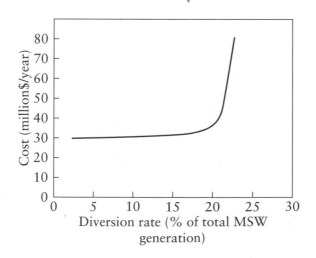

Figure 9-2 *Variation of cost with diversion rate, based on a life-cycle analysis.* Source: (3)

from several materials, including the plastic coating on the paper. They both burn well, although the foam cup produces 17,200 Btu/lb (40,000 kJ/kg), while the paper cup produces only 8600 Btu/lb (20,000 kJ/kg). In the landfill the paper cup degrades into CO_2 and CH_4, both greenhouse gases, while the foam cup is inert. Since it is inert, it will remain in the landfill for a very long time, while the paper cup will eventually (but very slowly!) decompose. If the landfill is considered a waste storage receptacle, then the foam cup is superior, since it does not participate in the reaction, while the paper cup produces gases and probably leachate. If, on the other hand, the landfill is thought of as a treatment facility, then the foam cup is highly detrimental since it does not biodegrade.

So which cup is better for the environment? If you wanted to do the right thing, which cup should you use? This question, like so many others in this book, is not an easy one to answer. Private individuals can, of course, practice pollution prevention by such a simple expedient as not using either plastic or paper disposable coffee cups but by using a refillable mug instead. The argument as to which kind of cup, plastic or paper, is better is then moot. It is better not to produce the waste in the first place. In addition the coffee tastes better from a mug! We win by doing the right thing.

Life Cycle Management

Once the life cycle of a material or product has been analyzed, the next (engineering) step is to manage the life cycle. If the objective is to use the least energy and cause the least damage to the environment, then much of the onus is on the manufacturers of these products. The users can have the best intentions, but if the products are manufactured in such a way as to make this impossible, then the fault is with the manufacturers. On the other hand, if the manufactured materials are easy to separate and recycle, then most likely energy is saved and the environment is protected. This process has become known as *pollution prevention*. There are numerous examples of how industrial firms have reduced emissions or other waste production, or have made it easy to recover waste products, and in the process saved money. Some automobile manufacturers, for example, are modularizing the engines so that junked parts can be easily reconditioned and reused. Printer cartridge manufacturers have found that refilling cartridges is far cheaper than remanufacturing them, and now offer trade-ins. All of the efforts by industry to reduce waste (and save money in the process) will influence the solid waste stream in the future.

Bad examples of industrial neglect of environmental concerns for the sake of short-term economic gain abound. The use and manufacture of nonrecyclable beverage containers, for example, is perhaps the most ubiquitous example. It might be instructive to establish some baseline of absolutely unconscionable industrial behavior in order to measure how far we have come in the pollution prevention process. The authors of this book recommend that there be an award established, called the *Trabi Award*, which could commemorate the worst, most environmentally unfriendly product ever manufactured. The Trablant, affectionately known as

the Trabi, was manufactured in East Germany during the 1970s and 1980s. This homely looking car (Figure 9-3) was designed to be the East German version of the Volkswagen (the People's Car). Its design objectives were to make it as cheap as possible, so the engineers used a two-stroke engine, which is a notoriously dirty engine and its wide use caused massive air pollution problems. All of the components were designed at the least cost and few survived normal use. Worst of all, the body was made of a fiberglass composite that was impossible to fix—except with duct tape! As far as the solid waste management was concerned, the Trabi had absolutely no recycling value since it could not be melted down or reprocessed in any other way, nor could it be burned in incinerators. After the reunification of Germany, thousands of Trabis were abandoned on the streets and had to be disposed of in landfills. The Trabi is the best example of engineering design when the sole objective is production cost and environmental concerns are nonexistent. The Trabi deserves to be immortalized as an exemplar of environmentally destructive design.

Figure 9-3   The East German Trablant, or "Trabi."

FLOW CONTROL

The San Diego Materials Recovery Facility is a perfect example of how nontechnical issues often override successful engineering. This 550,000-ton-per-year mixed waste materials recovery facility (dirty MRF) was built as a result of an agreement between a private company and the County of San Diego. When it went into operation, it met all of its performance requirements. Local communities, however, rather than send their municipal solid waste to the facility, elected to export their waste to remote landfills owned by their local haulers. Without sufficient waste to the MRF, the county had to close the facility after less than two years of operation.[5] The issue that caused the plant to close was the control of the solid waste stream, or *flow control.*

For several years Congress, states, and local communities struggled with the issue of flow control, the term used to describe the ability of the local municipality to direct its waste to a particular solid waste disposal or processing facility. This issue becomes critical when a community decides to finance a solid waste disposal facility. For example, a waste-to-energy facility might cost $200 million, and this represents a large investment by the community. Before such a facility can be financed, the owner must obtain "put-or-pay" commitments from participating communities who agree to send their solid waste to that facility or pay if the waste is not delivered. Such agreements provide the security to the bond holders that the facility will be able to pay the projected debt service during the life of the bonds.

Local communities are usually unable to direct private collectors to specific facilities. The courts have ruled that such requirements represent a restriction of the interstate commerce clause in the U.S. Constitution, which specifies that Congress shall make no laws to regulate commerce between states. Private waste collectors are free, therefore, to take their waste to other facilities, even in other states, regardless of any agreement that the local community may have made to finance the facilities.

To regain control of the waste, two options are available to local communities. The community could issue a franchise for the collection of solid waste, which includes the ability to direct solid waste to a specific facility. The second option is to replace the private collector with municipal collection. While both of these options are available, they may take as long as five years to implement. Thus municipalities have been looking to Congress for relief, but have not yet received it.

The reverse side of this issue is the ability to restrict solid waste from entering a local community. The United States Supreme Court issued a ruling in the case of the City of Philadelphia versus New Jersey in which Philadelphia tried to restrict the importation of waste from New Jersey. Again the Court ruled that the restriction of importation was unconstitutional on interstate commerce grounds. That ruling has been further refined so that a municipally owned landfill can restrict importation of waste, but a privately owned landfill cannot be restricted from receiving imported waste. In another case, the State of Oregon tried to place a surcharge of $3.00 per ton on waste from Washington; the courts overturned that action as well.

The courts have been unanimous in saying that laws restricting the flow of waste are unconstitutional. For example, in 1999 Virginia passed three laws to restrict the flow of waste into their state. Several large waste hauling companies filed suit, and in February 2000 the federal court overturned the laws. The judge stated in the decision that "The record demonstrates that Virginia acted to restrict the importation of municipal solid waste in a knee-jerk response to reports that increased levels of out-of-state MSW would soon be flowing into the Commonwealth, which—while perhaps advantageous politically, or commendable socially—is impermissible constitutionally."[6]

PUBLIC OR PRIVATE OWNERSHIP AND OPERATION

The debate between public and private ownership and operation is a long one and will continue into the foreseeable future. One thing can be said with certainty; there is no one right answer that applies to all communities.

Collection of solid waste is considered an essential public service, and the question is whether this service should be provided by private haulers or public employees. The private garbage company will claim it is more efficient, less wasteful, and more motivated. The public sector will claim it is less costly because it pays no tax and does not make a profit, is more responsive to the public, and provides a living wage for its employees. There is some truth to both claims. In the final analysis the municipality should consider cost, liability, and control when deciding on privatizing a solid waste system.[7]

A trend is to place collection services out to bid and let the municipality bid against the private companies. This approach is referred to as *competitive service delivery*. Unlike privatization, which turns the function over to the private sector, under competitive service delivery the public sector is allowed to compete for the project.[8]

While this approach appears to be good in principle, in reality it can be very difficult to implement. For example, once the public entity loses its collection system, it is almost impossible to get back into the business. The infrastructure necessary to provide the service—such as the personnel, equipment, and maintenance facilities—will no longer be available. Unlike private companies, government rules such as civil service and competitive procurement prevent local government from rapidly obtaining the resources necessary to provide collection services.

Another concern is the consolidation of the private sector. As companies purchase each other, the number of viable competitors continues to decrease. This has been described as the monopolization of solid waste through mergers and acquisitions.[9] With this approach companies can achieve *vertical integration*, whereby the company controls not only the collection of the waste, but also the transfer operation, the landfill, and the recycling operation. This decrease in competition may lead to an increase in costs for collection services.

Public or private ownership of landfills is another current solid waste issue. A landfill provides an essential public health function and is needed by every municipality. The public perception based on such well-publicized incidents as "The Garbage Barge" is that there is a landfill shortage. At the time of the "garbage barge" incident there was fear that landfill costs would soon exceed $100 per ton. In reality, no such shortage existed, and landfill prices dropped.

Public ownership does offer benefits to the local community. It can set its own rates, it can restrict the flow of MSW to the landfill, and it can decide on the level of service to be provided. On the other hand, a private landfill is free to compete in the open market and must offer competitive rates. One alternative that provides many benefits is to have the *operation* of a publicly owned landfill contracted out. This allows the local community to own an important asset while achieving the benefits of competition for the operation.

CONTRACTING FOR SOLID WASTE SERVICES

If a municipality does decide to contract with a private company for solid waste services, it will likely use a competitive bidding process to select the company. The municipality will prepare a *Request for Proposals* or *RFP*. In the RFP the municipality's level of solid waste service and terms and conditions for the contract or franchise agreement are described. While every RFP by its very nature is different, there are common elements that should be included. Those elements are described below:

Basic Service to Be Provided

The municipality must describe in detail the service included in the RFP. An example might be weekly curbside garbage collection of 12,000 residential customers with up to six 32-gallon cans of MSW per customer. This description specifies the following:

- collection frequency—weekly
- location of collection—curbside
- material collected—refuse
- number of customers—12,000
- type of customer—residential
- quantity of material—up to six 32-gallon cans
- container provider—customer

Each of these factors influences the cost of providing service and must be carefully spelled out in the RFP.

Options

In addition to the basic service to be provided, the municipality can include other options that may or may not be part of the proposal. For example, a municipality may request that the proposer bill the customer directly, or they may instead include the refuse collection bill with the water and sewer charges.

Terms of Agreement

The cost of providing the requested service is directly related to the term of the agreement. Start-up costs and capital costs are amortized over the life of the agreement. Thus the longer the term of the agreement, the lower the annual cost.

Cost of Service

Every RFP must specify how the proposer is to provide the cost information. Some RFPs ask for a very detailed breakdown of costs, while others ask only for the total cost. In addition the methodology used to adjust the costs during the term of the agreement should be specified.

Resources Provided

The RFP typically requires the proposer to provide a detailed description of the resources being provided. For example, the number and type of trucks being provided helps the municipality compare the proposals. In addition RFPs may require the names and resumes of the key personnel who will be assigned to the project. A classic example of needing to know this information was a project in Florida in which a private company was to assume the operation of a waste-to-energy project. One company proposed a plant manager who was currently under indictment for violating an order of the fire marshal at the waste-to-energy plant that he was currently operating.

Company Experience

A RFP usually includes a requirement that the proposer include a list of similar projects and references. This allows the selecting agency to determine if the proposer is qualified to provide the required service.

Record of Violations

Proposers are asked to provide a list of violations and judgments against the company and its officers. This information is used by the municipality to determine if the company meets their standard for contracting. For example, if the company has defaulted on its last three municipal contracts, or if the corporate officers have been convicted of bribing local officials, the company would probably not be selected for your contract. A record of past problems and poor performance may be sufficient reason for rejection. For example, the County of San Luis Obispo, California, recently rejected a large national firm because of its past practices. After being rejected the firm appealed to the courts to overturn the decision but lost the appeal.

Financial Resources

The municipality must be assured that the proposer has the financial resources to complete the project. Thus financial data are requested as part of the RFP. Examples might be audited annual financial statements or the most recent annual report.

Draft Agreement

A draft agreement should be included in the RFP. All proposers should be required to comment on the draft agreement and put any objections in writing prior to a selection being made. Finalizing the agreement with the selected company will be easier if the objections are known prior to the selection.

Standard Terms and Conditions

Most municipalities have a set of standard terms and conditions that are included in all RFPs. For example, the municipality has the right to reject all proposals, and the municipality will not reimburse any of the proposers for the cost of preparing their proposal.

In some cases a municipality will decide not to use a competitive process but instead to negotiate an extension with the existing hauler or service provider. If the existing contractor has provided good service and if the price is reasonable, it can be argued that the municipality will save time and money by not going through a selection process.

In 1999 the city of Escondido, California, decided not to go through a selection but instead to enter into a new 15-year agreement with the existing hauler. However, an initiative was placed on the ballot to force a competitive bid, and this initiative became the most expensive election in the history of the city. The existing hauler spent $379,515 and the city spent $107,493 to defeat the initiative, while the proponents of competitive bidding spent $352,547.[10] In the end the initiative was defeated, and the city entered into a 15-year agreement with the hauler as originally planned.

Agreement or Contract

If a municipality decides to contract for solid waste services, an *agreement* or *contract* will be required. Some agreements are as simple as several pages, while others require hundreds of pages. The agreement may be referred to as a *service agreement*, *franchise agreement*, or *contract* and may be either *exclusive* or *nonexclusive*. In an exclusive agreement, only the selected company can provide the specified service. Typically, residential garbage service is an exclusive service because only the contracted company can provide the service. In other cases, such as recycling, it may be exclusive under certain circumstances, but not under others. For example, the following is a section from the 1999 city of Arroyo Grande, California, recycling agreement:

> The Agreement for the Collection, processing and marketing of Recyclable Materials granted to Contractor shall be exclusive except as to the following categories of Recyclable Materials listed in this Section. The granting of this Agreement shall not preclude the categories of Recyclable Materials listed below from being delivered to and Collected and transported by others provided that nothing in this Agreement is intended to or shall be construed to excuse any person from obtaining any authorization from City which is otherwise required by law:
>
> A. Recyclable Materials separated from Solid Waste by the Waste Generator and for which Waste Generator sells or is otherwise compensated by a collector in a manner

resulting in a net payment to the Waste Generator for such Recycling or related services.

B. Recyclable Materials donated to a charitable, environmental or other non-profit organization.

C. Recyclable Materials which are separated at any Premises and which are transported by the owner or occupant of such Premises (or by his/her full-time employee) to a Facility;

D. Other Governmental Agencies within the City which can contract for separate solid waste and recycling services; and,

Contractor acknowledges and agrees that City may permit other Persons beside Contractor to Collect any or all types of the Recyclable Materials listed in this Section 4.2, without seeking or obtaining approval of Contractor under this Agreement.

This Agreement to Collect, transport, process, and market Recyclable Materials shall be interpreted to be consistent with state and federal laws, now and during the term of the Agreement, and the scope of this Agreement shall be limited by current and developing state and federal laws with regard to Recyclable Materials handling, Recyclable Materials flow control, and related doctrines. In the event that future interpretations of current law, enactment or developing legal trends limit the ability of the City to lawfully provide for the scope of services as specifically set forth herein, Contractor agrees that the scope of the Agreement will be limited to those services and materials which may be lawfully provided for under this Agreement, and that the City shall not be responsible for any lost profits and/or damages claimed by the Contractor as a result of changes in law.

Some municipalities that enter into franchise agreements with waste haulers will require that a *franchise fee* be paid by the waste hauler. This fee is in addition to the cost incurred by the selected company to provide the specified service. Franchise fees may be as little as one or two percent or in excess of 20 percent. In some cases the franchise fee is used to support related activities, such as recycling centers, while in other cases the fee goes to the general fund of the municipality.

FINANCING SOLID WASTE FACILITIES

Solid waste processing and disposal facilities can be either privately or publicly owned. In either case the owner has to minimize cost at a given level of service. Because solid waste facilities are most often long-term investments, the time value of money is important, and engineering economics plays a major role in deciding what kind of facility will be constructed. Funding of solid waste operations is similar to funding other utilities, such as water and sewerage services. Budgets consist of two general components: revenues and costs. Depending on the specifics of the solid waste system, such as ownership and contractual arrangements, the complexity of the financial system can vary. Some systems are complex and require financial experts, while others may require only a simple accounting system.

Solid waste operations receive revenue from the sale of services and goods. When the home or business pays its monthly garbage bill, these funds provide the revenue to cover the costs of the operation. Other sources of revenue are tipping fees at the landfill, the sale of recyclables, and the sale of electricity from a landfill gas turbine or waste-to-energy plant. In a publicly owned system the revenue should equal the cost. In a privately owned system the revenue should exceed the cost, resulting in a profit.

A central idea in engineering economics is the time value of money. A dollar today does not have the same value as a dollar a year from now. Ignoring inflation for now, a dollar today can be invested in an interest-bearing account and a year from now will be worth a dollar plus the interest earned. Thus a dollar today and a dollar a year from now cannot be added to get two dollars. They are as different as apples and oranges.

The initial cost of the facility is very important to the municipality or agency. Such a cost is known as a *capital cost* and is a one-time investment. Capital costs are paid from the proceeds of bank loans, general obligation bonds, and revenue bonds. It is common for a solid waste company to arrange for a bank loan to buy a garbage truck. The term of the loan is shorter than the life of the truck. The interest rate on the loan is based on the risk the lender perceives and includes such variables as the existence of a franchise agreement and the net worth of the company.

Municipal government can use *general obligation bonds* to finance capital projects. These bonds are based on the full faith and credit of the government. In addition the interest paid to the bond holder is usually tax exempt and thus the interest rate on these bonds is low. One problem with this type of bond is that they may require a vote of the citizens before they can be issued.

Municipal government can also use *revenue bonds* to finance capital projects. A revenue bond is guaranteed by the project. For example, a landfill may be funded by a revenue bond, and the revenue from the landfill (such as tipping fees) would be used to pay off the bond. Unlike a general obligation bond, these bonds have more risk because only the project revenue is pledged—not the full faith and credit of the municipal government. Thus these bonds have a higher interest rate and result in a higher cost to the municipality. The interest rate is generally still tax free and thus the rates are lower than normal bank loans.

One variation to the revenue bond is the ability of private companies to access revenue bond financing through government. This allows a private company to own the capital project while at the same time use tax exempt financing and receive a low interest rate. In addition the private company can depreciate the assets and reduce its tax bill. Many waste-to-energy facilities have been built using these bonds. The companies also receive accelerated depreciation and investment tax credits for the project. In 1984 the federal government limited the amount of private projects that could be financed using the tax exempt bonds; now solid waste projects must compete with other beneficial projects, such as low-cost housing, for this limited amount of funding.

Another complication is that public facilities require not only the capital cost for their construction, but also a yearly cost for their operation and maintenance (O&M). The latter are called the *O&M costs* and include such items as salary, replacement parts, service, fuel, and the many other costs incurred by the facilities.

Because of the time value of money, estimating the real cost of municipal facilities, such as solid waste collection and disposal operations, can be tricky. Economists and engineers use two techniques to "normalize" the dollars so that a true estimate of the cost of multiyear investment can be calculated. The first technique is to compare the costs of alternatives on the basis of *annual cost*, while the second calculates *present worth*. Both include the capital cost plus the O&M cost.

Calculating Annual Cost

The capital costs of competing facilities can be estimated by calculating the cost that the municipality or agency would incur if it were to pay interest on a loan of that amount. Calculating the annual cost of a capital investment is exactly like calculating the annual cost of a mortgage on a house. The owner (municipality or agency) borrows the money and then has to pay it back in a number of equal installments.

If the owner borrows X dollars and intends to pay back the loan in n installments at an interest rate of i, each installment can be calculated as

$$Y = \left[\frac{i(1+i)^n}{(1+i)^{(n-1)}} \right] X$$

where Y = the installment cost, $

i = annual interest rate, as a fraction

n = number of installments

X = the amount borrowed, $

The expression

$$\left[\frac{i(1+i)^n}{(1+i)^{(n-1)}} \right]$$

is known as the *capital recovery factor* or CRF. The capital recovery factor does not have to be calculated since it can be found in interest tables or is programmed into hand-held calculators. Table 9-1 shows the capital recovery factors if the interest is 6.125%.

EXAMPLE
9-1

A town wants to buy a refuse collection truck that has an expected life of 10 years. It wants to borrow the $150,000 cost of the truck and pay this back in 10 annual payments. The interest rate is 6.125%. How much are the annual installments on this capital expense?

From Table 9-1 the capital recovery factor (CRF) for $n = 10$ is 0.13667. The annual cost to the town would then be 0.13667 × $150,000 = $20,500. That is, the town would have to pay $20,500 each year for 10 years to pay back the loan on this truck. Note that this truck does not cost 10 × $20,500 = $205,000 because the dollars for each year are different and cannot be added.

Table 9-1 6.125% Compound Interest Factors

Years	CRF	PWF	SFF
1	1.06125	0.942	1.00000
2	0.54639	1.830	0.48514
3	0.37497	2.666	0.31372
4	0.28941	3.455	0.22816
5	0.23820	4.198	0.17695
6	0.20416	4.898	0.14219
7	0.17993	5.557	0.11868
8	0.16183	6.179	0.10058
9	0.14782	6.764	0.08657
10	0.13667	7.316	0.07452
11	0.12760	7.836	0.06635
12	0.12009	8.326	0.05884
13	0.11378	8.788	0.05253
14	0.10841	9.223	0.04716
15	0.10380	9.633	0.04255

*See following sections.

Calculating Present Worth

An alternative method for estimating the actual cost of a capital investment is to figure present worth: how much one would have to invest right now, Y dollars, at some interest rate i so that one could have available X dollars every year for n years.

$$Y = \left[\frac{1}{(1+i)^n} \right] X$$

where Y = the amount that has to be invested, $
i = annual interest rate
n = number of years
X = amount available every year, $

The term

$$\left[\frac{1}{(1+i)^n} \right]$$

is called the *present worth factor*, or PWF.

EXAMPLE 9-2

A town wants to invest money in an account drawing 6.125% interest so that it can withdraw $20,500 every year for the next 10 years. How much must be invested?

From Table 9-1 the present worth factor (PWF) for $n = 10$ is 7.316. Thus the money required is $7.316 \times \$20,500 = \$150,000$.

Calculating Sinking Funds

In some situations the municipality or agency must save money by investing it so that at some later date it would have available some specified sum. In solid waste engineering this most often occurs when the landfill owner must invest money during the active life of the landfill so that when the landfill is full, there are sufficient funds to place the final cover on the landfill. Such funds are known as *sinking funds*. It is necessary to calculate the funds Y necessary to be invested in an account that draws i percent interest so that at the end of n years the fund has X dollars in it. The calculation is

$$Y = \left[\frac{1}{(1+i)^n - 1} \right] X$$

and the term

$$\left[\frac{1}{(1+i)^n - 1} \right]$$

is known as the *sinking fund factor* (SFF).

EXAMPLE 9-3

A community wants to have $500,000 available at the end of 10 years by investing annually into an interest-bearing fund that yields 6.125% interest. How much would they have to invest annually?

From Table 9-1 the sinking fund factor (SFF) at 10 years is 0.07452, and the required annual investment is therefore $0.07452 \times \$500,000 = \$37,260$.

Note again that $10 \times \$37,260 = \$372,600$, which is considerably less than $500,000. The reason is that the investments during the early years are drawing interest and adding to the sum available.

Calculating Capital Plus O&M Costs

Capital investments require not only the payment of the loan in regular install-ments, but also the upkeep and repair of the facility. The total cost to the commu-nity is the sum of the annual payback of the capital cost plus the operating and maintenance cost.

EXAMPLE
9-4

A community wants to buy a refuse collection vehicle that has an expected life of 10 years and costs $150,000. They choose to pay back the loan in 10 annual install-ments at an interest rate of 6.125%. The cost of operating the truck (gas, oil, serv-ice) is $20,000 per year. How much will this vehicle cost the community every year?

From Table 9-1 the capital recovery factor for $n = 10$ is 0.13667, so the annual cost of the capital investment is 0.13667 × $150,000 = $20,500. The operating cost is $20,000, so the total annual cost to the community is $20,500 + $20,000 = $40,500.

Comparing Alternatives

One of the main uses of engineering economics is to fund the lowest-cost solution to a community need. The above analysis for a single truck can be applied equally well to alternative vehicles, and the annual costs (capital plus O&M) can be compared.

EXAMPLE
9-5

As an alternative to the truck analyzed in Example 9-4, suppose the community can purchase another truck that costs $200,000 initially, but has a lower O&M cost of $12,000. Which truck will be the most economical for the community?

The annualized capital cost for this truck is 0.13667 × $200,000 = $27,300. Adding the O&M cost of $12,000, the total annual cost to the community is $39,300. Comparing this to the total cost calculated in Example 9-4, it appears that the $200,000 truck is actually less expensive for the community to own and operate.

HAZARDOUS MATERIALS

Increased technological complexity and population densities have resulted in the identification of a new type of pollutant, commonly called a *hazardous substance*. The reported incidence of damage to the environment and to people by these mate-rials has increased markedly in the last few years. The EPA maintains a list of such incidents, and some of the better documented ones have been published.[11]

The term *hazardous substance* or *hazardous waste* is difficult to define, and yet a clear definition is necessary if specialized disposal standards are to be applied to such materials. A legal definition[12] of a hazardous waste suggested by the EPA is

> any waste or combination of wastes of a solid, liquid, contained gaseous, or semisolid form which because of its quantity, concentration, or physical, chemical, or infectious characteristics, may (1) cause or significantly contribute to an increase in mortality or an increase in serious irreversible or incapacitating reversible illness; or (2) pose a substantial present or potential hazard to human health or the environment when improperly treated, stored, transported or disposed of, or otherwise managed.

When deciding whether or not a specific substance is hazardous, it is useful to use a set of criteria against which the properties of the material in question can be judged. The EPA has defined hazardous materials in two ways: (1) the chemical is *listed* as a hazardous material, or (2) the material fails one of six tests that then define it as a hazardous material. The listing includes specific chemicals such as chlorinated pesticides, organic solvents, and over 50,000 others. The process used is generally that the EPA lists the material and then waits for a challenge from an interested party. If the interested party conducts tests that show the material not failing any of the six tests, it is *delisted*, making it possible to dispose of the material in ordinary landfills. All listed materials have to go to specially constructed hazardous waste landfills or other treatment centers, at many times the cost of nonhazardous disposal.

The six tests used to define a hazardous material are:

- Radioactivity. The stipulation being that the levels of radioactivity not exceed maximum permissible concentration levels as set by the Nuclear Regulatory Commission.
- Bioconcentration. This criterion captures many chemicals such as chlorinated hydrocarbon pesticides.
- Flammability. This standard is based on the National Fire Protection Association test for how easily a certain substance will catch fire and sustain combustion.
- Reactivity. Some chemicals, such as sodium, are extremely reactive if brought into contact with water.
- Toxicity. The criterion for toxicity is based on LD_{50} (lethal dose 50), or that dose at which 50% of the test species (e.g., rats) die when exposed to the chemical through a route other than respiration. Inhalation and dermal toxicity is next, where LC_{50} is the lethal concentration resulting in 50% mortality during an exposure time of 4 h. Dermal irritation is measured on a Federal Drug Administration scale of 1 to 10. A grade 8 irritant causes necrosis of the skin when a 1% solution is applied. Aquatic toxicity is measured by the 96-h mean toxic limit of less than 1000 mg/liter, although the present EPA criterion for aquatic toxicity is set at 500 ppm.[13] Phytotoxicity is the ability to cause poisonous or toxic reactions in plants based on the mean inhibitory limit of 1000 ppm or less. Since the publication of these criteria, the EPA has lowered its definition of phytotoxic poisons to 100 ppm.
- Genetic, carcinogenic, mutagenic, and teratogenic potential. These are all measured by tests developed by the National Cancer Institute.

Note that with this system of defining what is and is not hazardous the actual quantity of the material is not specified. This is unreasonable when disposal schemes are to be evaluated.

The most environmentally sound disposal scheme for hazardous materials (at any quantity) is destruction and conversion to nonhazardous substances. In many cases, however, this is either extremely expensive (e.g., dilute heavy metal and pesticide wastes) or technically impossible (e.g., some radioactive waste materials). Alternative disposal schemes are thus necessary.

By far the most widely used method of disposing of hazardous waste substances is the hazardous waste landfill, which is used to provide complete long-term protection for the quality of surface and subsurface waters from hazardous wastes deposited in the landfill, as well as to prevent other public health and environmental problems. The hazardous waste landfill differs from the ordinary sanitary landfill primarily in the degree of care taken to ensure minimal environmental impact. Clay liners, monitoring wells, and groundwater barriers are some of the techniques used in such landfills. The overall philosophy is strict segregation from the environment. It should be noted that just as is the case with aboveground storage, such landfills are usually not strictly disposal schemes, but rather holding operations. True disposal is still willed to future generations.

THE ROLE OF THE SOLID WASTE ENGINEER

The "good old days" in solid waste management included open burning dumps that were routinely set on fire. These dumps polluted the groundwater, provided no landfill gas control, and had unsafe working conditions resulting in numerous injuries, but all at low cost. While the public liked the low cost for garbage disposal, they were not aware of the many problems these practices caused. The solid waste engineer is responsible for transforming the industry into a professional field with best practices. Today open dumps have been replaced by sanitary landfills with gas-control leachate-collection systems. Garbage incinerators have been closed and modern waste-to-energy plants with state-of-the-art air-pollution-control equipment have replaced them. Collection has evolved from putting the garbage into a can to a series of recycling bins, yard waste cans, and waste oil containers. Household hazardous waste is managed separately from refuse. While all of these changes have been beneficial to society and the environment, they have resulted in an increased cost to the public. Thus on one hand, the solid waste engineer is hailed for bringing improvements to solid waste management; on the other hand, the engineer is blamed for increasing the cost of solid waste management.

To be effective, the solid waste engineer must be competent in three areas: technology, regulations, and public communications. These three areas are like the three legs on a stool. Without all three the stool will fall over.

Technology in the solid waste field is constantly evolving. For example, new landfill liners are being developed for different applications. Materials recovery facilities are being re-engineered to process different types of waste streams. Air-pollution-control equipment has become so extensive that it can be larger and more complex than the actual furnace in a waste-to-energy facility.

Regulations become more stringent and complex every year. Regulations pertaining to solid waste management are issued by federal, state, regional, and local agencies. In many cases regulations from various agencies conflict with each other. For example, landfills must have caps that restrict the infiltration of water into the landfill, but at the same time landfills must have gas extraction wells that penetrate the cap and could allow for the infiltration of surface water into the landfill.

Communication is integrally important in solid waste engineering. The engineer must be able to communicate with the public, and at the same time have the technical and regulatory knowledge to develop effective solid waste systems. If the engineer cannot convey that information to the public, including such decision makers as elected officials, then projects will not be implemented. For example, an engineer must be able to explain what it means when a health risk assessment determines that the waste-to-energy plant results in a 37 in one million increased risk of cancer.[14]

Integrated solid waste management services and supporting facilities, by their very nature, are public facilities and services. The public will use them, see them, and be affected by them. Because of the public nature of these facilities and services, the solid waste engineer cannot forget that it is the elected officials who represent the public. Thus those decisions that impact the public are best made by an informed elected official. More than one solid waste engineer has had a short career because the engineer forgot whom the public elected to make the ultimate decision.

FINAL THOUGHTS

The civility of our society depends on all of us agreeing to abide by good manners, high moral standards, and the law of the land. Given the choice, we all would want to live in a society where everyone agrees to abide by these conventions, recognizing that this benefits everyone. This argument is applicable to professional engineering as well. To help maintain a viable engineering profession, we should demonstrate good professional manners ourselves, and if the occasion requires, admonish others for boorish behavior. We should act as role models in conducting engineering on a high moral level and promote such behavior in others. And without doubt we should not become criminals. In short, we all have a responsibility to uphold the honor of professional engineering and to create a culture we all desire and in which we can all flourish.

But there is a larger question of why any of us should *act* in such a way. That is, if we find that having bad manners, or acting immorally, or even breaking a law is advantageous to us individually, why should we, at any given moment, not *act* in a manner that we would not necessarily want others to emulate?

The obvious answer to that question is that we don't want to get caught and suffer the consequences. Bad manners would subject us to ridicule; immoral conduct might cause us to be ostracized by others, or we might lose clients and business; and, of course, breaking a law might result in a fine or jail time. But consider now the possibility where we would not get caught and could not suffer any adverse consequences. Of course, we never know for sure that we will not get caught, but for the sake of argument, let's assume this extreme case. We have it all figured out, and it simply is impossible for us to get caught being ill-mannered, immoral, or illegal. Why should we, all things considered, still act as honorable engineers, especially if this might involve financial cost to us or in some other way cause us harm? The answer comes in three parts.

First, we are all members of a larger community, in this case the engineering community, and we all benefit from this association. Acting in a manner that brings harm or discredit to this community cannot, in the long run, be beneficial to us. Granted, the destruction of professional engineering may be far into the future and our small anti-social act would not be enough to destroy the profession, but we, along with all our contemporaries, have an obligation to uphold the integrity of the profession, eventually for our own good. Engineers should act honorably because the profession depends on them to do so.

Second, the anti-social act, even though we might get away with it, takes something out of us. There are circumstances where it clearly is better to lie than to tell the truth (such as saving an innocent life, for example), but all lies come at a cost to the teller.[15] There is, as it were, a reservoir of good in each human, and this can be nibbled away one justified lie at a time until the person is incapable of differentiating between lying and being truthful. This would also be true for bad manners and illegal acts. Every time we get away with something, we reduce our own standing as honorable human beings. Engaging in untruthful engineering reduces our own standing as professionals.

Finally, the reason for not being anti-social, even assuming we could get away with it, is that eventually our conscience would not stand for it. We all have a conscience within us that tells us the difference between right and wrong. Most of us, when we do anti-social things, *know* we are behaving badly and eventually regret such actions.[16]

But if, by telling lies and becoming a scoundrel one gains materially, why would this be an undesirable result? If an engineer ran an engineering practice where she made it a habit to lie to clients, why is this detrimental? Why will telling lies (that one continues to get away with) necessarily be a bad thing?

The truth is that engineers who behave without regard to manners, morals, or laws will eventually cause harm to befall upon themselves. They *will* lose clients, and their untruths *will* cause their works to fail, and they *will* have a bad conscience that will bother them. They will eventually think poorly of their own standing in the profession and regret their self-serving actions that may have been ill-mannered, immoral, or illegal.

So why be a good engineer? We might get caught if we don't; we have a common responsibility to the professional engineering community; we lose something of our own integrity when we behave badly; and we have a conscience. But what if, in the face of these arguments, we still are not convinced? There appear to be no knock-down ethical arguments available to make us change the mind of a person set on behaving badly. We have the human option to act in any way we wish. But if we have bad manners, act immorally, or break the laws, we are not behaving honorably, and eventually we will be harmed by our regrets for acting in this manner. That is, such behavior will always result in harm to us as professional engineers.

While the Viking society of northern Europe was in many ways cruel and crude, they had a very simple code of honor. Their goal was to live their lives so that when they died, others would say "He was a good man." The definition of what they meant by a "good man" might be quite different by contemporary standards, but the principle is important. If we live our professional engineering lives so as to uphold the exemplary values of engineering, the greatest professional honor any of us could receive would be to be remembered as a *good* engineer.

EPILOGUE

Well, dear reader, you have made it to the end of the book. The authors, with over 60 years of experience in solid waste management among them, have tried to share with you their collective knowledge. For some of you, the only future solid waste practice will be taking out the garbage. If that is the case, at least you will know that it does not just disappear when it goes into that truck. For others, this introduction may lead to a career in some aspect of solid waste management. For those readers, your journey is just beginning.

REFERENCES

1. Brunner, P. H., and W. R. Ernst. 1986. "Alternative Methods for the Analysis of Municipal Solid Waste." *Waste Management and Research* 4: 155.

2. "Life Cycle Analysis Measures Greenness, but Results May Not Be Black and White." 1991. *Wall Street Journal* (28 February).

3. Solano, E., R. D. Dumas, K. W. Harrison, S. Ranjithan, M. A. Barlaz, and E. D. Brill. "Integrated Solid Waste Management Using a Life-Cycle Methodology for Considering Cost, Energy, and Environmental Emissions—2. Illustrative Applications." Department of Civil Engineering, North Carolina State University.

4. Hocking, M. B. 1991. "Paper versus Polystyrene: A Complex Choice." *Science* 251, 1 (February).

5. Worrell, W. A. 1999. *The San Diego, California Mixed Waste Materials Recovery Facility: Technological Success, Political Failure.* R'99 Recovery, Recycling, Re-Integration Proceeding, Volume 1. Gallon, Switzerland: EMPA.

6. *Waste News* 5, Issue 38 (February 7, 2000).

7. *County of San Diego Privatization Study.* 1991. Deloitte and Touche and R.W. Beck (October 16).

8. *The Local Government Guide to Solid Waste Competitive Service Delivery.* 1995. Public Technology Inc.

9. Biering, R. A. 1999. "The Art of Saying "No" or Bambi meets Godzilla." *Proceedings* WasteCon 1999. Solid Waste Association of North America.

10. *The San Diego Union-Tribune*, February 12, 2000.

11. Wentz, C. A. 1989. *Hazardous Waste Management.* New York: McGraw-Hill Publishing Co.

12. Blackman, W. C. 1993. *Basic Hazardous Waste Management.* Boca Raton, Fla.: Lewis Publishers.

13. LaGrega, M. C., P. L. Buckingham, and J. C. Evans. 1994. *Hazardous Waste Management.* New York: McGraw-Hill Publishing Co.

14. *Health Risk Assessment for the Resource Recovery Facility Boiler No. 1.* 1989. Department of Solid Waste Management, Dade County, Florida. Malcome Pirnie (May).

15. Bok, S. 1978. *Lying: Moral Choice in Public and Private Life.* New York: Pantheon.

16. Pritchard, M. 1991. *On Being Responsible.* Lawrence, Kan.: University Press of Kansas.

ABBREVIATIONS USED IN THIS CHAPTER

CFC = chlorofluorocarbon
CRF = capital recovery factor
EPA = Environmental Protection Agency
MSW = municipal solid waste

O&M = operation and maintenance
PWF = present worth factor
RFP = request for proposal
SFF = sinking fund factor

PROBLEMS

9-1. The town of Recycleville has 30,000 citizens. Each home in Recycleville generates 1.1 tons of solid waste per year and has an average of 2.4 residents per service customer. The town has no commercial businesses or schools. The local landfill has space for 100,000 tons and charges $10 per ton. Once the landfill is full, the waste will need to be sent to a remote landfill at a cost of $90 per ton.

    a. What is the 10-year average disposal cost for the town, assuming a rate of inflation of 5% per year?

    b. What will be the cost per home?

9-2. A town near Recycleville (Problem 9-1), Trashville, generates about 40,000 tons per year of solid waste. Their landfill will be full in one year. Assume that after Trashville's landfill is full, their solid waste will come to Recycleville.

    a. If Recycleville wants to run its landfill at no cost to the citizens of Recycleville, what tipping fee should it charge Trashville?

    b. How much sooner would the landfill have to close if it accepted Trashville's refuse?

    c. If you are the mayor of Recycleville, what would you do about solid waste from Trashville?

9-3. If you are a private waste hauler trying to convince a city that you are less costly than public sector collection, how would you respond to the claim that you pay taxes, therefore your costs are higher than the public sector?

9-4. You are preparing the Request for Proposals for garbage collection service. What factors would automatically disqualify you as a bidder responding to the RFP?

9-5. You have been hired to develop an integrated solid waste management plan for the city of Confusion. You present your plan to the city council, and they decide that they will build the new landfill but not implement your recycling component. After the meeting the press asks you what you thought of the city council's decision. What do you tell the press?

9-6. A private company wants to construct a waste-to-energy plant that will have a capital cost of $20,000,000 and an operating cost of $300,000 per year. They expect to receive 300 tons of refuse per day, and the facility is to have a useful life of 15 years. The community insists that they will not pay more than $40 per ton to the company for disposal of their solid waste. What interest rate must the company obtain on their $20,000,000 loan just to break even? (You will need a set of interest tables, or a calculator with interest tables, to solve this problem.)

9-7. If you didn't have the garbage collection franchise in a city, what arguments would you use to sway the city council to put the collection agreement out to bid? If you already had the agreement, what arguments would you use to not put it out to bid?

9-8. Determine the current regulatory and legal status of flow control.

9-9. A community wants to construct a hazardous waste incinerator for household hazardous wastes. The capital cost of the facility is expected to be $5,000,000. The cost of operating this facility is estimated to be about $20,000 per year plus salaries of $100,000 annually. The insurance on the facility will be $5000 per year. What is the annual cost of this facility to the community? Assume an interest rate of 6.125% and a useful life of 15 years.

9-10. A town signs a contract with a dumpster manufacturer to provide 50 dumpsters at a cost of $5,000 each. The dumpsters are expected to last 5 years. A rival manufacturer insists that even though their dumpsters cost $8,000, the expected life of 10 years makes them a better deal to the community. Are they right?

9-11. A waste management agency decides to use some of the tipping fees from a landfill to buy extra land for expanding the landfill. The new land costs $2,000,000, and the existing landfill has an expected life of 8 years. How much money will the agency have to put into a sinking fund annually (at 6.125% interest) to have the $2,000,000 available to buy the land?

9-12. A community wants to construct a landfill that is expected to have a capital cost of $3,000,000 and an annual operating cost of $500,000. The landfill is expected to last 15 years, and the lowest-interest loan the town can get is at 6.125%. What must be the revenue stream to be able to finance this project?

APPENDIX A

The Lewisburg Solid Waste Problem

BACKGROUND

The borough manager, Nada Gray, of the Borough of Lewisburg has asked your firm to evaluate its options in solid waste management. She expects a preliminary study from your firm detailing the possible options for further study.

The impetus for this study is the expiration of a contract with Lycoming County Landfill, which is presently receiving all of Lewisburg's refuse under a grandfather contract of $40/ton. Lycoming County has told Lewisburg that it will begin to charge \$80/ton in the year $X + 2$ when the present contract expires. This potential threat has promoted the study you are conducting. (*Note:* X = the present year.)

The populations to be served and the sources and quantities of municipal solid waste (MSW) are estimated first. Unfortunately, there is no information on solid waste production in Lewisburg, and the best available information from other sources must be used. The first task is to estimate the solid waste production, both quantity and composition, generated in Lewisburg.

The second task involves the design of a collection system for the domestic waste, public institutional waste (e.g., schools), and the commercial solid waste. The industrial waste and the private institutional solid waste (such as Bucknell University) will continue to be collected by private haulers, and their contribution will be factored in only when the disposal options are considered.

The following disposal alternatives seem like reasonable possibilities to be investigated, if for no other reason than to be able to respond in a public hearing that you have done so. Composting of mixed waste is not included because you can demonstrate that this option is overwhelmingly expensive and should not even be considered for refuse.

1. Sanitary landfill
2. Recycling, with sanitary landfill
3. Materials Recovery Facility (MRF). This is a central refuse processing operation that includes the production of a refuse-derived fuel that will be accepted by the power company. Material not recovered will go to the landfill.
4. Waste-to-energy facility, with the production of steam and conversion to electricity that will be bought by the power company. A landfill for the unacceptable items and the incinerator ash is also necessary.

You have decided that your report should be organized as a series of chapters, which will be collected for the final report:

Chapter 1 Sources, Composition, and Quantities
Chapter 2 Recycling
Chapter 3 Collection
Chapter 4 Siting the Sanitary Landfill
Chapter 5 Design of the Sanitary Landfill
Chapter 6 Materials Recovery Facility
Chapter 7 Waste-to-Energy (Mass Burning)
Chapter 8 Technical, Economic, and Environmental Evaluation

That is, once you have developed some figures on the amount and type of solid waste generated, you will be comparing the various options for disposal. The plan is illustrated below:

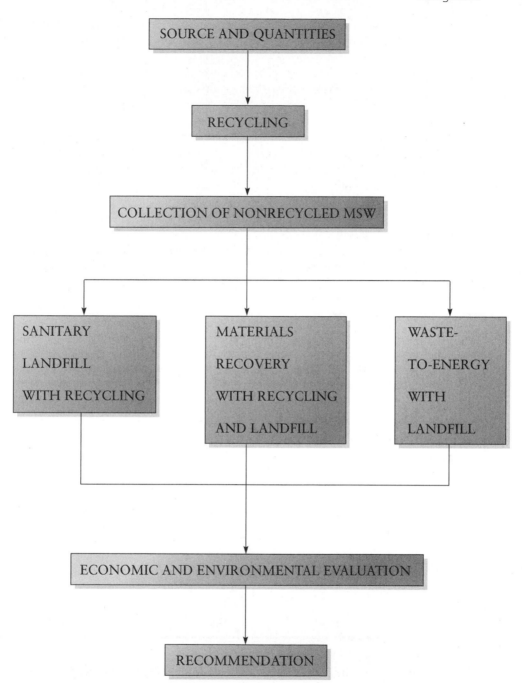

Remember that one of your alternatives is the acceptance of the high tipping fee by the Lycoming County Landfill and delivering all of the nonrecycled waste to them.

You will complete each chapter, developing the numerical and policy conclusions as necessary. You will work as groups, and all recommendations, calculations, and policy will be a group effort. However, **the final product must be your own writing**. Each chapter will have two sections: (1) the text: the written report and all graphs, tables or other illustrations necessary for the reader to understand your conclusions, and (2) the appendix: calculations and other supporting material, including working graphs. When the final report is prepared, the calculations will all go into a common appendix. Begin each chapter with the proper headings. You are encouraged to consult *Report Writing for Environmental Engineers and Scientists* for questions of style and format (available from Lakeshore Press, P. O. Box 92, Woodsville, NH 03785).

Individual chapters are due as indicated on the syllabus. A late penalty of 5% per day will be charged for any fraction of a day late. The day begins when the reports are collected at the beginning of the class period.

USEFUL DATA

Population of the Borough of Lewisburg

Year	Population
1900	4790
1910	4802
1920	4910
1930	5320
1940	5720
1950	5820
1960	5722
1970	5634
1980	5689
1990	5728
2000	5880

About 15% of the people in Lewisburg live in apartments served by dumpsters collected by the Borough of Lewisburg.

Transportation

Lewisburg is an old town with fairly narrow streets. Most streets are two ways with the exception of Fourth Street, which is one way from U.S. 15 to George Street. The town is located in the flood plain at the confluence of Buffalo Creek and the West Branch of the Susquehanna River. The main commercial street is Market Street, while U.S. Route 15 is the highway commercial route with significant truck traffic.

Industries in Lewisburg

Industry	Solid waste production (tons/year)	Composition
Silver Tip (metal fabricators)	200	80% aluminum 10% ferrous 5% mixed paper 5% other
Builders Supply	150	45% wood 15% ferrous 25% brick and mortar 15% other
Buffalo Valley Nursing Home	50	unknown

(*Note*: the residents of the nursing home are included in the census.)

All three industries are presently served by dumpsters collected by the borough.

Bucknell University

Bucknell University presently carries 1300 tons/year of refuse to the Lycoming County Landfill. In addition to the 1300 tons/year of refuse, Bucknell delivers recycled materials to the Lycoming County Landfill. The price paid to Bucknell for these materials and their quantities are as follows:

Clear glass, 30 tons/year, $10/ton
Aluminum, 10 tons/year, $600/ton
Mixed paper, 30 tons/year, no income/no cost

Existing Solid Waste Collection Program and Method of Disposal

The borough engineer estimates that at present there are 35 commercial dumpsters serviced by the borough, and that the residential collection, based on billing records, serves 1320 residences. The residences are serviced by alleyway collection for most of the borough, although the newer sections do not have alleys, and the residents are expected to carry their cans to the street for collection.

The borough does not have any idea how many tons of refuse are actually being disposed of because they are working under a grandfather clause with Lycoming County that allows the borough to deliver all of its refuse to the landfill at a fixed annual cost. This contract, however, is expected to expire by the year $X + 2$. The expected expiration of the (very favorable) contract is what prompted this study.

Existing Recycling Program

The Boy Scouts collect newsprint once a month. They take it to the Lycoming County Landfill and receive $10/ton. They collect about 50 tons/year.

The Kiwanis Club has a drop-off location for aluminum cans and clear glass. They transport the materials to the Lycoming County Landfill. Latest data show that they collect 15 tons of aluminum and 35 tons of clear glass per year. They pay for the drop-off containers and the transportation, and keep the income. Lycoming County Landfill pays $600 per ton for the aluminum and $10 per ton for the clear glass.

The borough has thus far not spent any money on developing its own recycling program although it is under mandate from the State of Pennsylvania to do so.

Borough Organization

The borough has an elected borough council, which elects its own chair, who then serves as the mayor. The council is elected by the citizens of Lewisburg only, and the council hires the town manager. She is the one who commissioned the study you are doing.

CHAPTER ONE

SOURCES, COMPOSITION, AND QUANTITIES

We assume here (as the worst case assumption) that there are no data on solid waste production or solid waste composition.

Three sources of solid waste generation data that may be useful to you are (1) national average (a crude and usually unacceptable estimate when applied to specific localities), (2) data from other nearby communities, and (3) data obtained for statewide surveys.

The municipal solid waste (MSW) can be identified as coming from distinct sources:

1. Residential
   Individual residences
   Apartments
2. Public institutional (public schools, for example)
3. Commercial
4. Industrial
5. Private institutional (such as Bucknell University)

Constuction and demolition debris is not considered in this study. Other than the C&D waste, all other components of MSW will need to be collected and disposed of, and therefore the quantities must be estimated.

You choose to use the population estimates for the year $X + 3$ in your calculations since this would be the earliest any one of the alternatives you evaluate would be placed into operation. Based on the census data (see USEFUL DATA), estimate the population to be served by the borough's solid waste collection and disposal program.

You can estimate that the public institutional waste is 10% of the total residential solid waste and the commercial waste is 20% of the sum of residential and public institutional solid waste. The commercial establishments in Lewisburg are served by the municipal collection system since there is no reasonably priced alternative. They are charged a fee for every dumpster-load collected.

National solid waste production averages are based on a sum of residential, public institutional, and commercial waste since these often cannot be distinguished during collection. Such figures are available in various publications (see EPA's *Decision Maker's Guide to Solid Waste Management*).

The collection of solid waste generated by Bucknell University is considered separately since it runs its own collection system. It may, however, contribute to the refuse destined for disposal.

The industries within the borough will use their own vehicles for transport or will contract privately to take their solid waste to the disposal site. The wastes generated must also be quantified.

For this chapter:

1. Estimate the population for the year $X + 3$. This is your design population. (*Note:* X = present year.)
2. Estimate the production and composition of the three different types of municipal solid waste for the year $X + 3$. The three types of solid waste are
   a. Residential, public institutional, commercial
   b. Industrial
   c. Private institutional
   Express the composition as percent by weight of various identifiable materials, such as newsprint, aluminum, etc. Assume that Bucknell University is the only source of private institutional waste.
3. Consider and select alternative tabular and/or graphical means to present your results. The graphics should be informative, accurate, and easy to read. You will

use the information from these figures for subsequent chapters. More than one graph may be useful and appropriate. This graphical presentation must include composition as well as quantities. The quality of the graphic will be a consideration in grading this report.

CHAPTER TWO

RECYCLING

The State of Pennsylvania has mandated that all towns with populations greater than 5000 must have a recycling program. (*Pennsylvania Act 101, Municipal Waste Planning, Recycling and Waste Reduction Act of 1988*) The percentages of materials recycled vary, but what is important is that an effort be made. Regardless of what other options are adopted by the borough, a recycling program is a must. It is unlikely that the borough will much longer be able to depend on the Boy Scouts and the Kiwanians, although they might be allowed to continue their programs. This is a problem that will complicate your recommendations.

A survey of potential markets for clean recycled materials indicates the following:

Material	Market and one-way distance (miles)	Price paid at the market site ($/ton)
Aluminum	Harrisburg (46 mi)	600
	Lycoming Landfill (20 mi)	600
Steel cans	Harrisburg (49 mi)	10
Glass		
Clear	Lycoming Landfill (20 mi)	10
Mixed	Harrisburg (45 mi)	0
Paper		
Corrugated	Lycoming Landfill (20 mi)	30
Newsprint	Danville (15 mi)	10
	Lycoming Landfill (20 mi)	10
Mixed	No market identified	
Plastic (No. 1 and 2)	Harrisburg (52 mi)	10

Transportation costs for the material from Lewisburg to its markets can be estimated as $0.25/mile/ton of materials. This includes the purchase of the trucks and salaries of the drivers.

The estimation of collection cost is very difficult since the costs will depend on the level of participation, the materials collected, and the collection systems used.

Estimate that collection will cost about $100/ton. This includes all recyclables, regardless of what is collected and what the participation rate is. Note that using a single collection cost for all materials is not a good assumption, of course, but it greatly simplifies our calculations.

For this chapter:

1. Assume initially that all of the materials listed above are candidates for recycling. For the year $X + 3$ estimate the fraction of each material produced by the town (from Chapter 1) that might be recycled (residential, commercial, and public institutional only). Estimate the quantities of each material to be collected, and sum these. Will the Boy Scouts and Kiwanians be allowed to continue their recycling programs?

2. Calculate the cost of collection and transport for the materials to be recycled by multiplying the total tonnage of expected separated material by $100/ton.

3. Bucknell University has a recycling program, and they want to deliver their collected material to the town materials handling facility to save on transportation costs. Estimate the quantity of recycled material produced by Bucknell University (again use the data developed in Chapter 1). Bucknell University wants some income for bringing the separated material to the borough materials handling facility. How much will you pay them? Remember that the borough will receive all of the income from the sale of the collected materials.

4. Calculate the net revenue received for each material if collected and sold to a scrap dealer, and the net cost of recycling each material (expressed as $/ton).

5. Based on the costs and quantities calculated above, decide what materials (if any!) you recommend be recycled, and calculate the tons/year that you expect to recycle. Remember the state mandate!

6. Based on your decision in paragraph 5, estimate the cost of a facility for collecting, storing, and shipping the separated material. Estimate this as costing $20/ton of material handled. This figure includes salaries and other operating expenses as well as the annualized capital cost of the building and equipment.

7. For the year $X + 3$ calculate the cost of the recycling program you recommend. This includes the cost of collection and transport, the fees paid to Bucknell University (if any), cost of the materials handling facility, minus the revenues received. This cost is represented by the symbol **H** and is used in the cost calculations for the next three chapters.

8. Based on your calculations in Chapter 1 and your conclusions from paragraph 5 above, calculate the total tons per year of MSW that still has to be disposed of. This figure is used in the next three chapters.

CHAPTER THREE

COLLECTION

The borough wants to continue collecting refuse using town-owned trucks and town employees. Commercial establishments will also be collected by town trucks and town employees. At the present time the borough services 35 dumpsters located in the commercial zone of Market Street (from Seventh to Front Street) and all along U.S. Route 15.[1] Bucknell University would continue to collect and truck their refuse in their own vehicles to whatever disposal facility the town constructs.

Refuse will be picked up from individual residences, but apartment buildings will be collected by dumpsters. Schools and other municipal facilities will need to have dumpsters for their refuse, as will commercial establishments. Two types of trucks—one type used for nonapartment house residential pickup and another type used for collecting apartments, institutions, and commercial establishments—must be purchased.

You may use the following cost data, or develop your own.

Collection vehicles for small containers (such as residential)

Size (yd^3)	Type	Capital cost ($)	Compaction ratio	Truck operating cost (not labor) ($/ton collected)
20	Rear load	150,000	1:4	1.60
16	Rear load	95,000	1:4	1.70
16	Side load	78,000	1:3	1.90

Collection vehicles for dumpsters (such as commercial)

20	Overhead	160,000	1:3	1.00
30	Overhead	200,000	1:3	1.00
40	Roll-off	140,000	1:1	0.70

Assume a crew of three persons per residential truck, each paid at $15/hour (including fringe benefits, insurance, etc.). Dumpster collection vehicles are managed by only one person. Expected vehicle life is 10 years. Assume salaries and benefits for one supervisor plus secretary at a total annual cost of $80,000.

Estimate that for individual households, one collection vehicle can service 80 residences per hour if the system is roll-out "green" cans placed on the curb by the residents, 30 per hour if the collectors have to go to the backyards to retrieve the green cans, 50 per hour if the residents set out ordinary nonrolling cans on the curb and take them back, and 30 per hour if the residents have backyard collection with nonrolling garbage cans.

If roll-out containers are to be purchased for individual households, estimate their cost as $50 per can. They have an expected life of five years.

Estimate that a full residential collection vehicle requires one hour of time to travel to and from the disposal site (wherever that will eventually be).

1. Maps are available at http://info.brookscole.com/Vesilind. Double click on each of the maps to download them to your hard drive.

Estimate that a dumpster has a capacity of 20 yd^3 and that it takes 30 minutes to empty a dumpster into the truck, including the travel to and from the dumpster site. Assume that each dumpster costs $8000 and that they have a useful life of 10 years. The commercial dumpsters will seldom be full and many more dumpsters will be needed than the minimum since commercial establishments want to have the dumpsters within a reasonable distance of their facilities. Estimate that it takes an hour for the dumpster truck to travel to and from the disposal facility.

For this chapter:

1. Design a plan for the collection of the nonrecycled MSW.
   a. Dumpsters and dumpster vehicles:
      What fraction of domestic solid waste will go into the dumpsters? Where will the dumpsters be placed to service the public institutions, the commercial establishments, and the apartments? How many commercial dumpsters will be provided? How many collection vehicles will be needed to service the dumpsters? Show maps with dumpster locations, including commercial establishments, apartment buildings, and public institutional establishments (excluding Bucknell University, which owns its own dumpsters).
   b. Routing of residential collection and vehicles:
      Establish a routing system for residential solid waste collection, including the type and number of trucks, the routes taken, frequency of collection, and number of personnel required (with job classifications). Recognize the existence of alleys in the older part of town, where the present collection occurs. If roll-out cans are to be used, who will pay for them? Who will pay for the replacement? Amortize any capital costs over the expected life.
2. Calculate the expected gross cost for the residential, commercial, and public institutional solid waste collection (remember that the industries and the college will collect their own) for the year $X + 3$. Include in your calculations the cost of the vehicles, the dumpsters (and the roll-out containers if you choose to use them) by annualizing the cost using the 6.125% interest table provided. (*Note*: If you borrow $1, you pay back $1.06125 the next year. If you borrow $1 and pay it back in 10 equal installments over the next 10 years, you pay $0.13667 every year.) The gross cost of collection for Lewisburg, as $/year, is represented by the symbol **G**.
3. Decide what you will charge commercial establishments for collecting the commercial dumpsters, as $/dumpster/year. How did you decide on this figure? Using the estimated number of dumpsters from paragraph 1, calculate the income to the town from charging commercial establishments, as $/year. This income is represented by the symbol **F**.
4. Calculate the net collection cost by subtracting the income from the fees charged to commercial establishments from the cost of collection, or

   Net cost of collection = **G** − **F**

5. Calculate the cost of collection on a tonnage basis by dividing the net cost of collection calculated above by the total quantity of refuse collected by the town (including commercial, but not including industrial or private institutional).

   We will assume that this collection cost ($/ton) is constant for all disposal options (with the exception of recycling, which will have its own collection costs). This cost, as $/ton, is used in the calculations in subsequent chapters and is represented by the symbol **C**.

CHAPTER FOUR

SITING THE SANITARY LANDFILL

The borough has decided that it might want to operate its own landfill (a very unrealistic assumption, of course). Many considerations must go into selecting a new site, including transportation, proximity to the town, social and political constraints, environmental constraints, and soil types.

For this chapter:

1. Estimate the solid waste production (minus recycled waste) for the Borough of Lewisburg for the years $X + 13$ and $X + 23$. (Make sure you explain to the reader why you have selected these two years.) You are welcome to make up whatever scenario you wish for the future of the industries and Bucknell University.

2. Set the expected life of the landfill. Too many years, and the land will be too expensive. Too few years, and the borough will be looking for a new landfill in a few years. Using the solid waste production figures from Chapter 1 (including the commercial and private institutional solid waste), as well as the estimates for the years $X +13$ and $X + 23$, estimate the land area needed. Assume a compacted refuse density of 1000 lb/yd^3. Estimate earth cover as 20% of the refuse volume. Make sure enough earth exists to provide the cover. If not enough is available on site, indicate where it will be obtained.

3. Select a site. Indicate its location on the USGS map.[1] Indicate roads that need to be constructed.

4. Anticipate the opposition to your site. Who will be against it, and why? How can this criticism be countered?

CHAPTER FIVE

DESIGN OF THE SANITARY LANDFILL

At the present time Pennsylvania law requires that all new landfills meet RCRA standards (Subtitle D landfills) and have, at a minimum, the following:

a. double liner (clay plus synthetic)
b. composit collection and treatment (or disposal)
c. gas venting or collection and use
d. a plan and fund for closure

---

1. Maps are available at http://info.brookscole.com/Vesilind. Double click on each of the maps to download them to your hard drive.

Assume land costs at between $10,000 and $5,000 per acre, depending on proximity to town, availability of transportation, and productivity. Assume $50,000 per mile for the construction of an unpaved road, and add $50,000 per mile for the paving. Assume site clearing, grading, fencing, etc. as $50,000 per acre. A scale costs $200,000 and is expected to last 20 years. Assume that we can lump the operating costs (landfill cell construction, including liners, pipelines for gas control, leachate-collection system, labor, and the annualized equipment costs) as $40 per ton of refuse landfilled. Assume closure costs as $100,000 per acre. This has to be handled as a sinking fund, and the cost attached to the cost of running the landfill. That is, a fund has to be established so that at the time the landfill is full, enough money is available in this fund to close the landfill at a cost of $100,000 per acre. You have to use the interest table to make this calculation.

For this chapter:

1. Using the land area established in Chapter 4, calculate the cost of the land and annualize this capital cost using the expected life of the facility at 6.125% interest.
2. Calculate the annualized cost of the scale.
3. Calculate the annual cost of the closure. (*Note:* You have to establish a sinking fund over the life of the landfill, and each year some funds will be placed into this fund. For example, if the interest rate is 6.125%, and you want to have $1 at the end of ten years, you have to invest $0.07542 every year.) You estimate the cost of closure, then calculate the annual contribution to this fund so that the money will be available when needed to close the landfill.
4. Estimate the annual operating cost of the landfill, using the figures for the first year the landfill would be in operation (year ___).
5. Add these four annual costs. This is your gross annual cost of the landfill for the year ___ . This cost is represented by the symbol **A** in the calculations below.
6. Establish the tipping fee charged to the industries and Bucknell University, expressed as $/ton of refuse received. How does this compare with your gross per ton cost calculated above? Should this tipping fee be lower or higher than the cost to the town? Why? If the fee is too high, the industries and Bucknell could take their refuse elsewhere. Remember that the Lycoming County Landfill, only 20 miles to the north of Lewisburg, charges a tipping fee of $40/ton for commercial and institutional customers. Calculate the total annual income to the town from this fee by multiplying the fee ($/ton) times the production of industrial plus private institutional solid waste for the first year of operation (year ____). This income is represented by the symbol **B** in the calculations below.
7. Calculate the net annual cost of the landfill by subtracting the annual income received (by charging the tipping fee) from the gross annual cost of the landfill, or

$$E = A - B$$

where **E** = net annual cost of landfill, $/year
     A = gross annual cost of landfill, $/year
     B = annual income derived from tipping fees, $/year

This is the net cost of the landfill, under this option, to the people of Lewisburg. Note that if **E** is negative, the borough is making money on the landfill.

8. Divide this net annual cost of the landfill (**E**) by the total tonnage of nonrecycled refuse (residential, commercial, and public institutional) expected during the first year of operation. This is the $/ton figure for the cost of landfilling to be used in subsequent chapters. Note that this cost per ton takes into account the tipping fees paid by the industries and institutional users. This cost of landfilling, as $/ton, is represented by the symbol **D** in subsequent chapters.

9. Calculate the cost of the landfilling option as the net annual cost of the landfill (paragraph 7 above) plus the cost of collection (from Chapter 3) plus the cost of the recycling program.

Annual cost of the landfill option with recycling ($/year) = **E** + **C**(**Q**) + **H**

where **E** = net annual cost of landfill, $/year
   C = cost of collection from Chapter 3, $/ton
   Q = total quantity of nonrecycled, nonindustrial, noninstitutional refuse collected by the town, tons/year
   H = cost of recycling program, $/year

This figure is used for comparative purposes in Chapter 8.

CHAPTER SIX

MATERIALS RECOVERY FACILITY

Assume that in this scenario all of the nonrecycled waste will be processed in a Materials Recovery Facility (MRF). The owner-separated recycled waste is also brought to the facility, but we assume that this will not add any cost to the operation (or reduce the cost of the recycling operation).

An understanding has been reached with the power company to burn a processed waste (so-called RDF-3) at the Shamokin Dam Power Plant. They will not, however, pay anything for the fuel. The transport distance is 12 miles one way, at a cost of $0.25/ton-mile. (That is, if 40 tons of RDF is sent to the power plant from Lewisburg, the transportation cost is 40 × 2 × 12 × 0.25 = $240 for that trip.)

You may use the following costs for estimating the cost of refuse processing:

Process	Cost per ton (including annualized capital and O&M)
Shredding	5.00
Air classifier	0.50
Magnet	0.15
Trommel screen	0.50
Any other process	0.40
Each conveyor	0.10

For the year X + 3 estimate the amortized annual cost of the building to house the facility as equal to the annual cost of processing (i.e., you simply double the cost of the processing). Assume that the processing facility will be at the site chosen for the landfill.

If you decide to use pickers, estimate their cost as $25,000 per person per year. The cost of operators and maintenance personnel is included in the above per ton figures.

For this chapter:

1. Design a solid waste processing facility. Specify the unit operations to be used and decide on the capacity of those unit operations. Draw a schematic of the process train and indicate what materials are to be recovered. (A schematic is a flow diagram, with the various processes identified by boxes and the flow of material shown by arrows.)
2. Draw a floor plan of the facility, including the refuse receiving area and the building. Be sure to indicate all storage and conveying equipment. Show how the material moves through your facility. If industrial solid waste is to be processed, how will it be received and how will it be processed? Will it be received at the same location, or will there be special receiving areas? The floor plan should be one step better than a schematic. It should include the various processing operations in rough size and space according to their operation. All material transfer operations must be identified.
3. Estimate the amount of RDF-3 to be produced and its fuel value.
4. Estimate the quantity of other materials to be recovered. Add to these the recycled materials. Use the prices listed in the previous chapter as market values (although this may not be accurate due to the low degree of purity) to calculate the expected income from the sale of the recovered materials.
5. Estimate the amount of waste destined for the landfill as all materials not recovered. Remember that some types of wastes cannot be processed through the MRF. Calculate the cost of landfilling using the prorated cost for the landfill as before.
6. Calculate the net cost of the central refuse processing option as

$$\text{Annual cost of the MRF option (\$/year)} = L - M + C(Q_{prod}) + D(Q_{rej}) + H$$

where L = annualized cost of equipment and building (capital plus O&M), \$/year
   M = annual income (or loss) from sale of recovered materials, including the cost of transport for the RDF-3, \$/year. Note that if M is negative, this must be treated as a cost.
   C = annual cost of collection of nonseparated materials, \$/ton (from Chapter 3)
   Q_{prod} = quantity of mixed refuse produced by the town and brought to the MRF, tons/year
   D = cost of landfilling rejected material, \$/ton
   Q_{rej} = quantity of materials not recovered (rejected) in MRF, tons/year
   H = cost of recycling operation, \$/year

This figure is used for comparative purposes in Chapter 8.

CHAPTER SEVEN

WASTE-TO-ENERGY (MASS BURNING)

The State of Pennsylvania considered waste-to-energy production as recycling. This option meets the State of Pennsylvania mandate for recycling, and it is no longer necessary to have the borough conduct a separate recycling program.

Mass burning is the combustion of unprocessed refuse. Burn units are sized on the basis of 24-hour operation. For example, if a furnace processes 20 tons per day and it is operated for 4 hours and shut down for the remainder of the day, the furnace is rated as a 120 ton/day facility. Only small modular units can be fired up on a daily basis, but these have been so troublesome that the manufacturers have discontinued their sale. The minimum waste-to-energy unit must receive 5 ton/day capacity on a 24-hour basis.[*] Such units cannot be shut down, except for cleaning and maintenance. If the amount of refuse is less than 5 tons/day, the refuse must be stored in a pit and fed continuously at a minimum rate of 5 tons/day (including weekends and holidays!). Typically, such units can operate as much as 300 days out of the year, with 65 days per year for repair and maintenance.

The annual cost can be estimated as follows:

Debt service (annualized)—$50,000/ton daily capacity
Operation (salaries, fuel, O&M)—$50,000/ton

For example, if the flow of waste is 10 tons/day and a furnace with a capacity of 15 tons/day is installed, the annualized capital cost is 15 × $50,000 = $750,000, and the operating cost is 10 × $50,000 = $500,000, for a total annual cost to the community of $1.25 million.

Assume linear capital cost extrapolation (no economies of scale—not a very good idea in real life). That is, a plant with a capacity of 100 tons/day, for example, has an annualized capital cost of $5 million/year, and assuming 24-hour operation, it has operational costs of $5 million/year. The debt service (annualized capital cost) obviously continues even when the plant is shut down for maintenance, as does some fraction of the operating cost (e.g., salaries), but we are ignoring these costs.

Estimate that a waste-to-energy plant will be able to reduce the volume of the refuse fed to it by 90%, but the remaining 10% of the waste (as ash) must go to the landfill. The density of the ash when placed in a landfill is about 2000 lb/cubic yard. During the time that the unit is down for repairs, all of the refuse must go to the landfill.

The facility is to produce steam and run a turbine, which will produce electricity to be sold to the power company at an income of $0.04/kWh. Assume that the plant will produce 500 kWh of electricity per ton of refuse delivered to the waste-to-energy facility.

For this chapter:

1. Select the type of mass-burn facility to be used. Draw a picture showing its major components. Note how the steam is produced and where the turbine is.
2. Specify what waste (if any) will be excluded from the facility.

---

[*] In real life there are no 5 ton/day mass combustors. This is far too little waste for a mass burn facility.

3. Estimate the quantity of refuse to be processed. This includes the materials that would otherwise have been recycled. Calculate the income derived from the sale of the electricity.

4. Estimate the amount of waste destined for the landfill as equal to the material not accepted at the mass-burn facility plus the ash produced. Calculate the cost of the landfilling using the figure from Chapter 5.

5. Estimate the cost of the mass-burn option as

Annual cost of the waste-to-energy option ($/year) =
$$N - P + D(Q_{ash}) + C(Q_{prod})$$

where

N = annualized cost of the mass-burn facility (capital plus O&M), $/year
P = annual income from the sale of electricity, $/year
D = cost of landfilling, $/ton
Q_{ash} = quantity of ash and unburned refuse landfilled, tons/year
C = collection costs, from Chapter 3, $/ton
Q_{prod} = quantity of refuse collected, tons/year

CHAPTER EIGHT

TECHNICAL, ECONOMIC, AND ENVIRONMENTAL EVALUATION

In this chapter collect all of the cost data generated in the earlier chapters and present comparisons in tabular or graphical form.

Discuss the technical considerations (dependability, reserve capacity, what future solid waste management systems might be available, industrial involvement, etc.). Present a brief discussion of the environmental concerns and ramifications of the various alternatives, listing each and responding to what are perceived to be the major public concerns. Remember that the borough is under a state mandate to have some type of recycling effort, so this must be included in your plan. Remember also that one of your alternatives is to recommend that the borough sign a long-term contract with the Lycoming County Landfill to send all of the refuse to them at a cost of $80/ton. Should the borough simply do that and save itself all the trouble of running a solid waste disposal program? What are the pros and cons of this?

Finally (!) choose one recommended alternative and justify your choice.

Bulk Densities of Refuse Components

This appendix is based on Appendix J of *Conducting a Diversion Study—A Guide for California Jurisdictions* published in March 2001 by California Integrated Waste Management Board. The table is a compilation of data from various sources listed at the end of this appendix.

Construction and Demolition

Material	Size	Study	LBS
Ashes, dry	1 cubic foot	FEECO	35–40
Ashes, wet	1 cubic foot	FEECO	45–50
Asphalt, crushed	1 cubic foot	FEECO	45
Asphalt/paving, crushed	1 cubic yard	Tellus	1,380
Asphalt/shingles, comp, loose	1 cubic yard	Tellus	418.5
Asphalt/tar roofing	1 cubic yard	Tellus	2,919
Bone meal, raw	1 cubic foot	FEECO	54.9
Brick, common hard	1 cubic foot	FEECO	112–125
Brick, whole	1 cubic yard	Tellus	3,024
Cement, bulk	1 cubic foot	FEECO	100
Cement, mortar	1 cubic foot	FEECO	145
Ceramic tile, loose 6″ × 6″	1 cubic yard	Tellus	1,214
Chalk, lumpy	1 cubic foot	FEECO	75–85
Charcoal	1 cubic foot	FEECO	15–30
Clay, kaolin	1 cubic foot	FEECO	22–33
Clay, potter's dry	1 cubic foot	FEECO	119
Concrete, cinder	1 cubic foot	FEECO	90–110
Concrete, scrap, loose	1 cubic yard	Tellus	1,855
Cork, dry	1 cubic foot	FEECO	15
Earth, common, dry	1 cubic foot	FEECO	70–80
Earth, loose	1 cubic foot	FEECO	76
Earth, moist, loose	1 cubic foot	FEECO	78
Earth, mud	1 cubic foot	FEECO	104–112

(continued)

Construction and Demolition, *continued*

Material	Size	Study	LBS
Earth, wet, containing clay	1 cubic foot	FEECO	100–110
Fiberglass insulation, loose	1 cubic yard	Tellus	17
Fines, loose	1 cubic yard	Tellus	2,700
Glass, broken	1 cubic foot	FEECO	80–100
Glass, plate	1 cubic foot	FEECO	172
Glass, window	1 cubic foot	FEECO	157
Granite, broken or crushed	1 cubic foot	FEECO	95–100
Granite, solid	1 cubic foot	FEECO	130–166
Gravel, dry	1 cubic foot	FEECO	100
Gravel, loose	1 cubic yard	Tellus	2,565
Gravel, wet	1 cubic foot	FEECO	100–120
Gypsum, pulverized	1 cubic foot	FEECO	60–80
Gypsum, solid	1 cubic foot	FEECO	142
Lime, hydrated	1 cubic foot	FEECO	30
Limestone, crushed	1 cubic foot	FEECO	85–90
Limestone, finely ground	1 cubic foot	FEECO	99.8
Limestone, solid	1 cubic foot	FEECO	165
Mortar, hardened	1 cubic foot	FEECO	100
Mortar, wet	1 cubic foot	FEECO	150
Mud, dry close	1 cubic foot	FEECO	110
Mud, wet fluid	1 cubic foot	FEECO	120
Pebbles	1 cubic foot	FEECO	90–100
Pumice, ground	1 cubic foot	FEECO	40–45
Pumice, stone	1 cubic foot	FEECO	39
Quartz, sand	1 cubic foot	FEECO	70–80
Quartz, solid	1 cubic foot	FEECO	165
Rock, loose	1 cubic yard	Tellus	2,570
Rock, soft	1 cubic foot	FEECO	100–110
Sand, dry	1 cubic foot	FEECO	90–110
Sand, loose	1 cubic yard	Tellus	2,441
Sand, moist	1 cubic foot	FEECO	100–110
Sand, wet	1 cubic foot	FEECO	110–130
Sewage, sludge	1 cubic foot	FEECO	40–50
Sewage, sludge dried	1 cubic foot	FEECO	35
Sheetrock scrap, loose	1 cubic yard	Tellus	393.5
Slag, crushed	1 cubic yard	Tellus	1,998
Slag, loose	1 cubic yard	Tellus	2,970
Slag, solid	1 cubic foot	FEECO	160–180
Slate, fine ground	1 cubic foot	FEECO	80–90
Slate, granulated	1 cubic foot	FEECO	95

(continued)

Construction and Demolition, *continued*

Material	Size	Study	LBS
Slate, solid	1 cubic foot	FEECO	165–175
Sludge, raw sewage	1 cubic foot	FEECO	64
Soap, chips	1 cubic foot	FEECO	15–25
Soap, powder	1 cubic foot	FEECO	20–25
Soap, solid	1 cubic foot	FEECO	50
Soil/sandy loam, loose	1 cubic yard	Tellus	2,392
Stone or gravel	1 cubic foot	FEECO	95–100
Stone, crushed	1 cubic foot	FEECO	100
Stone, crushed, size reduced	1 cubic yard	Tellus	2,700
Stone, large	1 cubic foot	FEECO	100
Wax	1 cubic foot	FEECO	60.5
Wood ashes	1 cubic foot	FEECO	48

Glass

Item	Amount	Study	LBS
Glass, broken	1 cubic foot	FEECO	80–100.00
Glass, broken	1 cubic yard	FEECO	2,160.00
Glass, crushed	1 cubic foot	FEECO	40–50.0
Glass, plate	1 cubic foot	FEECO	172
Glass, window	1 cubic foot	FEECO	157
Glass, 1 gallon jug	each	USEPA	2.10–2.80
Glass, beer bottle	each	USEPA	0.53
Glass, beverage—8 oz	1 bottle	USEPA	0.5
Glass, beverage—8 oz	1 case = 24 bottles	USEPA	12
Glass, beverage—12 oz	1 case = 24 bottles	USEPA	22
Glass, wine bottle	each	USEPA	1.08

Plastic

Item	Amount	Study	LBS
Film plastic/mixed, loose	1 cubic yard	Tellus	22.55
HDPE Film plastics, semi-compacted	1 cubic yard	Tellus	75.96
LDPE Film plastics, semi-compacted	1 cubic yard	Tellus	72.32

HDPE, common beverage containers

Item	Size	Study	LBS
Plastic, HDPE juice, 8 oz	8 oz.	USEPA	0.1
Plastic, 1 gallon HDPE jug	1 gallon	USEPA	0.33
Plastic, 1 gallon HDPE beverage cont.	milk/juice	USEPA	0.19–0.25

PETE, common beverage containers

Item	Size	Study	LBS
Plastic, 1 liter PETE beverage bottle w/o cap	1 liter	USEPA corr.	0.09
Plastic, PETE water bottle	50 oz. (over 1.5 liters)	USEPA	0.12
Plastic, PETE, 2 liter	1 bottle	USEPA	0.13

Plastic containers

Item	Size	Study	LBS
Plastic, 1 gallon container—mayo	1 gallon	USEPA	0.42
Plastic, 1/2 gallon plastic beverage cont.	1/2 gallon	USEPA	0.09
Plastic, beverage container	12 oz.	USEPA	0.05

Miscellaneous plastic items

Item	Amount	Study	LBS
Plastic, bubble wrap	33 gallons	USEPA	3
Plastic, bucket	25 gallons	USEPA	1.1
Plastic, bucket w/metal handle	5 gallons	USEPA	1.9
Plastic, cake decorator's boxes	each	USEPA	0.63
Plastic, grocery bag	100 bags	USEPA corr.	0.77
Plastic, HDPE 10–12 fluid oz	.	USEPA	0.05
Plastic, HDPE beverage case	.	USEPA	1.2
Plastic, HDPE bread case	.	USEPA	1.5
Plastic, HDPE gallon containers (not beverage)	1 gallon	USEPA	0.06
Plastic, HDPE (auto) oil container	1 quart size	USEPA	0.2
Plastic, pot	1 qt size	USEPA	0.25
Plastic, mixed HDPE & PETE	1 cubic yard	USEPA	32
Plastic, pallet, 48″ × 48″		USEPA	40
Plastic, sheeting	square yard	USEPA	1

(continued)

Miscellaneous plastic items, *continued*

Item	Size	Study	LBS
Plastic, whole, uncompacted PETE	1 cubic yard	USEPA	30–40
Polyethylene, resin pellets	1 cubic foot	FEECO	30–35
Polystyrene beads	1 cubic foot	FEECO	40
Polystyrene, packaging	33 gallon	USEPA	1.5
Styrofoam kernels	1 cubic yard	Tellus	6.27
Polystyrene, blown formed foam	1 cubic yard	Tellus	9.62
Polystyrene, rigid, whole	1 cubic yard	Tellus	21.76
PVC, loose	1 cubic yard	Tellus	341.12

Paper

Item	Size	Study	LBS
Books, hardback, loose	1 cubic yard	Tellus	529.29
Books, paperback, loose	1 cubic yard	Tellus	427.5
Egg flats	one dozen	USEPA	0.12
Egg flats	12″ × 12″	USEPA	0.5
Paper sacks	25# size	USEPA	0.5
Paper sacks	50# dry goods	USEPA	1
Calendars / Books	1 cubic foot	FEECO	50
Catalogs	100 pages ledger	USEPA	1
Computer printout, loose	1 cubic yard	USEPA	655
Mixed Paper, loose (construction, fax, manila, some chipboard)	1 cubic yard	USEPA	363.5
Mixed Paper, compacted (construction, fax, manila, some chipboard)	1 cubic yard	USEPA	755
Office Paper (white, color, CPO, junk mail)	13 gallon	USEPA	10.01
Office Paper (white, color, CPO, junk mail)	33 gallon	USEPA	25.41
Office Paper (white, color, CPO, junk mail)	55 gallon	USEPA	42.35
Shredded Paper	33 gallons	USEPA	8
White Ledger Paper	12″ stack	USEPA	12
White Ledger #20, 8.5″ × 11″	1 ream (500 sheets)	USEPA	5
White Ledger #30, 8.5″ × 14″	1 ream (500 sheets)	USEPA	6.4
White Ledger w/o CPO, loose	1 cubic yard	Tellus	363.51

(continued)

Paper, *continued*

Item	Size	Study	LBS
White Ledger, uncompacted stacked	1 cubic yard	USEPA	400
White Ledger, compacted stacked	1 cubic yard	USEPA	800
Colored message pads	1 carton (144 pads)	USEPA	22
Padded envelope	9" × 12"	USEPA	0.88
Magazines, 8.5" × 11"	10 units	USEPA	3
Manila envelope	1 cubic foot	FEECO	37
Newspapers	12" stack	USEPA	35
Newspapers	1 cubic foot	FEECO	38
Newspapers, loose	1 cubic yard	USEPA	400
Newspapers, stacked	1 cubic yard	USEPA	875
Phone book	Ventura	USEPA	4
Paper pulp, stock	1 cubic foot	FEECO	60–62.00
Tab cards, uncompacted	1 cubic yard	USEPA	605
Tab cards, compacted	1 cubic yard	USEPA	1,275.00
Yellow legal pads	1 case (72 pads)	USEPA	38

Chipboard

Item	Size	Study	LBS
Chipboard, beverage case	4 pack	USEPA	0.1
Chipboard, beverage case	6 pack	USEPA	0.2
Chipboard, cereal box	average	USEPA	0.15
Chipboard, fabric bolt		USEPA	0.69
Paperboard/Boxboard/ Chipboard whole	1 cubic yard	Tellus	21.5

OCC

Item	Size	Study	LBS
OCC, beverage case	4 six-packs full case	USEPA corr.	0.99
OCC, box, large	48"×48"×60"	USEPA	4
OCC, box, medium	24"×24"×30"	USEPA	2.2
OCC, box, small	12"×12"×15"	USEPA	1.1
OCC, flattened boxes, loose	1 cubic yard	Tellus	50.08
OCC, stacked	1 cubic yard	USEPA	50
OCC, whole boxes	1 cubic yard	Tellus	16.64
OCC, uncompacted	1 cubic yard	USEPA	100
OCC, compacted	1 cubic yard	USEPA	400

Organics

Item	Size	Study	LBS
Yard trims, mixed	1 cubic yard	USEPA	108
Yard trims, mixed	40 cubic yards	USEPA	4,320
Grass	33 gallons	USEPA	25
Grass	3 cubic yards	USEPA	840
Grass & leaves	3 cubic yards	USEPA	325
Large limbs & stumps	1 cubic yard	Tellus	1,080
Leaves, dry	1 cubic yard	Tellus	343.7
Leaves	33 gallons	USEPA	12
Leaves	3 cubic yards	USEPA	200–250
Pine needles, loose	1 cubic yard	Tellus	74.42
Prunings, dry	1 cubic yard	Tellus	36.9
Prunings, green	1 cubic yard	Tellus	46.69
Prunings, shredded	1 cubic yard	Tellus	527

Other Organic Material

Item	Size	Study	LBS
Hay, baled	1 cubic foot	FEECO	24
Hay, loose	1 cubic foot	FEECO	5
Straw, baled	1 cubic foot	FEECO	24
Straw, loose	1 cubic foot	FEECO	3
Compost	1 cubic foot	FEECO	30–50
Compost, loose	1 cubic foot	Tellus	463.39

Food

Item	Size	Study	LBS
Bread, bulk	1 cubic foot	FEECO	18
Fat	1 cubic foot	FEECO	57
Fats, solid/liquid (cooking oil)	1 gallon	USEPA	7.45
Fats, solid/liquid (cooking oil)	55 gallon drum	USEPA	410
Fish, scrap	1 cubic foot	FEECO	40–50
Meat, ground	1 cubic foot	FEECO	50–55
Oil, olive	1 cubic foot	FEECO	57.1
Oyster shells, whole	1 cubic foot	FEECO	75–80
Produce waste, mixed, loose	1 cubic foot	Tellus	1,443

Manure

Item	Size	Study	LBS
Manure	1 cubic foot	FEECO	25
Manure, cattle	1 cubic yard	Tellus	1,628

(continued)

Manure, *continued*

Item	Size	Study	LBS
Manure, dried poultry	1 cubic foot	FEECO	41.2
Manure, dried sheep & cattle	1 cubic foot	FEECO	24.3
Manure, horse	1 cubic yard	Tellus	1,252

Wood

Item	Size	Study	LBS
Cork, dry	1 cubic foot	FEECO	15
Pallet, wood or plastic	average 48″ × 48″	USEPA	40
Particle board, loose	1 cubic yard	Tellus	425.14
Plywood, sheet 2′ × 4′	1 cubic yard	Tellus	776.3
Roofing/shake shingle, bundle	1 cubic yard	Tellus	435.3
Sawdust, loose	1 cubic yard	Tellus	375
Shavings, loose	1 cubic yard	Tellus	440
Wood chips, shredded	1 cubic yard	USEPA	500
Wood scrap, loose	1 cubic yard	Tellus	329.5
Wood, bark, refuse	1 cubic foot	FEECO	30-50
Wood, pulp, moist	1 cubic foot	FEECO	45–65
Wood, shavings	1 cubic foot	FEECO	15

Miscellaneous

Item	Size	Study	LBS
Toner cartridge	.	USEPA	2.5

Rubber

Item	Size	Study	LBS
Tire, bus	.	USEPA	75
Tire, car	.	USEPA	20
Tire, truck	.	USEPA	60–100
Rubber, car bumper	.	USEPA	15
Rubber, manufactured	1 cubic foot	FEECO	95
Rubber, pelletized	1 cubic foot	FEECO	50–55

Textiles

Item	Size	Study	LBS
Clothing, used, mixed	cubic yard	Tellus	225
Fabric, canvas	square yard	USEPA	1
Leather, dry	1 cubic foot	FEECO	54
Leather, scrap, semi-compacted	1 cubic yard	Tellus	303
Rope	1 cubic foot	FEECO	42
String	yard	USEPA	1 gram
Used clothing, mixed, loose	1 cubic yard	Tellus	225
Used clothing, compacted	1 cubic yard	Tellus	540
Wool	1 cubic foot	FEECO	15–30
Carpet & padding, loose	1 cubic yard	Tellus	84.4

Metals

Aluminum

Item	Size	Study	LBS
Aluminum foil, loose	1 cubic yard	Tellus	48.1
Aluminum scrap, cubed	1 cubic yard	Tellus	424
Aluminum scrap, whole	1 cubic yard	Tellus	175
Aluminum cans, uncrushed	1 case = 24 cans	USEPA	0.89
Aluminum cans, crushed	13 gallons	USEPA corr.	7.02
Aluminum cans, crushed	33 gallons	USEPA	17.82
Aluminum cans, crushed	39 gallons	USEPA	31.06
Aluminum cans, crushed & uncrushed mix	1 cubic yard	Tellus	91.4
Aluminum cans, uncrushed	1 full grocery bag	USEPA	1.5
Aluminum cans, uncrushed	13 gallons	USEPA	2.21
Aluminum cans, uncrushed	33 gallons	USEPA	5.61
Aluminum cans, uncrushed	39 gallons	USEPA	6.63
Aluminum cans (whole)	1 cubic yard	USEPA	65
Aluminum, chips	1 cubic foot	FEECO	15–20
Aluminum/Tin cans commingled-uncrushed	33 gallon	USEPA	11.55

Ferrous

Item	Size	Study	LBS
Metal scrap	55 gallon	USEPA	226.5
Metal scrap	cubic yard	USEPA	906
Metal, car bumper	each	USEPA	40

(continued)

Ferrous, *continued*

Item	Size	Study	LBS
Paint can	5 gallon	USEPA	2.21
Radiator, ferrous	each	USEPA	20
Hanger (adult)	each	CIWMB	0.14
Hanger (child)	each	CIWMB	0.09
Tin can, ferrous	#2.5	USEPA	0.13
Tin can, ferrous	#5	USEPA	0.28
Tin can, ferrous	#10	USEPA	0.77
Tin coated steel cans	1 cubic yard	USEPA	850
Tin coated steel cans	1 case (6 #10 cans)	USEPA	22
Tin, tuna can (3/4 of #10)	each	USEPA	0.58
Tin, cat food can, ferrous	8 oz.	USEPA	0.14
Tin, dog food can, large	22 oz.	USEPA	0.22
Tin, dog food can, Vet	15.5 oz.	USEPA	0.11
Tin, cast	1 cubic foot	FEECO	455
Cast iron chips or borings	1 cubic foot	FEECO	130–200
Iron, cast ductile	1 cubic foot	FEECO	444
Iron, ore	1 cubic foot	FEECO	100–200
Iron, wrought	1 cubic foot	FEECO	480
Steel, shavings	1 cubic foot	FEECO	58–65
Steel, solid	1 cubic foot	FEECO	487
Steel, trimmings	1 cubic foot	FEECO	75–150
Brass, cast	1 cubic foot	FEECO	519
Brass, scrap	1 cubic yard	Tellus	906.43
Bronze	1 cubic foot	FEECO	552
Copper fittings, loose	1 cubic yard	Tellus	1,047.62
Copper pipe, whole	1 cubic yard	Tellus	210.94
Copper, cast	1 cubic foot	FEECO	542
Copper, ore	1 cubic foot	FEECO	120–150
Copper, scrap	1 cubic yard	Tellus	1,093.52
Copper, wire, whole	1 cubic yard	Tellus	337.5
Chrome ore (chromite)	1 cubic foot	FEECO	125–140
Lead, commercial	1 cubic foot	FEECO	710
Lead, ores	1 cubic foot	FEECO	200–270
Lead, scrap	1 cubic yard	Tellus	1,603.84
Nickel, ore	1 cubic foot	FEECO	150
Nickel, rolled	1 cubic foot	FEECO	541

Furniture

This data is from the U.S. EPA Business Users Guide Study.

Item	Type	Material	Size	LBS
Desk	Executive, single pedestal	Wood	72″ × 36″	246.33
Desk	Executive, double pedestal	Wood	.	345
Desk	Double pedestal	Laminate	72″ × 36″	299.5
Desk	Double pedestal	Laminate	60″ × 30″	231
Desk	Single pedestal	Laminate	72″ × 36″	250
Desk	Single pedestal	Laminate	42″ × 24″	146
Desk	Double pedestal	Metal	72″ × 36″	224.67
Desk	Double pedestal	Metal	60″ × 30″	184.75
Desk	Double pedestal	Metal	54″ × 24″	124
Desk	Single pedestal	Metal	72″ × 36″	189
Desk	Single pedestal	Metal	48″ × 30″	133.67
Desk	Single pedestal	Metal	42″ × 24″	146
Desk	Single pedestal	Metal	40″ × 20″	82
Desk	Small modular panel system	.	.	422
Desk	Large modular panel system	.	.	650
Workstation	with return	Laminate	60″ × 30″	329.33
Workstation	with return	Metal	60″ × 30″	230.67
Bridge	Executive	Wood	.	76.67
Bridge	.	Laminate	.	140
Credenza	.	Wood	.	250.78
Credenza	.	Laminate	.	230.14
Credenza	with knee space	Metal	60″ × 24″	156.67
Round conference table	.	Wood	42″ diameter	91.5
Bookcase	3 shelves	Wood	36″ wide	90
Bookcase	4 shelves	Wood	36″ wide	110.9
Bookcase	5 shelves	Wood	36″ wide	138.8
Bookcase	6 shelves	Wood	36″ wide	134.6
Bookcase	7 shelves	Wood	34″ wide	138.5
Bookcase	4 shelves	Laminate	.	85
Bookcase	5 shelves	Laminate	.	110
Bookcase	2 shelves	Metal	34″–36″	44.5
Bookcase	3 shelves	Metal	34″–36″	57.5
Bookcase	4 shelves	Metal	34″–36″	70.5
Bookcase	5 shelves	Metal	34″–36″	89
Bookcase	6 shelves	Metal	34″–36″	101
File Cabinet	2 drawer, lateral	Wood	.	155.14

(continued)

Furniture, *continued*

This data is from the U.S. EPA Business Users Guide Study.

Item	Type	Material	Size	LBS
File Cabinet	2 drawer, lateral	Laminate	.	171.5
File Cabinet	2 drawer, lateral	Metal	30"–42"	230.67
File Cabinet	4 drawer, lateral	Metal	36"	207.33
File Cabinet	2 drawer, vertical	Metal	Letter size	60.6
File Cabinet	4 drawer, vertical	Metal	Letter size	107.6
File Cabinet	2 drawer, vertical	Metal	Legal size	71.5
File Cabinet	4 drawer, vertical	Metal	Legal size	123.5
Chair	Executive desk	.	.	51.167
Chair	Guest arm	.	.	38.2
Chair	Swivel arm	.	.	45.25
Chair	Secretary with no arms	.	.	31.76
Chair	Stacking	.	.	15.83
Personal Computer	CPU (Central Processing Unit)	.	.	26
Computer monitor	.	.	.	30
Computer printer	.	.	.	25.33

Source Acronyms Used:
CIWMB   California Integrated Waste Management Board
FEECO   FEECO Incorporated
Tellus   Tellus Institute, Boston, Massachusetts
USEPA   United States Environmental Protection Agency
                 Business Users Guide

REFERENCES

EPA Municipal and Industrial Solid Waste, Washington, D.C. and University of California Los Angeles Extension Recycling and Municipal Solid Waste Management Program. 1996. *Business Waste Prevention Quantification Methodologies—Business Users Guide* (November). Los Angeles, Calif.

EPA. 1997. *Measuring Recycling: A Guide For State and Local Governments* (September). Publication number EPA530-R-97-011.

Cal Recovery Inc., Tellus Institute, and ACT...now. 1991. *Conversion Factors for Individual Material Types*. Submitted to California Integrated Waste Management Board (December).

FEECO International, Inc. *FEECO International Handbook*, 8th Printing (Section 22-45 to 22-510). Green Bay, Wis.

Harivandi, M. Ali, Victor A. Gibeault, and Trevor O'Shaughnessy. 1996. "Grasscycling in California." *California Turfgrass Culture* 46: 1–2.

Alameda County Waste Management Authority Home Composting Survey. 1992. *Source Reduction Through Home Composting*. Summary published in *Biocycle* (April).

APPENDIX C

Conversions

Multiply	by	to obtain
acres	0.404	ha
acres	43,560	ft^2
acres	4047	m^2
acres	4840	yd^2
acre-ft	1233	m^3
atmospheres	14.7	lb/in^2
atmospheres	29.95	in. mercury
atmospheres	33.9	ft of water
atmospheres	10,330	kg/m^2
Btu	252	cal
Btu	1.053	kJ
Btu	1,053	J
Btu	2.93×10^{-4}	kWh
Btu/ft^3	8,905	cal/m^3
Btu/lb	2.32	kJ/kg
Btu/lb	0.555	cal/g
Btu/s	1.05	kW
Btu/ton	278	cal/tonne
Btu/ton	0.00116	kJ/kg
calories	4.18	J
calories	0.0039	Btu
calories	1.16×10^{-6}	kWh
calories/g	1.80	Btu/lb
$calories/m^3$	0.000112	Btu/ft^3
calories/tonne	0.00360	Btu/ton
centimeters	0.393	in.
cubic ft	1728	in^3
cubic ft	7.48	gal
cubic ft	0.0283	m^3
cubic ft	28.3	L
cubic ft/lb	0.0623	m^3/kg
cubic ft/s	0.0283	m^3/s
cubic ft/s	449	gal/min
cubic ft of water	61.7	lb of water
cubic in. of water	0.0361	lb water
cubic m	35.3	ft^3
cubic m	264	gal

(continued)

Multiply	by	to obtain
cubic m	1.31	yd^3
cubic m/day	264	gal/day
cubic m/hr	4.4	gal/min
cubic m/hr	0.00638	million gal/day
cubic m/s	1	cumec
cubic m/s	35.31	ft^3/s
cubic m/s	15,850	gal/min
cubic m/s	22.8	mil gal/day
cumec	1	m^3/s
cubic yards	0.765	m^3
cubic yards	202	gal
C-ration	100	rations
decacards	52	cards
feet	0.305	m
feet/min	0.00508	m/s
feet/s	0.305	m/s
fish	10^{-6}	microfiche
foot lb (force)	1.357	J
foot lb (force)	1.357	Nm
gallons	0.00378	m^3
gallons	3.78	L
gallons/day	43.8 × 10^{-6}	L/s
gallons/day/ft^2	0.0407	m^3/day/m^2
gallons/min	0.00223	ft^3/s
gallons/min	0.0631	L/s
gallons/min	0.227	m^3/hr
gallons/min	6.31 × 10^{-5}	m^3/s
gallons/min/ft^2	2.42	m^3/hr/m^2
gallons of water	8.34	lb water
grams	0.0022	lb
grams/cm^3	1,000	kg/m^3
hectares	2.47	acre
hectares	1.076 × 10^5	ft^2
horsepower	0.745	kW
horsepower	33,000	ft-lb/min
inches	2.54	cm
inches of mercury	0.49	lb/in^2
inches of mercury	0.00338	N/m^2
inches of water	249	N/m^2
joules	0.239	cal
joules	9.48 × 10^{-4}	Btu
joules	0.738	ft-lb
joules	2.78 × 10^{-7}	kWh
joules	1	Nm
joules/g	0.430	Btu/lb
joules/s	1	W
kilocalories	3.968	Btu
kilocalories/kg	1.80	Btu/lb
kilograms	2.20	lb

(continued)

Multiply	by	to obtain
kilograms	0.0011	tons
kilograms/ha	0.893	lb/acre
kilograms/hr	2.2	lb/hr
kilograms/m^3	0.0624	lb/ft^3
kilograms/m^3	1.68	lb/yd^3
kilograms/m^3	1.69	lb/yd^3
kilograms/tonne	2.0	lb/ton
kilojules	9.49	Btu
kilojules/kg	0.431	Btu/lb
kilometers	0.622	mile
kilometer/hr	0.622	miles/hr
kilowatts	1.341	horsepower
kilowatt-hr	3600	kJ
lite year	365	days drinking low-calorie beer
liters	0.0353	ft^3
liters	0.264	gal
liters/s	15.8	gal/min
liters/s	0.0228	million gal/day
megaphone	10^{12}	microphone
meters	3.28	ft
meters	1.094	yd
meters/s	3.28	ft/s
meters/s	196.8	ft/min
microscope	10^{-6}	mouthwash
miles	1.61	km
miles/hr	0.447	m/s
miles/hr	88	ft/min
miles/hr	1.609	km/hr
milligrams/L	0.001	kg/m^3
million gal	3,785	m^3
million gal/day	43.8	L/s
million gal/day	3785	m^3/day
million gal/day	157	m^3/hr
million gal/day	0.0438	m^3/s
newtons	0.225	lb(force)
newtons/m^2	2.94×10^{-4}	in. mercury
newtons/m^2	1.4×10^{-4}	lb/in^2
newtons/m^2	10	poise
newton-m	1	J
pounds (mass)	0.454	kg
pounds (mass)	454	g
pounds (mass)/acre	1.12	kg/ha
pounds(mass)/ft^3	16.04	kg/m^3
pounds (mass)/ton	0.50	kg/tonne
pounds (mass)/yd^3	0.593	kg/m^3
pounds (force)	4.45	N
pounds (force)/in^2	0.068	atmospheres
pounds (force)/in^2	2.04	in. of mercury
pounds (force)/in^2	6895	N/m^2

(continued)

Multiply	by	to obtain
pounds (force)/in^2	6.89	kPa
pound cake	454	Graham crackers
pounds of water	0.01602	ft^3
pounds of water	27.68	in^3
pounds of water	0.1198	gal
square ft	0.0929	m^2
square m	10.74	ft^2
square m	1.196	yd^2
square miles	2.59	km^2
tons (2000 lb)	0.907	tonnes (1000 kg)
tons	907	kg
tons/acre	2.24	tonnes/ha
tonnes (1000 kg)	1.10	ton (2000 lb)
tonnes/ha	0.446	tons/acre
two kilomockingbirds	2000	mockingbird
watts	1	J/s
yard	0.914	m

Author Index

Abert, J. G. 277
AbuGhararah, Z. H. 333
Adams, S. 24
Ahlert, R. 167
Ali Khan, M. Z. A. 63, 333
Allison, T. 168
Alter, H. 219, 276, 360
Anderson, R. 63
Andreottola, G. 167
Aquino, J. T. 334
Augenstein, D. 167
Austin, L. G. 219, 220
Azar, S. 166
Barber, E. B. 361
Barktoll, A. W. 360
Barlaz, M. A. 334, 384
Bates, L. 219
Beardsley, J. B. 276
Bell, P. M. 63, 360
Bennett, J. G. 219
Berthouex, P. M. 219
Biaciardi, C. 102
Biering, R. A. 384
Blackman, W. C. 384
Blazquez, R. 167
Boettcher, R. A. 276
Bok, S. 384
Boley, G. L. 333
Bonaparte, A. 166
Bond, F. C. 199, 220
Bordne, E. F. 166
Boughey, A. S. 26
Bowerman, F. 360
Boyd, G. 63
Bridle, T. L. 334
Brill, E. D. 384
Broadbent, S. R. 220
Brown, H. 26
Bruner, D. R. 168
Brunner, C. 333
Brunner, P. H. 384
Buckingham, P. L. 384
Buney, F. 63
Burgess, R. L. 102
Cailas, M, D. 63

Calcott, T. C. 220
Callahan, C. 361
Cannas, P. 167
Carnot, N. 296
Carruth, D. 63
Chadwick, E. 2
Chang, N-B. 334
Chesner, W. H. 333
Chian, E. S. 167
Cole, M. A. 361
Coleman, E. 361
Collins, R. J. 333
Commoner, B. 5
Conn, W. D. 26
Constable, T. W. 334
Converse, A. O. 102
Cookson, A. 361
Cooper. C. D. 167
Corbo, P. 167
Cossu, R. 167
Coulson, J. M. 220
Cross, J. F. 26
Crutcher, A. J. 167
Daigger, G. 26
Dair, F. R. 26
Davidson, G. R. 63
Dayal, G. E. 63
DeCesare, R. S. 276
DeGeare, T. V. 166
Dessanti, D. J. 167
DeWalle, F. 167
DeWeese, A. 102
Diaz, L. F. 220, 360, 361
Dickens, C. 2
Dilbert 24
Doorns, M. 168
Dorairaj, R. 102
Drobney, N. L. 220, 277
DuLong 283
Dumas, R. D. 334, 384
Dunlop, C. E. 361
Eimers, J. L. 333
Elliott, S. J. 334
Engdahl, R. B. 276
Englebrecht, R. 166

Epstein, B. 220
Erdincler, A. 333
Erlich, P. 5
Ernst, W. R. 381
Euler, L. 76
Evans, I. 219
Evans, J. C. 384
Everett, J. 88, 89, 102
Farrell, A. 361
Fenn, D. G. 166
Finnie, W. C. 102
Forrester, J. 5
Forrester, K. 334
Foust, A. S. 276
Franconeri, P. 219
Froeisdorph, G. 361
Fryberger, W. L. 276
Fung, T. 333
Fungaroli, A. 166
Gardner, H. P. 220
Gaudin, A. M. 192, 219
Gawalpanchi, R. R. 219
Genereux, J. 166
Genereux, M. M. 166
Georgescu-Roegen, N. 102
Glenn, J. 26
Goldberg, D. 26
Goldstein, I. S. 361
Golueke, C. 348
Graham, L. 361
Greely, S. A. 276
Greenberg, M. R. 102
Grosh, C. J. 166
Gross, B. A. 166
Grossman, D. 63
Gullett, B. 333
Gulueke, C. G. 360, 361
Hagerty, D. J. 360
Hajny, G. L. 361
Ham, R. 183, 219
Hammurabi 358
Hanley, K. J. 166
Harrelson, L. xviii
Harris, C. C. 166, 219
Harris, J. M. 167

Harrison, K. W. 334
Harrison, S. 384
Hart, F. T. 102
Hasselriis, F. 250, 272, 330, 276, 334
Hawkins, M. 63
Heer, J. E. 360
Hendricks, S. L. 63
Henstock, M. 276
Herckert, J. R. 361
Hering, R. 276
Herschfield, S. 166
Hickman, L. 173, 219
Hille, S. J. 360
Hockett, D. 63
Hocking, M. B. 384
Hope, R. F. 361
Houser, V. L. 168
Howard, D. 63
Huang, S-H. 334
Hudson, F. 63
Hull, H. E. 220, 277
Hummel, P. 168
Izumin, S. 276
Johnke, B. 334
Jones, K. H. 360
Kelly, D. J. 168
Kermit the Frog 25
Kerstetter, J. D. 333
Kerzee, R. 63
Kesler, S. E. 26
Kick 219
King, D. 167
Klee, A. 63
Kruglack, W. 276
Kurz, F. 360
Kwan, K. 77, 102
LaGrega, M. C. 384
Lawrence, A. W. 166, 360
Lema, J. M. 167
Leopold, A. 59, 63
Levy, S. J. 333
Lewis, K. 166
Liddell, C. L. 276
Liebman, J. C. 102, 360
Liu, J. 333
Liu, X. 361
Lloyd, C. M. 166
Lober, D. 26, 63
Lossin, R. D. 360
Lyons, J. K. 333
Madison, R. D. 219
Makar, H. V. 276
Male, J. W. 102
Malloy, M. G. 277
Malthus, T. 4
Mandel, M. 361
Maratha, R. 102
Markanya, A. 361
Marks, D. H. 63
Martello, W. P. 333
Martin, M. W. 168

Mather, R. J. 166
Matthews, C. W. 276
McAbee, M. K. 361
McAdams, C. L. 334
McBean, E. A. 167
McCarty, P. L. 360
McEnroe, B. M. 167
McGuinnes, J. 166
McKinley, V. 361
Meadows, D. 5, 26
Medeiros, J. 361
Melosi, M. V. 26
Meloy, T. P. 219
Mendez, R. 167
Michaels, E. L. 276
Miller, C. 63, 102
Mitchell, J. R. 219
Monteith, H. D. 360
Mumford, L. 26
Mureebe, A. 167
Murphy, C. 63
Murray, D. L. 276
Nagano, I. 276
Nelson, A. C. 166
Nelson, W. 276
Newton, I. 241
Niessen, W. R. 63, 333
Nishtala, S. R. 334
Nixon, R. 6
Nordstedt, R. A. 360
Nyns, E. J. 167
Obeng, D. M. 220
O'Brien, J. 63, 166
Oosternoek, J. 361
Othman, M. A. 166
Overton, M. 276
Pacey, J. 167, 168
Paode, R. D. 333
Parker, B. L. 276
Pas, E. I. 166, 220
Pavoni, J. L. 360
Pearce, D. A. 361
Perry, J. H. 220
Peter, H. W. 167
Pheffer, J. 360
Phelps, E. 332
Phillips, C. A. 168
Pilgrim, K. 63
Pohland, F. G. 166, 167
Pomeroy, C. D. 219
Porteus, A. 361
Pritchard, M. 385
Purcell, T. C. 361
Purchewitz, D. 167
Rammler, E. 197, 219
Ranjithan, S. 384
Rathje, W. 63
Rees, J. F. 166
Reese, E. T. 361
Reindl, J. 166
Reinhart, D. 63, 166, 167, 168, 219
Remson, I. A. 166

Resnick, W. 219
Rhyner, C. R. 63
Ricardo, D. 4
Richard, T. L. 361
Richardson, G. 167
Richardson, J. F. 220, 276
Rietema, K. 198, 276
Riley, P. 102
Rimer, A. 63, 220
Rittinger, T. 198
Rodrique, D. S. 360
Rodrique, J. O. 360
Rogers, C. J. 361
Rose, H. E. 276
Rosin, P. 193, 219
Rovers, F. A. 167
Ruf, J. A. 63, 219
Ruiz, N. A. 167
Ryan, B. 63
Salton, K. 276
Sandelli, G. J. 168
Savage, G. 63, 194, 219, 220, 276, 361
Sawell, S. E. 334
Schert, J. 102
Schinzinger, R. 168
Schroeder, P. R. 166
Schur, D. A. 102
Schwartz, S. 333
Schwegler, R. E. 26
Scoparius, P. V. 361
Senden, M. M. G. 276
Sfeir, H. 63
Sharp-Hansen, S. 168
Sherman, S. 26
Shumatz, L. A. 26
Shuster, K. A. 102
Simpson, B. 220
Singh, P. 63
Sino, D. F. 361
Skinner, J. H. xvi
Sleats, R. 166
Smit, J. P. 361
Smith, A. 4
Snow, J. 2
Solano. P. 384
Span, L. A. 361
Spence, P. 277
Stahl, J. 168
Stanley, E. S. 333
Stegman, R. 167
Stelzner, E. 334
Stessel, R. I. v, 228, 276
Stevenson, J. P. 360
Stokes, W. 241, 257, 258, 259
Stratton, F. E. 219
Sullivan, M. E. 276
Swager, R. 63
Sweeney, P. J. 276
Taggart, A. F. 219, 274, 277
Tanzer, E. K. 219
Tchobanaglous, G. H. 63, 167

Tels, M. 276
Tertin, R. F. 220, 277
Thalenberg, S. 168
Theisen, H. 63, 167
Thoreau, D. 5
Thorneloe, S. A. 167
Thornthwaite, C. W. 166
Tiezzi, E. 102
Townsend, T. 167, 168
Travers, C. 168
Trezek, G. 63, 194, 219, 220, 276, 360, 361
Ulgiati, S. 102
Upadhyay, R. 63
Venkataremani, E. S. 167

Vesilind, P. A. 63, 102, 166, 220, 276, 333, 360
Vigil, U. 63, 167
Viney, I. 166
Walker, J. M. 361
Walter, R. 360
Waner, J. A. 168
Waring, G. 2, 229
Warren, D. 166
Wathne, M. 102
Weand, B. L. 168
Webster, N. A. 168
Wen, C. Y. 333
Wenz, C. A. 384
Wheless, E. 168

Wilbey, S. J. 360
Wilson, D. L. 333
Wilson, G. B. 361
Winebreak, J. 361
Wong, M. 168
Woodbury, P. 361
Woodruff, K. L. 276
Worrell, W. 220, 228, 276, 384
Yadav, R. P. 63
Young, A. 166
Zaki, W. N. 276
Zappi, P. A. 166
Zavetski, S. 277
Zhang, L. 361
Zhao, A. 167

Subject Index

A

abrasiveness, of refuse 174
acid formers 339
acid hydrolysis 355
acid rain 320
adherence separators 268
aerated static pile 349, 351
aeration in composting 350
aerodynamic diameter 241, 243
Agent Orange 329
agreement (contract) 372
agricultural waste 24, 33
Alabama xvii
Altoona, PA 355
air classification 15
air classifiers 15
    baffled 254, 255
    constricted 254, 255
    effectiveness 249
    pulsed flow 254, 255
    vibrating 247, 249
    zig-zag 254, 255, 256
air pollutants 319
    gaseous 319
    particulates 319
    primary 319
    secondary 319
air pollution control 319
    gaseous air pollution 319
    particulate air pollution 319
air:solids ratio 253
airports 115, 116
airspace, in landfills 157
albedo 322
allocation models 79
aluminum 9, 265
American Society of Civil Engineers
    (ASCE) 3, 164, 358
American Society for Testing and
    Materials (ASTM) 40, 284,
    312
anaerobic bioreactors 142
anaerobic degradation 119, 337
anaerobic digestion 337, 342
angle of friction 151

angle of nip 208
angle of repose 174
annual cost 374
Arroyo Grande Landfill, CA 372
ash 23, 316, 318
Athens 1

B

backyard collection 67
baffled air classifiers 254, 255
bag filters 323, 325
Baltimore 5
beer cans 4
belt magnets 261
binary separation 224
biochemical processes 336
bioconcentration 379
biodegradability of MSW compo-
    nents 57, 58
biodegradation 118
bioreactors 110, 142
Black Death 1
blue bag system 87
boiler 300, 304
bomb calorimeter 289
Bond Law 196
Bond work index 199, 204, 205
bonds 374
bottom ash 316
bounce separators 268
brownfields 354
Bruntland Report 358
bulk density 18, 45, 54, 55, 111,
    112, 174, 182, 184
burden depth, under magnets 262

C

cadmium 321
California 74, 124, 189, 372
calorimeter 49, 153, 289
Cambridge, MA 161
can flatteners 210

can snatchers 70, 73
Canada 329
canyon landfills 157
cap failure 154
capillary barrier 153
capital costs 374
capital recovery factor 375
caps for landfills 126, 152
carbon:nitrogen ratio 346, 348
Carnot heat engine 296
cascading 232
cataracting 232
cellulose 336, 355
centrifuging 233
characteristic size 193, 197
Charleston, VW 2
chemical composition of MSW 49
Chicago 3, 161
Chinese Postman problem 77
chlorofluorocarbons 365
Cincinnati 125
Clean Air Act 7, 322
clean materials recovery facilities 86,
    224, 271
Club of Rome 5
code 13
coding 224, 229, 239
coffee cup debate 365
coliforms 21
collection 67
    backyard 67
    curbside 67, 88
    fully automated 70
    history 3
    macro-routing 76
    micro-routing 76
    plastic bags 71
    semi-automated 70
Columbia 151
combustion 21, 283
    air 293
    efficiency 295
    grates 303
    mass-burn 310
    modular 307, 308

overfire air 305
semi-suspended 310
starved air 307
suspension 310
waste-to-energy 307
waterwall 305, 306
combustors
mass-burn combustors 310
modular 307, 308
refractory 307
rotary kiln 305
thermal balances 293, 297
waste-to-energy 307
water-wall 305, 306
commercial waste 80
commingled waste 86
common law 6
compaction, in landfills 155, 156
compactors 86, 89, 111, 184
competitive service delivery 369
compositional analysis 285
composting 39, 344
in-vessel 349
pathogens, destruction of 345
rotating 353
windrows 349, 350
Comprehensive Environmental
Response Compensation and
Recovery Act (CERCLA) 7
compression 54, 56, 184
coning 40
constricted air classifier 254, 255
construction and demolition (C&D)
waste 31, 33
consumers 8
contracting 370
contracts 372
control of gases 328
control of particulates 322
conveying 176
conveyors 176
live bottom hoppers 178
metal pan 176
pneumatic 178, 180
rubber-belted 176
screw 180
cooling towers 316, 317
corrugated cardboard 4
covers for landfills 117
crew size 74, 90
critical speed 234
crushing 183
cubetter 252
curbside collection 67, 88
curbside recycling 88
cutting 183
cyclone 247, 255, 323, 324

D

Dade County, FL 206
daily cover 158

deadheading 76
degradation (in landfills) 118, 119,
337
delisted hazardous materials 379
denox system 328
density of materials 55
density of refuse 18, 45, 54, 55. 111,
112,174, 182
desiccation 154
design standards 118
diameter of particles 48
aerodynamic 241, 243
effective 242
diapers, disposable 364
digesters 337, 338
digestion, anaerobic 337, 342
Dilbert 24
dioxin 329
direct osmosis 140
dirty materials recovery facilities 86,
224, 269
disc screens 236, 238
dismal science 4
districting 76
diversion 32
Dow Chemical 4
drag chain conveyors 182
drag chain unloading 87
drop test 248, 251
drum magnet 261
dry scrubbers 328
DuLong formula 283
dumps 6, 17
dumpsters 80, 81
DuPont Chemical Corporation 11
Durham, NC 6, 84
dust 200

E

Eco-fuel 252
economics 4
eddy current separators 263, 264
effective diameter 242
effectiveness of air classification 249
effectiveness of separation 227
Rietema Efficiency 228
Worrell-Stessel Efficiency 228
efficiency of combustion 295
electric power generation 149
electromechanical separation 260
electrostatic precipitators 325, 327
electrostatic separators 265
endangered species 115
energy conversion 21
energy recovery 283
energy use in shredders 204
environmental impact 116
environmental impact statements 6,
116
environmental justice 117
enzymatic hydrolysis 355

equipment for landfills 155
Escondido CA 372
Euler's tour 77
European Union 109
evaporation 140
evapotranspiration 123, 124, 125
evapotranspiration cover 153
expediency, principle of 332
explosions in shredders 200
extracellular enzymes 339
extract 226, 230, 248, 268

F

fabric filters 323
fatal flaw analysis 115
fermentation 357
field capacity 121, 123, 142
filling sequence for landfills 155
final cap 126, 152
financing of solid waste facilities 373
fireplaces 331
flail mill 189
flammability 379
flaring 148
float/sink separators 238
floating velocity 178
flood plains 115
Florida 49, 1131, 138, 148, 163,
189, 207
Florida, University of 95
flotation 260
flow control 368
fluff 312, 313
fly ash 316
franchise agreement 372
franchise fee 373
Fresh Kills Landfill, NY 95
fuel cells 149
fully automated collection 70

G

garbage grinders 336
gases, from landfills 20, 38, 109, 143
collection systems 143, 147
combustion 148
electric power generation 149
extraction 144, 162
flaring 148
fuel cells 149
monitoring 160
production 128
quality of landfill gases 130
quantities of landfill gases 128
synthetic fuel 149
use 147
vehicle fuel 148
wells 144, 162
gases, from anaerobic digesters 340
gasification 309

gaseous air pollution 319
Gaudin model 192
Gaudin-Melloy model 197
general obligation bond 374
generation of municipal solid waste
        33, 36, 40, 318
    input method 33
    output method 40
geomembranes 132, 133
geonets 137
geosynthetic drainage 137
Germany 367
global warming 322
glucose production 355
Governor's Island, NY 2
granulating 210
Grassroots Recycling Network 86
grate spacing, in shredders 203, 204
grates, in combustion 303
Great Britain 17
Great Sanitary Awakening 2
green-can-on-wheels 70, 72
greenhouse gases 131, 322
Greensboro, NC 11
grinding 15, 183, 185, 354
gross calorific energy 53
groundwater 116, 122, 160

H

hammer wear 201, 203
hammermill 186
    horizontal 186
    vertical 186, 188, 197
Hammurabi, Code of 358
hand sorting 229
Harz jig 247
Hawaii 355
hazardous materials 378
hazardous waste 24, 31, 38
hazardous waste landfills 380
health and safety 69, 200
heat value 49, 52, 53, 283, 286, 292,
        312
    higher heat value (HHV) 33, 292
    lower heat value (LHV) 53
    moisture-free 52
    moisture- and ash-free 52
heavy fraction 248
heavy liquid separator 259
heavy media separator 259
heuristic routing 78
high density polyethylene (HDPE)
        13, 14, 15, 132, 134, 207,
        269
history of solid waste management 1
home scrap 10
horizontal hammermill 186, 187
Hydrologic Evaluation of Landfill
        Perforce (HELP) 126

I

incineration 2, 300
inclined table 266
Indiana 161
industrial waste 33
injuries in collection 69
input method of estimating solid
        waste generation 33
Integrated Solid Waste Management
        23, 24
interest tables 376
Interior, U. S. Department of 6
in-vessel composting 349
invisible hand 4
involuntary risk 331
iron law of wages 4
Italy 329

J

jigs 246

L

land treatment of leachate 141
landfills 17, 151
    airspace 157
    angle of friction 151
    canyon 157
    caps 152,154
    daily cover 158
    design 132
    equipment 155
    filling sequence 155
    gases 344
    history 3
    liners 20,132, 134
    mining 163
    moisture balance 125
    monitoring 158
    monofills 23
    operational layers 155
    settlement 162
    side slopes 151
    stability 151
    water balance 122
LandGEM 129, 130
latent heat of vaporization 292
law of population 4
laissez-faire 4
leachate 20, 109, 120
    collection 133, 134, 159
    detection 134
    disposal 138
    land treatment 141
    production 121
    quality 127
    recirculation 141
    storage 134
    treatment 138
    use 143

lead 321
leaves 32, 354
legal restriction to recycling 14
legislation 6
    Clean Air Act 7, 322
    Comprehensive Environmental
        Response Compensation and
        Recovery Act (CERCLA) 7
    National Environmental Policy
        Act (NEPA) 6
    Occupational Safety and Health
        Act 69, 200
    Resource Conservation and
        Recovery Act (RCRA) 7, 109,
        110, 115, 117, 133, 134, 152,
        160, 275
    Rivers and Harbors Act 6
    Solid Waste Disposal Act 6
life cycle analysis 3363, 365
life cycle assessment 363
life cycle management 366
light fraction 248
liners for landfills 20, 132, 134
link 77
listed hazardous wastes 379
litter 91, 92
live bottom feeders 178
live bottom unloading 84
Los Angeles 88, 124
Love Canal 7
low density polyethylene (LDPE) 13,
        14, 153
lower heat value (LHV) 53, 292

M

macro-routing 76
magnetic separation 15
magnets 260
    belt 261
    drum 261
    suspended 261
Manchester, England 2
markets 14, 15
mass-burn combustors 310
Massachusetts 161
Massachusetts Institute of
        Technology 5
materials balances 293
materials density 55
materials flow 8
materials recovery facilities 13, 15,
        224, 368
    clean 86
    dirty 86, 224, 269
materials separation systems 269
mechanical composting 352, 353
mechanical properties 56
mechanical vapor compression 141
medial waste 24
mercury 321
metal pan conveyors 176

methane formers 339
methane from landfills 343
methane generation 337
micro-routing 76
Middle Ages 1
mining landfills 162
mining waste 33
Mobro xvii
modular combustors 307, 308
moisture balance in landfills 125
moisture content 45, 46, 287, 345
moisture-free heat value 52
moisture- and ash-free heat value `
    52
monitoring landfills 158
monofills 23
municipal solid waste (MSW)
    characterization 38
    chemical composition 49
    collection 32
    defined 30
    generation 33, 36, 39
    mechanical properties 56
    moisture content 45, 46, 287
    processing 173

N

National Environmental Policy Act
    (NEPA) 6
National Pollution Discharge
    Elimination System (NPDES)
    139
negative sorting 229
net calorific energy 53
New England 49
New Jersey 75, 368
New York xvii, 2, 84, 109, 138, 229
New Zealand 148
newsprint 18
Newton's Law 241
node 77
noise 200
North Carolina 24, 37, 38, 117
Norway 320

O

Occupational Safety and Health Act
    69, 200
ocean disposal 1, 109
odor 185
Ohio 320
operating cost 374
operational layers in landfills 155
optical sorting 266, 267
Oregon 368
organic fraction 109
Orlando, FL 125

output method of estimating solid
    waste generation 40
overfire air 305
overflow 248
ownership of solid waste operations
    369
ozone 320

P

packers 69
    dual compartment 87
paint 11
pan scraper 155, 156
paper 15, 16
    as fuel 311
    flow in Switzerland 364
    heat value 312
    newsprint 18
paper cups 4
particle size 48, 60, 173, 191, 215,
    216, 253
particulate air pollution 319
pathogens, destruction in composting
    345
pellets 312, 313
Pennsylvania 7
percolation 122, 125
performance standards 118
permitting 110, 117
pH in composting 348
Philadelphia 368
photogrammetry 40, 44
photosynthesis 293
physical characteristics of refuse 173
pi breakage theorem 211
pickers 15
picking 15, 229
planning 110
plastic bags used in collection 71
plastics 13, 14, 34, 132, 207, 269
plunger jig 246
pneumatic conveying 178, 180
pollution prevention 11, 366
polychlorinated dibenzodioxins
    (PCDD) 329
polychlorinated dibenzofurans
    (PCDF) 329
polyethylene
    high density (HDPE) 13, 14,15,
        132, 207, 269
    low density (LDPE) 13, 14
polyethylene terephthalate (PETE) 4,
    13, 14, 15, 207, 269
polynary separation 224, 226
polypropylene (PP) 13, 14
polystyrene (PS) 13, 14
polyvinyl chloride (PVC) 13, 14
Pompano Beach, FL 49
population densities 36, 37
porosity 182

positive sorting 229
power requirements
    for shredders 197
    for trommel screens 236, 237
precedence, in law 6
precision in sampling 41
primary air pollutants 319
primary settlement 162
principle of expediency 332
private ownership of solid waste
    facilities 376
product 226
prompt industrial scrap 10
property values 116, 117
proximate analysis 49, 51, 288
public ownership of solid waste
    facilities 369
Publicly Owned Wastewater
    Treatment (POTW) 138
pulping 206
pulsed flow air classifiers 254, 255
push blade unloading 84
pyrolysis 307

Q

quartering 40
quick condition 247

R

radio pills 175
radioactivity 379
radiological tagging 175
ram feed 308
rats 185
reactivity 379
rebound 184
reciprocating grates 303
reciprocating screens 236
recirculation of leachate 141
recovery 8, 10
recycling 7, 8, 10, 12
recyclables 67, 71
    collection of 85
reduction 10
refuse
    abrasiveness 174
    angle of repose 174
    as collected 32
    as generated 32
    bulk density 174
    defined 31
    diverted 32
    moisture 175
    physical characteristics 173
    variation in generation 36
refuse-derived fuel (RDF) 15, 22, 49,
    53, 185, 250, 252, 310, 311,
    312, 313

Eco-fuel 252
 fluff 312, 313
 pellets 312, 313
regulations 6
rejects 226, 230, 248, 268
request for proposals (RFP) 370
Resource Conservation and Recovery
    Act (RCRA) 7, 109, 110, 115,
    117, 133, 134, 152, 160, 275
retail sales 37, 38
reuse 8, 10, 11
revenue bond 374
reverse osmosis 140
Rietema Efficiency 228
risk 331
Rittinger's Law 198
Rivers and Harbors Act 6
roll crushers 207
roll mill 198
roll-off containers 82
Rome 1
rotary kiln 305
rotating composting system 353
rotation failure of landfill caps 154
Rosin-Rammler paper 195, 196
Rosin-Rammler particle size distribu-
    tion 193, 194, 204
routing of trucks 76
    heuristic 78
    macro-routing 76
    micro-routing 76
rubber-belted conveyor 176
runoff 123, 125

S

sacred lands 116
safety 69, 200
sampling MSW 40
San Diego 368
sanitary landfills 17, 18, 151
    (see landfills)
scavengers 3
schools 115
scrapers 155
screens 230
    disc 236, 238
    reciprocating 236
    trommel 231
screw conveyors 180
scrubbers 323
    dry scrubbers 328
    wet scrubbers 328
Seattle 84, 96
Seattle Stomp 96
secondary air pollutants 319
secondary settlement 162
seismic faults 115
semi-automated collection 70
semi-suspended combustion 310
separation of materials 224
service agreement 372
settlement 162

settling chambers 323
shaking tables 266
shear 189, 190
shear failure, in landfills 154
shear shredder 189, 190
shearing 183
shredding 183
    energy use 204
    explosions 200
    hammermills 180
side slopes, on landfills 158
sinking fund factor 377
sinking funds 377
siting of waste facilities 7, 110, 114
size reduction 15, 180, 183, 189,
    190, 200, 204
sludge 24, 33, 355
Solid Waste Disposal Act 6
sorting 15, 229
    negative 229
    positive 229
source reduction 11
special wastes 24
split 231, 272
stabilization, in landfills 119, 120,
    151
starved air combustion 307
steam production in waste-to-energy
    307
street cleanliness 91
stereophotogrammetry 175
stoichiometric air 294
stoichiometric oxygen 294
Stokes Law 241, 257, 258, 259
stoners 265
storage 175
storage bin 314
storage pits 300
stormwater management 151
styrofoam 4
Subtitle D landfills 6, 109, 117, 133,
    134, 139, 152, 160
sulfur oxides 319
sulfuric acid 320
Superfund 3
suppression of explosions 201
suspended magnets 261
suspension combustion 310
sustainable development 358
Svesco, Italy 329
swales 152
switching 13, 224, 229, 239
Switzerland 364
synthetic fuel 149
systems analysis 97

T

Tampa, FL 148
terminal velocity 245
thermal balances in combustion 293,
    297
thermogram 289

3M Corporation 11
time, in collection 74
tipping 17
tipping fees 18, 37, 38
tipping floors 175
tips 17
tires 24
toxicity 379
transfer of moisture 45, 47
transfer stations 82, 83
travel time 75, 90
traveling grates 303
Trabi Award 366
Trablant 366, 367
triboelectric charging 265
TV dinners 4

U

ultimate analysis 49, 51, 283, 284
underfire air 302
underflow 248
unicoursal network 77
uniformity coefficient 48
upflow separator 260
U. S. Army Corps of Engineers 126
U. S. Bureau of Mines 195
U. S. Department of Interior 6
U. S. Environmental Protection
    Agency 3, 6, 7, 11, 23, 32,
    39, 49, 66, 117, 118, 129,
    140, 153, 318, 319, 320, 329,
    330, 365, 378, 379
U. S. Public Health Service 6

V

Valley of the Drums 7
vapor compression distillation 140
vehicle fuel 148
venting 201
vertical hammer mill 186, 188, 197
vertically integrated markets 15, 17,
    369
vibrating air classifier 247, 249
vibrating feeder 178, 180
Viet Nam War 329
Vikings 383
Virginia 84, 369
void ratio 182
volatile organic carbon (VOC) 11,
    131
volume-based fee system 68
volume reduction 112
voluntary risk 331

W

walking floor unloading 84, 179
waste heat 344
waste-to-energy 22, 297, 300
water balance 122

waterwall combustion 305, 306
weight-based fee system 68
wet scrubber 328
wetlands 115
Williams Patent Crusher 196
windrows 349, 350

winnowing 239, 240
Wisconsin 34, 35
Wisconsin, University of 183
work index 199, 204, 205
Worrell-Stessel Efficiency 228

XYZ

X-rays 175
yard waste 71, 88, 354
    composition of 354
zero waste 86
zig-zag air classifiers 254, 255, 256